METHODS IN MOLECULAR BIOLOGY

For further volumes:
http://www.springer.com/series/7651

For over 35 years, biological scientists have come to rely on the research protocols and methodologies in the critically acclaimed *Methods in Molecular Biology* series. The series was the first to introduce the step-by-step protocols approach that has become the standard in all biomedical protocol publishing. Each protocol is provided in readily-reproducible step-bystep fashion, opening with an introductory overview, a list of the materials and reagents needed to complete the experiment, and followed by a detailed procedure that is supported with a helpful notes section offering tips and tricks of the trade as well as troubleshooting advice. These hallmark features were introduced by series editor Dr. John Walker and constitute the key ingredient in each and every volume of the *Methods in Molecular Biology* series. Tested and trusted, comprehensive and reliable, all protocols from the series are indexed in Pub Med.

Antiviral Resistance in Plants

Methods and Protocols

Edited by

Kappei Kobayashi and Masamichi Nishiguchi

Faculty of Agriculture, Ehime University, Matsuyama, Ehime, Japan

Humana Press

Editors
Kappei Kobayashi
Faculty of Agriculture
Ehime University
Matsuyama, Ehime, Japan

Masamichi Nishiguchi
Faculty of Agriculture
Ehime University
Matsuyama, Ehime, Japan

ISSN 1064-3745 ISSN 1940-6029 (electronic)
Methods in Molecular Biology
ISBN 978-1-4939-9634-6 ISBN 978-1-4939-9635-3 (eBook)
https://doi.org/10.1007/978-1-4939-9635-3

This Humana imprint is published by the registered company Springer Science+Business Media, LLC, part of Springer Nature.
The registered company address is: 233 Spring Street, New York, NY 10013, U.S.A.

Preface

Plant virus diseases have long been serious threats for agricultural production. Different strategies have been developed to combat plant virus infection, which can roughly be categorized in two ways: one is to prevent viruses from infecting plants by shutting off the infection pathways, such as vector insects, and the other is to make plants combat virus infection, e.g., the use of resistant cultivars introgressed with natural virus resistance genes or those engineered by recombinant DNA technologies, or to use something like vaccines.

The first five chapters describe the techniques for identifying and studying the virus resistance gene involved in plant innate immunity. Tomita et al. briefly overview the studies of the most common group of plant virus resistance genes encoding NB-LRR class resistance proteins and describe the techniques for functional analysis of the resistance proteins. Kobayashi et al. describe a way to insert an epitope tag into the middle of the coding region for NB-LRR protein, which is a prerequisite for protein-level studies of plant virus resistance. Sato and Takahashi provide protocols for the analyses of plant virus resistance mechanisms and signal transduction pathways therein together with a comprehensive list of *Arabidopsis* mutants useful for the analyses. Hashimoto et al. describe protocols of virus-induced gene silencing for the analysis of antiviral signaling in plants, which, together with Sato and Takahashi, provide a detailed protocol for the antiviral resistance signaling in two major model plants, *Arabidopsis thaliana* and *Nicotiana benthamiana*. Idehara et al. describe a method to introduce random mutations into a virus gene, which is useful for analyzing a potential breakdown of the resistance.

The next five chapters describe techniques related to novel mechanisms of plant virus resistance. Ishibashi and Ishikawa provide the protocol for their original cell-free virus replication system and its use for the identification of the new type of plant virus resistance protein. Sasaki et al. describe the techniques for the identification of a new type of plant virus resistance protein that hampers virus cell-to-cell movement. Hosoe et al. describe the method for protoplast transfection in *Arabidopsis thaliana*, which is useful for studying the mode of plant virus resistance. Osmani et al. describe the protocol for the targeted mutagenesis of plant virus resistance-related genes in potato plants, which is also useful for the breeding of plant virus resistance in plants. Ali et al. describe a method for the new engineered resistance to plant DNA viruses.

The next four chapters describe methods for the analysis and practical use of RNA silencing. Donaire and Llave describe a bioinformatics method for small RNA profiling, which is fundamental for understanding RNA silencing-mediated virus resistance. Guo et al. describe the protocol for the genetic screen for exploring the underlying mechanisms of antiviral RNA silencing. Carbonell and Daròs provide the protocol for artificial small RNAs, which are expected to provide powerful tools to combat plant viruses in genetically engineered plants. Nishiguchi et al. describe the protocol for a new strategy of engineered plant virus resistance, in which the host genes essential for virus replication are targeted.

The final two chapters describe methods for the development of plant viral vaccines. Kubota and Tomitaka describe a method for large-scale virus inoculation and quick methods for virus detection, which are useful for developing and evaluating naturally occurred attenuated virus strains. Li et al. describe a method for an engineered plant virus vaccine using an efficient virus-based system for RNA silencing.

We are grateful to all the authors for their excellent contribution and to Prof. John M. Walker for his patient guidance to complete the book.

Matsuyama, Ehime, Japan

Kappei Kobayashi
Masamichi Nishiguchi

Contents

Preface ... *v*
Contributors ... *ix*

1 Identification and Functional Analysis of NB-LRR-Type Virus Resistance Genes: Overview and Functional Analysis of Candidate Genes ... 1
Reiko Tomita, Ken-Taro Sekine, Chika Tateda, and Kappei Kobayashi

2 Insertion of Epitope Tag into NB-LRR Class Plant Virus Resistance Protein ... 11
Kappei Kobayashi, Reiko Tomita, and Ken-Taro Sekine

3 Reverse Genetic Analysis of Antiviral Resistance Signaling and the Resistance Mechanism in *Arabidopsis thaliana* ... 27
Yukiyo Sato and Hideki Takahashi

4 Analysis of Antiviral Resistance Signaling Pathways by Virus-Induced Gene Silencing in *Nicotiana benthamiana* ... 85
Masayoshi Hashimoto, Yasuyuki Yamaji, and Ken Komatsu

5 Random Mutagenesis of Virus Gene for the Experimental Evaluation of the Durability of NB-LRR Class Plant Virus Resistance Gene ... 97
Kengo Idehara, Reiko Tomita, Ken-Taro Sekine, Masamichi Nishiguchi, and Kappei Kobayashi

6 A Cell-Free Replication System for Positive-Strand RNA Viruses for Identification and Characterization of Plant Resistance Gene Products ... 115
Kazuhiro Ishibashi and Masayuki Ishikawa

7 Plant Protein-Mediated Inhibition of Virus Cell-to-Cell Movement: Far-Western Screening and Biological Analysis of a Plant Protein Interacting with a Viral Movement Protein ... 123
Nobumitsu Sasaki, Yasuhiko Matsushita, and Hiroshi Nyunoya

8 Transfection of Protoplasts Prepared from *Arabidopsis thaliana* Leaves for Plant Virus Research ... 145
Naoi Hosoe, Takuya Keima, Yuji Fujimoto, Yuka Hagiwara-Komoda, Masayoshi Hashimoto, Kensaku Maejima, Shigetou Namba, and Yasuyuki Yamaji

9 CRISPR/Cas9-Mediated Editing of Genes Encoding rgs-CaM-like Proteins in Transgenic Potato Plants ... 153
Zhila Osmani, Shinnosuke Jin, Masafumi Mikami, Masaki Endo, Hiroki Atarashi, Kaien Fujino, Tetsuya Yamada, and Kenji S. Nakahara

10 A Simplified Method to Engineer CRISPR/Cas9-Mediated Geminivirus Resistance in Plants ... 167
Zahir Ali, Syed Shan-e-Ali Zaidi, Manal Tashkandi, and Magdy M. Mahfouz

11 Computational Workflow for Small RNA Profiling in Virus-Infected Plants 185
Livia Donaire and César Llave

12 A Sensitized Genetic Screen to Identify Novel Components and Regulators of the Host Antiviral RNA Interference Pathway 215
Zhongxin Guo, Xian-Bing Wang, Wan-Xiang Li, and Shou-Wei Ding

13 Design, Synthesis, and Functional Analysis of Highly Specific Artificial Small RNAs with Antiviral Activity in Plants 231
Alberto Carbonell and José-Antonio Daròs

14 Resistance Breeding Through RNA Silencing of Host Factors Involved in Virus Replication 247
Masamichi Nishiguchi, Emran Md. Ali, Hui Chen, Masayuki Ishikawa, and Kappei Kobayashi

15 Large-Scale Inoculation and Evaluation Methods for Attenuated Plant Viruses 261
Kenji Kubota and Yasuhiro Tomitaka

16 RNA Silencing-Mediated Apple Latent Spherical Virus Vaccine in Plants 273
Chunjiang Li, Noriko Yamagishi, and Nobuyuki Yoshikawa

Index 289

Contributors

ZAHIR ALI • *Laboratory for Genome Engineering, Division of Environmental and Biological Sciences and Engineering, King Abdullah University of Science and Technology, Thuwal, Saudi Arabia*

EMRAN MD. ALI • *Faculty of Agriculture, Ehime University, Matsuyama, Ehime, Japan; Department of Plant Pathology, University of Georgia, Tifton, GA, USA*

HIROKI ATARASHI • *Research and Development Organization, Kikkoman Corporation, Noda, Japan*

ALBERTO CARBONELL • *Instituto de Biología Molecular y Celular de Plantas, (Consejo Superior de Investigaciones Científicas-Universitat Politècnica de València), Valencia, Spain*

HUI CHEN • *Faculty of Agriculture, Ehime University, Matsuyama, Ehime, Japan; Department of Agronomy, Kansas State University, Manhattan, KS, USA*

JOSÉ-ANTONIO DARÒS • *Instituto de Biología Molecular y Celular de Plantas, (Consejo Superior de Investigaciones Científicas-Universitat Politècnica de València), Valencia, Spain*

SHOU-WEI DING • *Department of Microbiology and Plant Pathology and Center for Plant Cell Biology, University of California, Riverside, CA, USA*

LIVIA DONAIRE • *Departamento de Biología del Estrés y Patología Vegetal, Centro de Edafología y Biología Aplicada del Segura (CEBAS), Consejo Superior de Investigaciones Científicas (CSIC), Murcia, Spain*

MASAKI ENDO • *Institute of Agrobiological Sciences, National Agriculture and Food Research Organization, Tsukuba, Japan*

YUJI FUJIMOTO • *Graduate School of Agricultural and Life Sciences, The University of Tokyo, Tokyo, Japan*

KAIEN FUJINO • *Research Faculty of Agriculture, Hokkaido University, Sapporo, Japan*

ZHONGXIN GUO • *Vector-Borne Virus Research Center, Haixia Institute of Science and Technology, College of Plant Protection, Fujian Agriculture and Forestry University, Fuzhou, Fujian, People's Republic of China*

YUKA HAGIWARA-KOMODA • *Graduate School of Agricultural and Life Sciences, The University of Tokyo, Tokyo, Japan*

MASAYOSHI HASHIMOTO • *Graduate School of Agricultural and Life Sciences, The University of Tokyo, Tokyo, Japan*

NAOI HOSOE • *Graduate School of Agricultural and Life Sciences, The University of Tokyo, Tokyo, Japan*

KENGO IDEHARA • *Faculty of Agriculture, Ehime University, Matsuyama, Ehime, Japan*

KAZUHIRO ISHIBASHI • *Plant and Microbial Research Unit, Division of Plant and Microbial Sciences, Institute of Agrobiological Sciences, National Agriculture and Food Research Organization, Tsukuba, Ibaraki, Japan*

MASAYUKI ISHIKAWA • *Institute of Agrobiological Sciences, National Agriculture and Food Research Organization, Tsukuba, Ibaraki, Japan*

SHINNOSUKE JIN • *Research Faculty of Agriculture, Hokkaido University, Sapporo, Japan*
TAKUYA KEIMA • *Graduate School of Agricultural and Life Sciences, The University of Tokyo, Tokyo, Japan*
KAPPEI KOBAYASHI • *Faculty of Agriculture, Ehime University, Matsuyama, Ehime, Japan*
KEN KOMATSU • *Laboratory of Plant Pathology, Graduate School of Agriculture, Tokyo University of Agriculture and Technology, Fuchu, Tokyo, Japan*
KENJI KUBOTA • *Central Region Agricultural Research Center, National Agriculture and Food Research Organization, Tsukuba, Ibaraki, Japan*
CHUNJIANG LI • *Faculty of Agriculture, Iwate University, Morioka, Japan*
WAN-XIANG LI • *Department of Microbiology and Plant Pathology and Center for Plant Cell Biology, University of California, Riverside, CA, USA*
CÉSAR LLAVE • *Departamento de Biología Medioambiental, Centro de Investigaciones Biológicas (CIB), Consejo Superior de Investigaciones Científicas (CSIC), Madrid, Spain*
KENSAKU MAEJIMA • *Graduate School of Agricultural and Life Sciences, The University of Tokyo, Tokyo, Japan*
MAGDY M. MAHFOUZ • *Laboratory for Genome Engineering, Division of Environmental and Biological Sciences and Engineering, King Abdullah University of Science and Technology, Thuwal, Saudi Arabia*
YASUHIKO MATSUSHITA • *Gene Research Center, Tokyo University of Agriculture and Technology, Tokyo, Japan*
MASAFUMI MIKAMI • *Graduate School of Nanobioscience, Yokohama City University, Yokohama, Japan*
KENJI S. NAKAHARA • *Research Faculty of Agriculture, Hokkaido University, Sapporo, Japan*
SHIGETOU NAMBA • *Graduate School of Agricultural and Life Sciences, The University of Tokyo, Tokyo, Japan*
MASAMICHI NISHIGUCHI • *Faculty of Agriculture, Ehime University, Matsuyama, Ehime, Japan*
HIROSHI NYUNOYA • *Gene Research Center, Tokyo University of Agriculture and Technology, Tokyo, Japan; Faculty of Science and Engineering, Waseda University, Tokyo, Japan*
ZHILA OSMANI • *Research Faculty of Agriculture, Hokkaido University, Sapporo, Japan; Faculty of Agriculture, Tarbiat Modares University, Tehran, Iran*
NOBUMITSU SASAKI • *Gene Research Center, Tokyo University of Agriculture and Technology, Tokyo, Japan*
YUKIYO SATO • *Graduate School of Agricultural Science, Tohoku University, Sendai, Japan*
KEN-TARO SEKINE • *Faculty of Agriculture, University of the Ryukyus, Nishihara, Okinawa, Japan*
HIDEKI TAKAHASHI • *Graduate School of Agricultural Science, Tohoku University, Sendai, Japan*
MANAL TASHKANDI • *Laboratory for Genome Engineering, Division of Environmental and Biological Sciences and Engineering, King Abdullah University of Science and Technology, Thuwal, Saudi Arabia*
CHIKA TATEDA • *Iwate Biotechnology Research Center, Kitakami, Iwate, Japan*
REIKO TOMITA • *Faculty of Agriculture, University of the Ryukyus, Nishihara, Okinawa, Japan*

YASUHIRO TOMITAKA • *Kyushu Okinawa Agricultural Research Center, National Agriculture and Food Research Organization, Koshi, Kumamoto, Japan*
XIAN-BING WANG • *State Key Laboratory of Agro-Biotechnology, College of Biological Sciences, China Agricultural University, Beijing, People's Republic of China*
TETSUYA YAMADA • *Research Faculty of Agriculture, Hokkaido University, Sapporo, Japan*
NORIKO YAMAGISHI • *Faculty of Agriculture, Iwate University, Morioka, Japan; Agri-Innovation Center, Iwate University, Morioka, Japan*
YASUYUKI YAMAJI • *Graduate School of Agricultural and Life Sciences, The University of Tokyo, Tokyo, Japan*
NOBUYUKI YOSHIKAWA • *Faculty of Agriculture, Iwate University, Morioka, Japan; Agri-Innovation Center, Iwate University, Morioka, Japan*
SYED SHAN-E-ALI ZAIDI • *National Institute for Biotechnology and Genetic Engineering (NIBGE), Faisalabad, Pakistan*

Chapter 1

Identification and Functional Analysis of NB-LRR-Type Virus Resistance Genes: Overview and Functional Analysis of Candidate Genes

Reiko Tomita, Ken-Taro Sekine, Chika Tateda, and Kappei Kobayashi

Abstract

Coexpression of a plant NB-LRR-type resistance (*R*) gene and corresponding viral avirulent *(Avr)* gene introduced in *Nicotiana benthamiana* using *Agrobacterium tumefaciens* confers hypersensitive response (HR). Such *Agrobacterium*-mediated transient gene expression methods have contributed to the identification of new plant *R* genes and facilitated the analysis of their functions. Here we describe a model method, by which several tobamovirus *R* genes from Solanaceous plants have been successfully identified and characterized molecularly.

Key words Resistance protein, Avirulent effector, Transient expression, Agroinfiltration, *R* gene

1 Introduction

Plants have evolved various resistance (*R*) genes operating in a gene-for-gene-specific manner, in which the R protein limits the invasion of viruses, bacteria, oomycetes, fungi, or pests by activating a host-specific defense response, depending on the recognition of pathogen-derived elicitors [1, 2]. The *R* gene-mediated defense involves the production of antimicrobial compounds, accumulation of reactive oxygen species, and hypersensitive response (HR) accompanying a local necrotic lesion [3]. In the past decade, many *R* genes have been identified based on positional cloning [4]. Since the isolation of *Tobacco mosaic virus* resistance gene *N* in 1994 [5], over 20 *R* genes against viruses have been identified from various plants [6]. Most of the *R* genes known encode immune receptor proteins consisting of a nucleotide-binding domain (NB), and a leucine-rich repeat domain (LRR). An NB-LRR-type R protein functions as a receptor of the corresponding avirulent effector protein (Avr) and as a molecular switch for activating intercellular defense signaling. These activities are tightly mediated by

Kappei Kobayashi and Masamichi Nishiguchi (eds.), *Antiviral Resistance in Plants: Methods and Protocols*, Methods in Molecular Biology, vol. 2028, https://doi.org/10.1007/978-1-4939-9635-3_1,

interdomain and intradomain interactions [4, 7]. Whole genome sequencing can help us find a candidate *R* gene based on sequence homology [8–11]. However, after identification, the validation of a candidate gene, that is, the transformation of candidate *R* genes to produce a resistant phenotype in a susceptible plant is effort-intensive [5, 12–14].

To analyze the function of R protein, *Agrobacterium* infiltration (agroinfiltration) in the leaves of *Nicotiana* plants has been used for transient gene expression[15]. R-Avr interaction reproduced transiently via *Agrobacterium* infection can develop a lesion owing to HR in *Nicotiana* plants, even if the *R* gene was isolated from a different species. Such transient systems have revealed several functions of R proteins including characterization of domains related to the recognition and activation, and intermolecular and intramolecular interactions [16–18]. Moreover, they have contributed to the identification of cofactors that interact with R and/or Avr proteins and signaling transductions for resistance responses. Here we describe a model method of agroinfiltration-mediated transient gene expression for characterizing candidate *R* genes, in which a combination of transiently expressed R and Avr proteins causes visible lesions on *Nicotiana benthamiana* leaves. This method has been utilized successfully for identifying tobamovirus resistance genes in Solanaceae (pepper and tobacco) and analyzing their functions based on mutational analyses [19–21]. Our method can facilitate the isolation and characterization of a novel NB-LRR-type *R* gene against virus.

2 Materials

2.1 Plasmid DNAs for Transformation of Agrobacterium Tumefaciens

1. Plant expression vector pBA: a derivative of pBI121 for *R* gene expression (*see* **Note 1**).
2. Plant expression vector pGT: a derivative of pGreen0000 for *Avr* gene expression (*see* **Note 2**).
3. pBTPIW: an expression vector of *Tobamovirus* genome (*see* **Note 2**).
4. *Escherichia coli* DH5α for constructing the binary vectors and obtaining the plasmid DNAs.
5. Vectors for cloning such as pCR4 (Thermo Fisher Scientific, Waltham, MA, USA) are useful to modify the expressing gene sequences in the steps of sub-cloning.
6. Epitope tag sequences such as hemagglutinin (HA), c-Myc, or FLAG added to coding sequence of *R* and *Avr* in the 5′ or 3′ terminals, or in the intermediate legion for the purpose of confirming protein expressed after agroinfiltration in *Nicotiana benthamiana* (*see* **Note 3**).

2.2 Preparation for Culture of Agrobacterium tumefaciens

1. Purified plasmid DNAs using a standard procedure or a plasmid DNA purification kit such as LaboPass Mini (Cosmo Genetech, Seoul, Korea).
2. *Agrobacterium tumefaciens* LBA4404 Electro-Cells (Takara, Osaka, Japan) (*see* **Note 4**).
3. A modular electroporation system such as Gene Pulser (Bio-Rad, Hercules, CA, USA) and electroporation cuvettes (Bio-Rad).
4. L-broth medium: 10% (w/v) Bacto tryptone, 0.5% (w/v) Bacto yeast extract, 0.5% (w/v) NaCl, 0.1% (w/v) glucose. Autoclave in aliquots. For solid (plate) medium, add 1.5% agar before autoclaving.
5. 50 mg/mL kanamycin dissolved in water. Store at −25 °C.
6. 20 mg/mL rifampicin dissolved in methanol. Store at −25 °C.
7. 50% (v/v) glycerol filter-sterilized.
8. Sterilized cryogenic tubes.
9. Infiltration Buffer: 10 mM $MgCl_2$, and 10 mM MES pH 5.6.
10. 150 mM acetosyringone dissolved in dimethyl sulfoxide. Store at −25 °C.

2.3 Growth Condition of Nicotiana benthamiana

1. *Nicotiana benthamiana* are grown on a commercial soil mixture (Sakata Seed Co., Yokohama, Japan) in glasshouse at 25 ± 5 °C.
2. *N. benthamiana* plants at the six- to nine-leaf stages infiltrated with *Agrobacterium*.
3. After agroinfiltration, transfer the plants in a growth chamber under a 22 °C cycle of 16-h days and 8-h nights.

2.4 Preparation for Western Blot

1. Screw-cap tubes for tissue disruption. 2 mL microtubes (catalog number 72.693, Salstedt, Nümbrecht, Germany) are suitable for the Micro Smash.
2. Stainless steel beads: 2 beads/tube, 5 mm diameter.
3. Liquid nitrogen.
4. A tissue disruptor such as a Micro Smash system (TOMY, Tokyo, Japan).
5. Extraction Buffer: 10 mM sodium phosphate (pH 8.0). Proteinase inhibitor such as cOmplete™ (Roche, Basel, Switzerland) is added at a final concentration of 0.1 M immediately before use. Store at 4 °C.
6. 2-Mercaptoethanol.
7. Protein molecular marker such as SeeBlue® Plus2 Pre-stained Protein Standard (Thermo Fisher Scientific).
8. NuPAGE® LDS Sample Buffer (Thermo Fisher Scientific).

9. NuPAGE® 3 to 8% Tris-Acetate Gels (Thermo Fisher Scientific).
10. NuPAGE® Tris-Acetate SDS Running Buffer (Thermo Fisher Scientific).
11. NuPAGE® 12% Bis-Tris Gels (Thermo Fisher Scientific).
12. MES SDS Running Buffer (Thermo Fisher Scientific).
13. PVDF membrane (Thermo Fisher Scientific).
14. NuPAGE® Transfer Buffer (Thermo Fisher Scientific).
15. ECL prime Western blotting reagent pack (GE healthcare Japan, Tokyo, Japan).
16. Antibodies (*see* **Note 5**).
17. Amido Black-staining solution: 0.1% (w/v) Amido Black in 45% (v/v) methanol and 10% (v/v) acetic acid.
18. Destaining solution: 90% (v/v) methanol and 2% (v/v) acetic acid.

3 Methods

3.1 Construction of Binary Vector Plasmid DNA for Expressing R or Avr Genes

1. Insert the coding sequence of the *R*-gene candidate into the pBA vector under the *Arabidopsis thaliana* ACT2 promoter along the multicloning site (*see* **Note 1**). This plasmid product is referred to as pBA-R. For example, the coding sequence of the tobamovirus resistance gene L^3 or N' was inserted using *Sse*8387I and *Sal*I or *Sse*8387I and *Kpn*I in pBA binary vector (pBA-L^3, pBA-N′, respectively) (*see* **Note 6**).
2. Insert the coding sequence of the viral *Avr* gene into the pGT vector (*see* **Note 2**). This plasmid product is referred to as pGT-Avr. For example, the coding sequence of the tobamovirus coat protein gene of the *Tobacco mosaic virus* (TMV), *Pepper mottle virus* (PMMoV) IW isolate, or PMMoV RK isolate was inserted in the pGT binary vector (pGT-TMV-CP, pGT-PIW-CP, and pGT-PMRK, respectively) (*see* **Note 7**).

3.2 Culture of Agrobacterium Harboring a Binary Vector Plasmid DNA

1. Transform *Agrobacterium tumefaciens* LBA4404 using pBA-R or pGT-Avr through electroporation. Inoculate the L-broth plate medium containing 50 mg/L kanamycin and 20 mg/L rifampicin. Incubate for 2 days at 28 °C. Inoculate 2 mL of the L-broth liquid medium containing 50 mg/L kanamycin and 20 mg/L rifampicin with a single colony of the transformant and culture for 1 day at 28 °C with vigorous shaking. Dilute the culture in a fresh 5 mL of L-broth containing 50 mg/L kanamycin and 20 mg/L rifampicin. Culture overnight at 28 °C with vigorous shaking until the optical density reaches an absorbance of 0.5 at 600 nm [A_{600}].
2. Collect the bacterial cells by centrifugation for 15 min at 5000 × *g* and suspend in 1 mL of Infiltration Buffer.

3. Transfer 700 μL of growth culture in the L-broth media into a cryogenic tube with 300 μL glycerol to make a stock of *Agrobacterium*. Stock at −80 °C. Repeat, but this time inoculate the L-broth liquid/plate medium with the frozen cell stocks.
4. Dilute 100 μL of the suspended cells in 900 μL of water, and check the A_{600} and estimate the cell density.
5. Adjust the suspended cells to $A_{600} = 1.0$ using the Infiltration Buffer.
6. Mix the *Agrobacterium* suspensions carrying pBA-R or pGT-Avr to 1:1 (final concentration ratio of $A_{600} = 0.5{:}0.5$), (*see* **Note 8**).
7. Add 1 μL of 150 mM acetosyringone to 1 mL of infiltration mixture, mix gently and leave at 25 °C for 2 h before agroinfiltration.

3.3 Agroinfiltration and Validation of Conferring a Hypersensitive Response

1. For coinfiltration of *Agrobacterium* cultures carrying *R* and *Avr* protein genes, equal volumes of cultures were mixed prior to agroinfiltration (*see* **Note 8**).
2. Use a needle-less syringe to infiltrate the *Agrobacterium* cultures into the underside of a *Nicotiana benthamiana* leaf.
3. Observe the infiltrated leaf for 1–4 days after infiltration. Cell death lesions may develop where the leaf was infiltrated. Look for the expression of a combination of R and its corresponding Avr proteins (Fig. 1; *see* **Note 9**).

	R	CP	Virurence
A	; L^3	TMV	(Avr)
B	; L^3	PMMoV ($P_{1,2}$)	(Avr)
C	; L^3	PMMoV ($P_{1,2,3,4}$)	(Vir)
D	; N′	PMMoV ($P_{1,2,3,4}$)	(Avr)
E	; N′	PMMoV ($P_{1,2}$)	(Avr)
F	; N′	TMV	(Vir)

Fig. 1 Hypersensitive response (HR) lesions in *N. benthamiana* leaf infiltrated with *Agrobacterium*. HR lesions were induced by coexpression of the L^3/N′ gene and coat protein (CP) gene of tobamovirus. Experimental settings are shown in the left diagram, in which six combinations of R and CP were coexpressed by infiltration of *Agrobacterium*. A = L^3 and CP of *Tobacco mosaic virus* (TMV), B and C = L^3 and CP of *Pepper mild mottle virus* (PMMoV) pathotypes, D and E = N′ and CP of PMMoV pathotypes, F = N′ and CP of TMV. Four days after agroinfiltration, the leaf was photographed. The dashed circles indicate the infiltrated areas

4. Collect leaf tissues to detect expressed proteins 2 days after infiltration (*see* **Note 10**). Place the plant tissue sample (20 mg) (*see* **Note 11**) and two stainless steel beads into a 2-mL screw-cap microtube. Tightly seal the cap and place the tube into liquid nitrogen.

3.4 Detection of Expressed Proteins

To confirm the expression of the transgenes in *Nicotiana benthamiana* leaves, use western blotting to detect epitope-tagged proteins.

1. Place the plant tissue sample (20 mg) (*see* **Note 11**) and two stainless steel beads into a 2-mL screw-cap microtube. Tightly seal the cap and place the tube in liquid nitrogen. Grind the sample to a fine powder using a cell disrupter (e.g., Micro Smash). Work quickly to ensure that the tissue powder remains frozen.
2. Add 50 μL of Extraction Buffer, and mix thoroughly by vigorous shaking.
3. Centrifuge at 8000 × *g* for 5 min at 4 °C.
4. Collect 21 μL of the supernatant into a new tube. Add 7.5 μL NuPAGE® LDS Sample Buffer and 1 μL 2-mercaptoethanol. Mix gently and incubate for 5 min at 70 °C.
5. SDS-PAGE using NuPAGE® gels performed as per the manufacturer's protocol (*see* **Note 12**). Applied sample volumes in each well of the NuPAGE gels are 15 μL or 5 μL for detecting R protein or Avr protein, respectively (*see* **Note 13**).
6. Transfer the proteins to PVDF membranes using the NuPAGE® Transfer Buffer according to the manufacturer's protocol.
7. Perform western blotting using the standard protocol to detect epitope-tagged protein, (Fig. 2) (*see* **Note 14**).

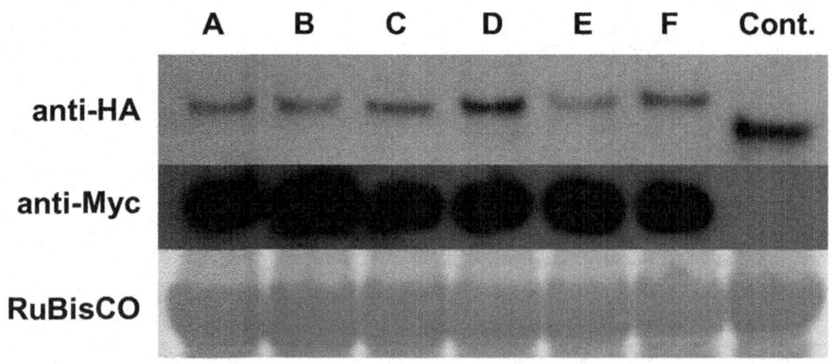

Fig. 2 Western blot of proteins extracted from *N. benthamiana* leaves transiently expressing R and Avr proteins. Proteins were detected using anti-HA antibody (top box) or anti-Myc antibody (middle box). In the bottom box, the RuBisCO large subunit protein stained with Amido Black is shown as a loading control. A = L^3-T3HA and 7 × c-Myc-CP of *Tobacco mosaic virus* (TMV), B and C = L^3-T3HA and 7 × c-Myc-CP of *Pepper mild mottle virus* (PMMoV) pathotypes, D and E = N′-T3HA and 7 × c-Myc-CP of PMMoV pathotypes, F = N′-T3HA and 7 × c-Myc-CP of TMV. Control (Cont.) = L/N′ homolog PIHX-HA and nontagged viral CP

8. As an internal control, stain posttransfer membranes with Amido Black-staining solution for a minute and destain the membranes with destaining solution for three minutes. Photograph the bands of the large RuBisCO subunit.

4 Notes

1. In case the *L³* gene was cloned into pBI121 in the downstream of the 35S promoter, *Agrobacterium* transformed with this binary plasmid could not be obtained. Probably because the 35S promoter is leaky to express the *L³* gene in *Agrobacterium*, and the L³ protein may be toxic to *Agrobacterium*. That is why the alternative promoter should be chosen. The modified binary plasmid vector pBA was generated from pBI121 [19], in which the 35S promoter was replaced with the *Arabidopsis thaliana* actin 2 promoter region and a multicloning site (comprising *Xba*I, *Sse*8387I, *Kpn*I, *Sal*I, and *BamH*I restriction sites) in place of the GUS gene expression unit. pBA is available from K.T. Sekine (k-sekine@agr.u-ryukyu.ac.jp) or K. Kobayashi (kappei@ehime-u.ac.jp).
2. The vector pGT vector was generated from pGreen0000 [19, 22]. For expression of tobamovirus coat protein genes, pBI121 can also be used. The vector pBTPIW can be used for the expression of tobamovirus coat protein genes with PMMoV genome under the control of a tandem 35S promoter [19]. pGT and pBTPIW are available from K.T. Sekine (k-sekine@agr.u-ryukyu.ac.jp) or K. Kobayashi (kappei@ehime-u.ac.jp).
3. Although the cucumber mosaic virus resistance gene, *RCY1*, for example, can be fused with epitope tags at its N- and C-terminal without the loss of function [14, 23], the N/C--terminal epitope tags compromised the function of L³ and its homologs [20]. For detection of L³ protein, we constructed the pBA-L³-T3HA vector, in which a HA tag sequence was inserted in the predicted turn structure between CC and NB regions (for details, *see* next chapter). Similarly, pBA-N′-T3HA vector was designed for the expression of N′ fused with a HA tag. For detection of Avr proteins, the tobamovirus CP can be detected using commercial antibodies against tobamovirus spp. We also constructed an expression vector of CP fused with 7 × c-Myc tag in its N-terminal based on the pBI121 vector (*see* **Note 7**).
4. For the transformation of *Agrobacterium* with the pGT vector, LBA4404 carrying pSoup plasmid must be used.
5. Suitable primary/secondary antibodies should be used depending on the expressed protein or the fused epitope tag.

6. To evaluate the R protein accumulation after the agroinfiltration, fuse an epitope tag sequence to the *R* gene coding sequence as an option. For validation of the function of the candidate R protein, use the natural coding sequence without any epitope tag at first.
7. To evaluate the Avr protein accumulation after the agroinfiltration, fuse an epitope tag sequence to the *Avr* gene coding sequence as an option. Although antibodies against tobamovirus CP are also useable for this purpose, each CPs in different tobamovirus species can be detected by same antibody if an epitope tag is fused. We constructed pBI-7Myc-TMCP, pBI-7Myc-PIWCP, and pBI-7Myc-PMRKCP, in which CPs of TMV, PMMoV IW isolate, and PMMoV RK isolate, respectively, fused with tandem 7 × c-Myc in its N-terminal and were cloned under control of the 35S promoter in pBI121.
8. When CP is expressed from pBTPIW vector, *Agrobacterium* culture can be reduced. Mix the *Agrobacterium* suspensions carrying either pBA-R or pBTPIW at 1:0.01.
9. In combination of R and the corresponding Avr coexpressed by the transient expression assay, HR confers in the area infiltrated with *Agrobacterium*. L^3 recognizes TMV and PMMoV IW isolates but PMMoV RK isolate. On the other hand, N′ recognizes PMMoV IW and RK isolates but TMV. In Fig. 1, cell-death lesions were observed in these combinations of R-Avr coexpression. Timing and intensity of cell death depends on accumulation amounts of R and Avr proteins and also the plant growth condition. Also take notice that an epitope tag such as 7 × c-Myc in the expressed protein sometimes disturbs the R–Avr interaction resulting in conferring weaken HR.
10. It is better to collect leaf tissues before appearance of visible HR.
11. 20 mg is the smallest weight for performing western blot to detect R and Avr proteins.
12. We use the NuPAGE® 3 to 8% Tris-Acetate Gel and NuPAGE® 12% Bis-Tris Gel for detecting R and CP proteins, respectively.
13. R proteins such as L^3 and N′ are hard to detect because the accumulated protein amounts are lower than CP proteins.
14. In western blot detecting the HA-tagged R proteins, we use anti-HA antibody (Roche) as a primary antibody and anti-rat antibody HRP conjugated (Bio-Rad) as a secondary antibody. For the c-Myc-tagged CP proteins, we use anti-c-Myc antibody (Santa Cruz Biotechnology, TX, USA) as a primary antibody and anti-mouse antibody HRP conjugated (Bio-Rad) as a secondary antibody.

Acknowledgments

This work was supported in part by JSPS KAKENHI grant numbers 24780043 and 17K15230.

References

1. Jones JD, Dangl JL (2006) The plant immune system. Nature 444:323–329
2. Cook DE, Mesarich CH, Thomma BP (2015) Understanding plant immunity as a surveillance system to detect invasion. Annu Rev Phytopathol 53:541–563
3. Heath MC (2000) Hypersensitive response-related death. Plant Mol Biol 44:321–334
4. Collier SM, Moffett P (2009) NB-LRRs work a "bait and switch" on pathogens. Trends Plant Sci 14:521–529
5. Whitham S, Dinesh-Kumar SP, Choi D, Hehl R, Corr C, Baker B (1994) The product of the tobacco mosaic virus resistance gene *N*: similarity to toll and the interleukin-1 receptor. Cell 78:1101–1115
6. de Ronde D, Butterbach P, Kormelink R (2014) Dominant resistance against plant viruses. Front Plant Sci 5:307
7. Takken FL, Tameling WI (2009) To nibble at plant resistance proteins. Science 324: 744–746
8. Meyers BC, Kozik A, Griego A, Kuang H, Michelmore RW (2003) Genome-wide analysis of NBS-LRR-encoding genes in *Arabidopsis*. Plant Cell 15:809–834
9. Zhou T, Wang Y, Chen JQ, Araki H, Jing Z, Jiang K, Shen J, Tian D (2004) Genome-wide identification of NBS genes in *japonica* rice reveals significant expansion of divergent non-TIR NBS-LRR genes. Mol Gen Genomics 271:402–415
10. Kohler A, Rinaldi C, Duplessis S, Baucher M, Geelen D, Duchaussoy F, Meyers BC, Boerjan W, Martin F (2008) Genome-wide identification of *NBS* resistance genes in *Populus trichocarpa*. Plant Mol Biol 66:619–636
11. Jupe F, Pritchard L, Etherington GJ, Mackenzie K, Cock PJ, Wright F, Sharma SK, Bolser D, Bryan GJ, Jones JD, Hein I (2012) Identification and localisation of the NB-LRR gene family within the potato genome. BMC Genomics 13:75
12. Whitham S, McCormick S, Baker B (1996) The *N* gene of tobacco confers resistance to tobacco mosaic virus in transgenic tomato. Proc Natl Acad Sci U S A 93:8776–8781
13. Takahashi H, Miller J, Nozaki Y, Takeda M, Shah J, Hase S, Ikegami M, Ehara Y, Dinesh-Kumar SP (2002) *RCY1*, an *Arabidopsis thaliana RPP8/HRT* family resistance gene, conferring resistance to cucumber mosaic virus requires salicylic acid, ethylene and a novel signal transduction mechanism. Plant J 32: 655–667
14. Sekine KT, Kawakami S, Hase S, Kubota M, Ichinose Y, Shah J, Kang HG, Klessig DF, Takahashi H (2008) High level expression of a virus resistance gene, *RCY1*, confers extreme resistance to *Cucumber mosaic virus* in *Arabidopsis thaliana*. Mol Plant-Microbe Interact 21:1398–1407
15. Bendahmane A, Querci M, Kanyuka K, Baulcombe DC (2000) *Agrobacterium* transient expression system as a tool for the isolation of disease resistance genes: application to the *Rx2* locus in potato. Plant J 21:73–81
16. Mestre P, Baulcombe DC (2006) Elicitor-mediated oligomerization of the tobacco N disease resistance protein. Plant Cell 18: 491–501
17. Rairdan GJ, Moffett P (2006) Distinct domains in the ARC region of the potato resistance protein Rx mediate LRR binding and inhibition of activation. Plant Cell 18:2082–2093
18. Sacco MA, Mansoor S, Moffett P (2007) A RanGAP protein physically interacts with the NB-LRR protein Rx, and is required for Rx-mediated viral resistance. Plant J 52: 82–93
19. Tomita R, Sekine KT, Mizumoto H, Sakamoto M, Murai J, Kiba A, Hikichi Y, Suzuki K, Kobayashi K (2011) Genetic basis for the hierarchical interaction between *Tobamovirus* spp. and *L* resistance gene alleles from different pepper species. Mol Plant-Microbe Interact 24:108–117
20. Sekine KT, Tomita R, Takeuchi S, Atsumi G, Saitoh H, Mizumoto H, Kiba A, Yamaoka N, Nishiguchi M, Hikichi Y, Kobayashi K (2012) Functional differentiation in the Leucine-rich repeat domains of closely related plant virus-resistance that recognize common Avr proteins. Mol Plant-Microbe Interact 25: 1219–1229

21. Hamel LP, Sekine KT, Wallon T, Sugiwaka Y, Kobayashi K, Moffett P (2016) The chloroplastic protein THF1 interacts with the coiled-coil domain of the disease resistance protein N′ and regulates light-dependent cell death. Plant Physiol 171:658–674

22. Hellens RP, Edwards EA, Leyland NR, Bean S, Mullineaux PM (2000) pGreen: a versatile and flexible binary Ti vector for *Agrobacterium*-mediated plant transformation. Plant Mol Biol 42:819–832

23. Takahashi H, Shoji H, Ando S, Kanayama Y, Kusano T, Takeshita M, Suzuki M, Masuta C (2012) *RCY1*-mediated resistance to *Cucumber mosaic virus* is regulated by LRR domain-mediated interaction with CMV(Y) following degradation of RCY1. Mol Plant-Microbe Interact 25:1171–1185

Chapter 2

Insertion of Epitope Tag into NB-LRR Class Plant Virus Resistance Protein

Kappei Kobayashi, Reiko Tomita, and Ken-Taro Sekine

Abstract

It is prerequisite to detect plant disease resistance proteins for studying the function of the proteins. Numerous studies have used epitope tags fused to either N- or C-terminus for the detection of resistance proteins. However, some resistance proteins do not tolerate the terminal fusions of epitope tags. In this chapter, we provide a protocol for searching the protein regions in which the inserted epitope tag does not affect the protein function. In the protocol, we first perform an in silico search to select the insertion site candidates and then insert there a short sequence containing restriction sites to find out the sites, in which the insertion does not affect the protein function. Epitope tags are inserted into the experimentally selected sites to produce a functional protein with an epitope tag.

Key words Agroinfiltration, Epitope tag, Protein function, Protein structure, Resistance protein

1 Introduction

In plant–virus interaction research, the study of virus resistance genes has been playing a pivotal role. The study of *N* gene, which is the first cloned plant virus resistance gene, has provided us with the numerous novel knowledge not only on the structure of resistance protein [1] but also the identification of viral avirulence protein [2] and downstream signal transduction pathways [3–7], the perception of viral effector [8], and the importance of alternative splicing [9] and oligomerization of resistance protein [10] in the resistance function and the signal transduction, respectively. In addition to the *N* gene, several other virus resistance genes have contributed to improving our understanding of virus resistance mechanisms. In potato Potexvirus resistance Rx protein, studies have shown that intramolecular interactions within coiled-coil (CC), nucleotide binding-ARC (Apoptosis, *R*-gene products, and CED4), and leucine-rich repeat (LRR) domains [11–13], and that between Rx and host protein RanGAP have crucial roles in

Kappei Kobayashi and Masamichi Nishiguchi (eds.), *Antiviral Resistance in Plants: Methods and Protocols*, Methods in Molecular Biology, vol. 2028, https://doi.org/10.1007/978-1-4939-9635-3_2,

pathogen perception by the resistance protein [14, 15]. In RCY1, a *Cucumber mosaic virus* resistance protein of *Arabidopsis thaliana*, the importance of resistance protein degradation has been documented in the downstream signaling after pathogen perception [16]. In these pioneering research works, especially those which demonstrated the importance of protein–protein interactions and accumulation of resistance proteins, it was essential that the proteins and their interaction partners can be epitope-tagged without affecting their function.

As carefully discussed in the review by Jarvik and Telmer [17], many proteins accept epitope tags at their termini without severe loss of function. Indeed, the aforementioned resistance proteins are epitope-tagged at either of their termini. When we cloned pepper *L* gene alleles, however, our trial to tag them at either terminus were not successful at all as briefly described in another paper reporting the cloning and characterization of *Nicotiana sylvestris* *N′* gene [18] (*see* Chapter 1). In the aforementioned review [17], the insertion of short peptides or epitope tags into random sites within the polypeptides does not necessarily affect the protein function. The literature searches 20 years ago found that more than a half, in average in 16 different proteins, of randomly inserted short peptides did not severely affect the protein function. In the case of *Escherichia coli* β-galactosidase [19] and lac repressor [19], however, most of the insertion mutant proteins severely lost their functions. These studies clearly demonstrated that the tolerance of proteins to short insertions depends on the nature of proteins and that we need to test different sites when we want to insert an epitope tag into a protein, which is affected by both N- and C-terminal fusions of epitope tags.

In this chapter, we provide an example of a protocol for inserting an epitope tag into L resistance protein. The protocol comprises (1) selection of insertion site candidates through a secondary structure prediction; (2) insertion of 18-nt (6-aa) sequence containing two restriction sites separated with 6-nt stuffer; (3) functional analysis of resistance proteins with 6-aa insertion; (4) insertion of epitope tag sequences; (5) functional analysis of epitope-tagged resistance proteins. In the selection of insertion sites, we avoided α-helixes and β-sheets because these structures play pivotal roles in protein structure and thus in protein functioning. Two out of 12 candidates tested were found to be tolerant to the insertion of epitope tags. Although the success rate was not very high, the protocol would provide a wide range of investigators, who want to study the function, accumulation, degradation, and protein–protein interactions of proteins involved in the virus resistance as well as other proteins, with a clue to solve the problem of nonfunctional terminally tagged proteins.

2 Materials

2.1 Designing and Construction of Resistance Gene with Inserted Restriction Sites

1. Computer system installed with a protein secondary structure prediction software (*see* **Note 1**).
2. pBAL3: A binary vector construct of L^3 gene [20] (*see* **Note 2**).
3. Oligonucleotide primers for site-directed insertion.

L3-S5F	GGCGCTGAAAACTTAATGGA
L3-S8R	TCGGCAGAGTGCTAATAGGC
L3-S13F	AGTGATGCGGACAATTCACA
PIHc-SalR	gtcgacTCACAGGCATTCACAGTCAA (restriction site in lowercase)
LTC1-R	cccgggactagacctaggATCACTCAAGTTCAGTTTGC*
LTC1-F	cctaggtctagtcccgggGATTATTTTCTCGACATAAAG*
LTC2-R	cccgggactagacctaggTGTTCTAGTTTCTAGTTTCT*
LTC2-F	cctaggtctagtcccgggTCAACTTCTTTGGTCGATGA*
LTC3-R	cccgggactagacctaggATCACTTGACAATAAACGGT*
LTC3-F	cctaggtctagtcccgggTCAAATGGAGAAAATCTGAC*
LTC4-R	cccgggactagacctaggATGCAACTGTTGTACCAGAC*
LTC4-F	cctaggtctagtcccgggCATTCTGGTAACCAATACTT*
LTC5-R	cccgggactagacctaggACCGTATCTTTCAGAAGACT*
LTC5-F	cctaggtctagtcccgggGGGAAATTCTTAATGCATGA*
LTC6-R	cccgggactagacctaggTGAAGATGCAATTTGGGCCA*
LTC6-F	cctaggtctagtcccgggAAACTTTGTGTCAGGTTGGA*
LTC7-R	cccgggactagacctaggTTGGCACTCTTCCAACCTGA*
LTC7-F	cctaggtctagtcccggggGATCTCATATATTGGAACA*
LTC8-R	cccgggactagacctaggCATTGAATACGATGTGTGCC*
LTC8-F	cctaggtctagtcccgggAGAGATGGTGACTTTGAGAAA*
LTC9-R	cccgggactagacctaggCTTAGAGAGCGGTTTCAATT*
LTC9-F	cctaggtctagtcccgggTCAGAGCAGCTGAGGACATT*
LTC10-R	cccgggactagacctaggACCATAGAATTCTTCCGTCA*
LTC10-F	cctaggtctagtcccgggAGTCCGTCCTCTGAAAAGCC*
LTC11-R	cccgggactagacctaggATCTGGTGCCTCCAATTTCA*
LTC11-F	cctaggtctagtcccgggAGTAGTAGGATGATTTCAGA*
LTC12-R	cccgggactagacctaggTCCATCAGGAAAGGACTCTA*
LTC12-F	cctaggtctagtcccgggGGATTGCCCTTCAATTTACA*

*Restriction sites and stuffer sequence are shown in lowercase

4. High fidelity PCR enzyme: PrimeSTAR GXL DNA polymerase (Takara Bio) or equivalent. Supplied with buffer and dNTP mixture.
5. Gel purification kit.
6. PCR cloning kit: TOPO Zero blunt PCR cloning kit (Thermo) or equivalent.
7. Competent *E. coli*.
8. LB medium: 1% (w/v) tryptone, 0.5% (w/v) yeast extract, 1% (w/v) NaCl. Adjust pH to 7.0 with 1 N NaOH. Sterilize by autoclave.
9. Kanamycin: 50 mg/mL stock solution.
10. LB-K50 plates: Add 1.5% (w/v) agar to LB medium and sterilize by autoclave. After cooling down to 55 °C, add 50 mg/L (final concentration) kanamycin and pour into plastic petri dishes.
11. Incubator at 37 °C.
12. Shaking incubator at 37 °C.
13. Miniprep kit.
14. Restriction enzymes: BglII, BlnI, EcoRI, SacI, SalI, SmaI, SpeI, XhoI.
15. Sequencing facility.
16. Sequencing primers (Nucleotide positions in the L^3 gene sequence, accession No. AB523370).

L3-S1R	TCACAGGCATTCACAGTCAAAC	(3987–3966)
L3-S2R	TTGGATGGGTGTCTCCAAAT	(2733–2714)
L3-S3R	GCAGTGACCGAATTTGAGGT	(3550–3531)
L3-S4R	TAGACGCCACTCCTTTCGAC	(3345–3326)
L3-S6F	AAAGCAAATCGGTGACCTTG	(435–454)
L3-S7F	TTGTCCCTATTGTTGGAATGG	(611–631)
L3-S9R	CAGGCATCTCTGCAAATTCA	(2566–2547)
L3-S10F	TATGCCGCAAATCAGAAGTG	(1163–1182)
L3-S11F	CACCTGTGGATTGCTAATGG	(1357–1376)
L3-S12F	CAAGAGGGTGCTGCATAACA	(1707–1726)
L3-S13F	AGTGATGCGGACAATTCACA	(2233–2252)
L3-S14F	TCTGGTTCTTCTAAGGCTGGA	(2761–2781)
L3-S15F	TGGCAGTGACGAAGAGATTG	(3387–3406)
L3-S16R	AGCACTTGGAGGCTTTTCAA	(2039–2020)
L3-S17R	AAGACTCTGGGACCCTTTCG	(1459–1440)
L3-S18R	TCATCCCACTCATCACAGTCA	(902–882)

17. Alkaline phosphatase: Molecular biology grade. Alkaline Phosphatase (Shrimp) (Takara Bio) or equivalent.
18. Ligation kit: Ligation kit ver. 2.1 (Takara Bio) or equivalent.
19. 1% agarose gel in TAE buffer.
20. Agarose gel electrophoresis apparatus.

2.2 Functional Analysis of Resistance Gene Clones with Inserted Restriction Sites

1. Agrobacterium: *Agrobacterium tumefaciens* GV3101 strain.
2. Viral agroinfection clones: pBTPIW and pBTPIW-ToCP [18] (*see* **Note 2**).
3. L-broth: 1% (w/v) tryptone, 0.5% (w/v) yeast extract, 0.5% (w/v) NaCl, 0.1% (w/v) glucose. Sterilize by autoclave.
4. Antibiotics: 50 mg/mL kanamycin, 20 mg/mL rifampicin, 20 mg/mL gentamycin.
5. L broth R20G20 plates: Add 1.5% (w/v) agar to L-broth and sterilize by autoclave. After cooling down to 55 °C, add 20 mg/L (final concentration) rifampicin and 20 mg/L (final concentration) gentamycin, and pour into plastic petri dishes.
6. L broth R20G20: Add 20 mg/L (final concentration) rifampicin and 20 mg/L (final concentration) gentamycin.
7. 500 mL Erlenmeyer flask, sterilized.
8. 10% glycerol: Sterilize by autoclave. Chill on ice before use.
9. Electroporation cuvettes: 2 mm gap between electrodes.
10. Electroporation apparatus.
11. L broth R20G20K50 plates: Add 1.5% (w/v) agar to L-broth and sterilize by autoclave. After cooling down to 55 °C, add 50 mg/L (final concentration) kanamycin, 20 mg/L (final concentration) rifampicin, and 20 mg/L (final concentration) gentamycin, and pour into plastic petri dishes.
12. L broth R20G20K50: Add 50 mg/L (final concentration) kanamycin, 20 mg/L (final concentration) rifampicin, and 20 mg/L (final concentration) gentamycin to L broth.
13. 45% glycerol: Dilute glycerol with deionized water and sterilize by autoclave.
14. Seeds: *Nicotiana benthamiana* seeds.
15. Plastic petri dish: 90 mm in diameter.
16. Filter papers: Round sheets of qualitative filter paper, 70 mm in diameter.
17. Planting pots: Small plastic pots containing soil (*see* **Note 3**).
18. Liquid fertilizer: Hyponex or equivalent.
19. Spectrophotometer.
20. Disposable plastic cuvettes.

21. Infiltration buffer: 10 mM K-MES pH 5.5 (2-(*N*-morpholino) ethanesulfonic acid, adjusted to pH 5.5 with KOH), 10 mM $MgCl_2$. Sterilize by autoclave.
22. 20 mg/mL acetosyringone: Dissolved in dimethyl sulfoxide.
23. 1 mL syringe: Needles are not required.

2.3 Insertion of Epitope-Tag to Resistance Gene Coding Sequence and Its Functional Analysis

1. Oligonucleotides with 5′-phosphate group: BglII sites partially overlapping with XbaI cohesive ends (compatible with BlnI cohesive end) are shown in lowercase (*see* **Note 4**).

HA-U	CTagatctTATCCATATGATGTTCCAGATTATGC TAGTCCG
HA-N	CGGACTAGCATAATCTGGAACATCATATGGATAagat
Myc-U	CTagatctGAGCAGAAGTTGATCTCAGAGGAGGAC TTGAGTCCG
Myc-N	CGGACTCAAGTCCTCCTCTGAGATCAACTTCTG CTCagat
FLAG-U	CTagatctGACTACAAGGATGACGATGACAAGAGTCCG
FLAG-N	CGGACTCTTGTCATCGTCATCCTTGTAGTCagat

2. Annealing buffer: Ten times concentrated. 100 mM, Tris–HCl (pH 7.5), 100 mM $MgCl_2$, 10 mM dithiothreitol, 500 mM NaCl.
3. Cloning reagents as in Subheading 2.1, **items 6–17**.
4. PCR purification kit.
5. Agroinfiltration reagents as in Subheading 2.2, **items 1–23**.

3 Methods

3.1 Designing and Construction of Resistance Gene with Inserted Restriction Sites

3.1.1 Selection of Insertion Site Candidates in the Coding Region of a Resistance Gene

In contrast to the aforementioned classic studies inserted short peptides into random sites, we selected several insertion site candidates based on the predicted secondary structure to minimize the possible effect of insertion on the resistance protein function.

1. Identify protein motifs and domains in the resistance protein of interest (e.g., L^3 protein in this example) (Fig. 1) through the search for conserved motifs.
2. Predict the secondary structure of L^3 protein using some software and/or web services (Fig. 2) (*see* **Note 1**).
3. Select insertion site candidates (Fig. 3a) based on the following criteria.
   (a) Short stretches of hydrophilic amino acids.
   (b) Short stretches of amino acids locating within putative linker regions that connect distinct domains.

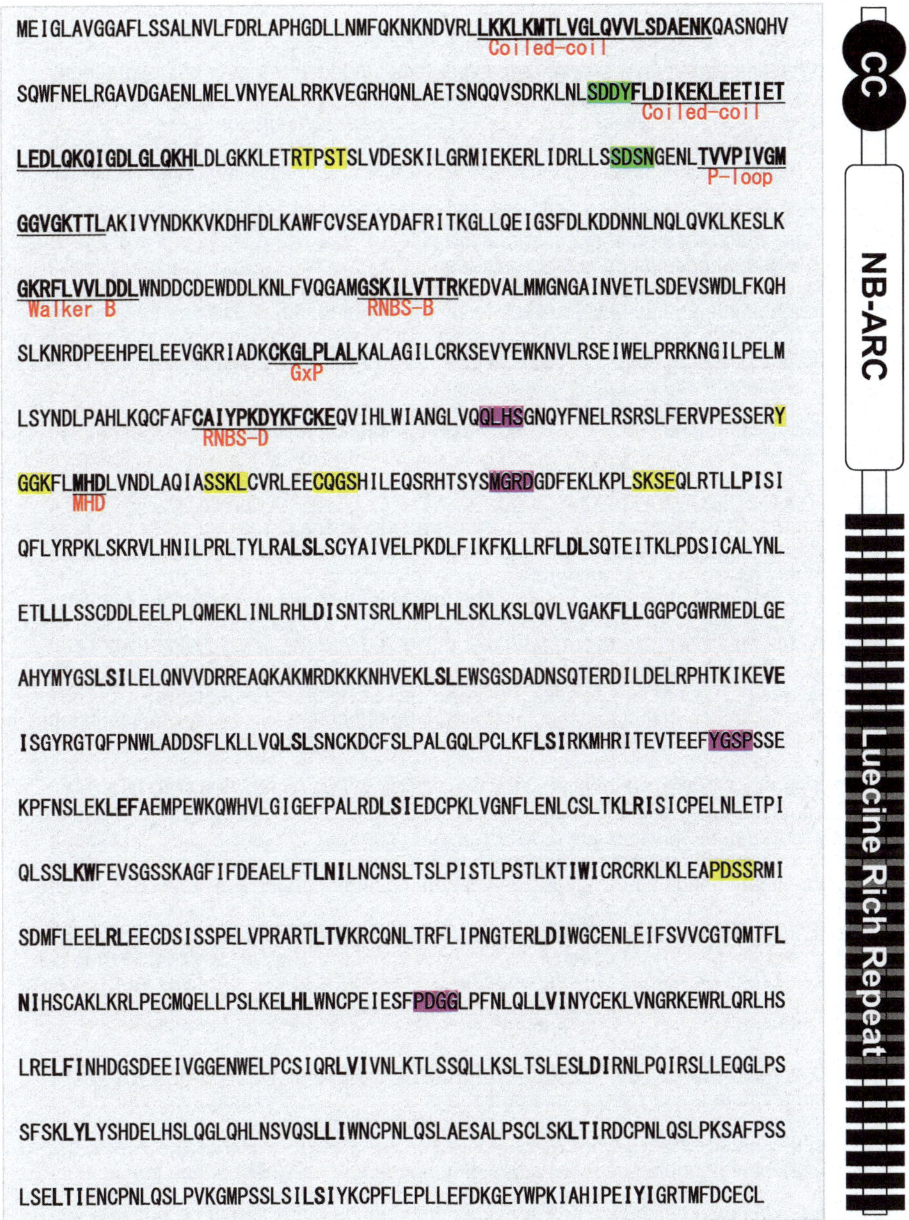

Fig. 1 Domain structure, amino acid sequence, functional motifs, and insertion site candidates in L[3] tobamovirus resistance protein. Domains of L[3] resistance protein: CC, coiled-coil domain; NB-ARC, Nucleotide binding-ARC domain; LRR, leucine-rich repeat domain. Amino acids sequence motifs are as described previously [20]. Two amino acids each in upstream and downstream of the insertion site candidates are highlighted: green, those which tolerate the insertion of epitope tags; yellow, those which tolerate the insertion of six amino acids but not epitope tags; magenta, those which do not tolerate even the six amino acids insertion

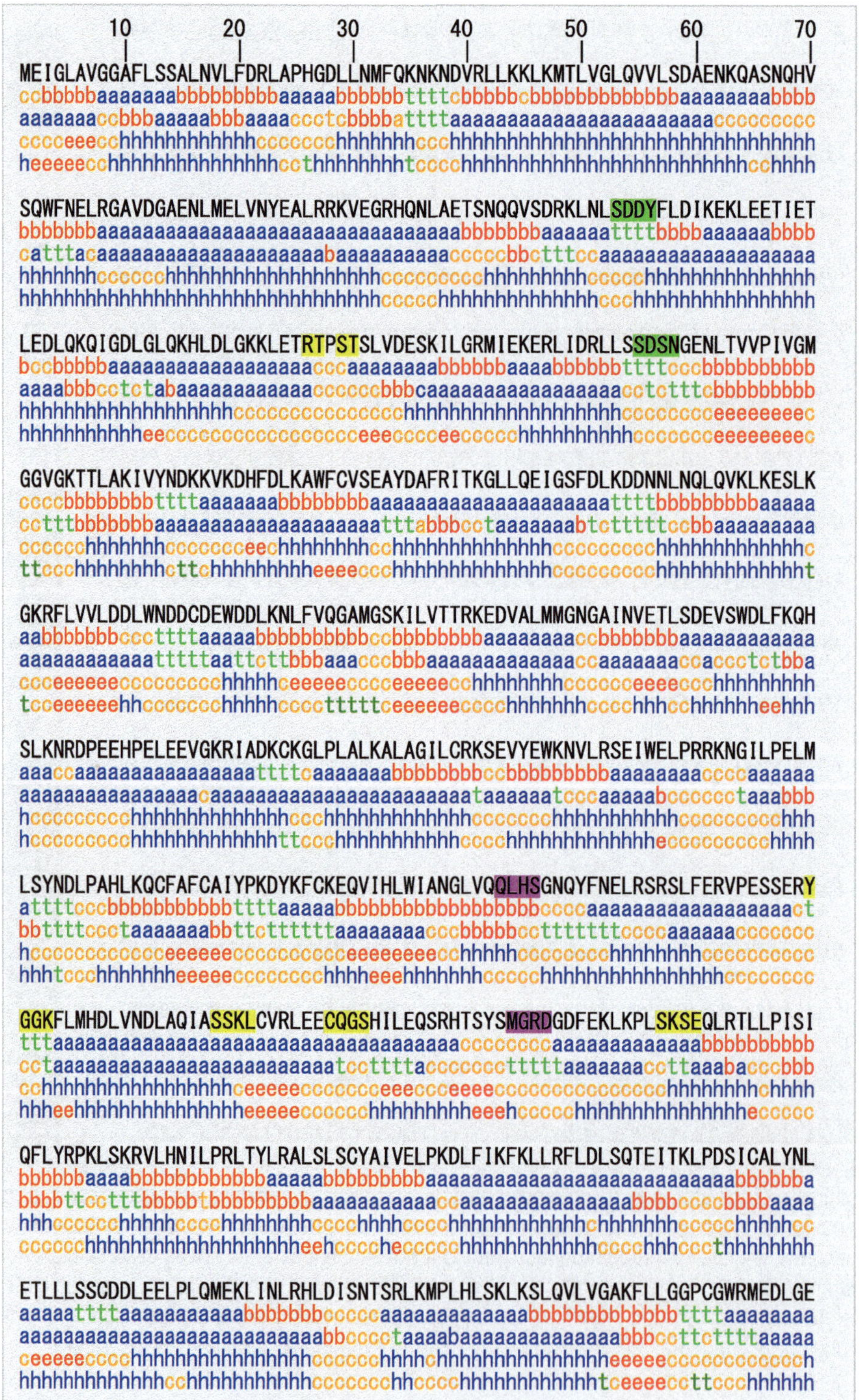

Fig. 2 Predicted secondary structures and insertion sites in L^3 tobamovirus resistance protein. Amino acid sequences are shown in black letters on the top, followed by Chou-Fasman prediction in GENETYX, Robson prediction in GENETYX, sopma prediction and GOR4 prediction (*see* **Note 1**). The predicted secondary structures are shown: t in green, turn structure; c in orange, random coil structure; b or e in red, beta-sheet structure; a or h in blue, alpha-helix structure. The insertion sites are shown as in Fig. 1

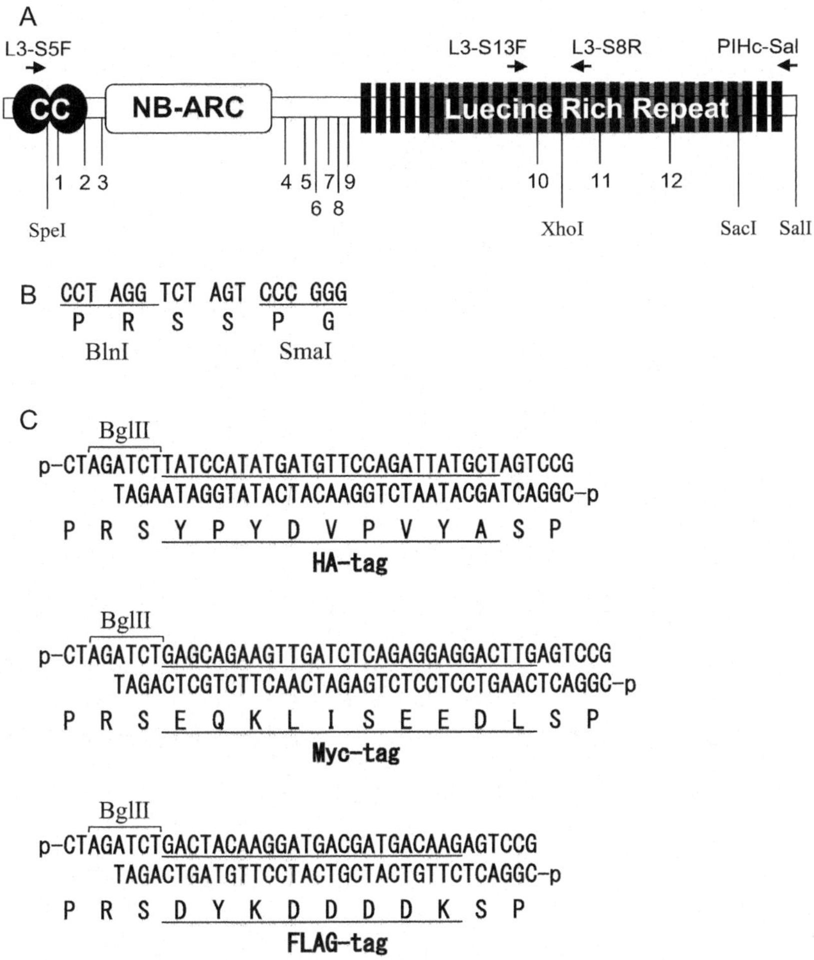

Fig. 3 Schematic representation of insertion sites, PCR primers and restriction sites (**a**), and the structures of insertion site (**b**) and the epitope tags (**c**)

(c) Short stretches of amino acids in which random coil or turn structures are predicted rather than α-helixes and β-sheets.

4. Design primers as follows:

   Forward primers, insertion sequence (Fig. 3b) + 20-nt downstream sequence. Reverse primers, complementary insertion sequence +20-nt complementary upstream sequence. In this example, the primers were named LTC (L protein Tagging Candidates) + (serial number of candidates)-F or -R (Forward and Reverse) (*see* Subheading 2.1, **item 3**).

3.1.2 Construction of Expression Vector for Resistance Genes with Inserted Sequence

1. Set up PCR reaction with pBAL3 as a template and high fidelity PCR enzyme according to manufacturer's protocol. The outer primer pairs (L3-S5F and L3-S8R, or L3-S13F and PIHc-SalR) (Fig. 3) should be able to amplify a DNA fragments containing two unique restriction sites. The insertion site primers should be used with the outer primer pairs to amplify a DNA fragment, in which an insertion site locates between two unique restriction sites. For example, the first PCR reactions for the upstream and downstream fragments of the LTC1 insertion site contain L3-S5F/LTC1-R and LTC1-F/L3-S8R, respectively. The PCR reaction can be carried out under the following condition: 94 °C, 2 min; 25 cycles of 95 °C for 10 s, 55 °C for 10 s, and 68 °C for 1 min; and 68 °C for 3 min, but should be optimized for the PCR enzyme and primers.
2. Check the amplification on a 1% agarose gel, gel-purify the amplification products using a gel purification kit.
3. Mix approximately the same molar amount of the upstream and downstream fragments of each insertion site. Set up the second PCR reaction using the mixture as a template, the outer primer pairs used in the first PCR, and high fidelity PCR enzyme according to manufacturer's protocol. The PCR reaction can be carried out under the following condition: 94 °C, 2 min; 95 °C for 10 s, 50 °C for 1 min, and 68 °C for 1 min; 24 cycles of 95 °C for 10 s, 50 °C for 10 s, and 68 °C for 1 min; and 68 °C for 3 min.
4. Check the amplification on a 1% agarose gel, and clone the fragments using a PCR cloning kit. Transform *E. coli* and pick up some colonies and culture them overnight at 37 °C in 2 mL LB medium containing 50 mg/L Kanamycin for plasmid DNA extraction.
5. Prepare plasmid DNA from overnight cultures using a miniprep kit according to the manufacturer's protocol and confirm the insert by restriction analysis with EcoRI (when using the TOPO PCR cloning kit).
6. Confirm the sequence of the clones using appropriate sequencing primers.
7. Digest 1 μg each of plasmid DNA of selected clones and pBAL3 with SpeI/XhoI and XhoI/SacI for the insertion sites introduced using outer primer pairs L3-S5F/LTC1-R and LTC1-F/L3-S8R, respectively. Include alkaline phosphatase in the reaction for pBAL3.
8. Inactivate the alkaline phosphatase, confirm the digestion on an agarose gel, and gel purify the DNA fragments that contain the insertion sites.

9. Ligate the gel purified DNA fragments with an appropriate vector. Transform *E. coli* and pick up some colonies and culture them overnight at 37 °C in 3 mL LB medium containing 50 mg/L kanamycin for plasmid DNA extraction.
10. Prepare plasmid DNA from overnight cultures using a mini-prep kit according to the manufacturer's protocol and confirm the insertion sites by the double digestion with SalI and BnlI, which locates only in the insertion sites.

3.2 Functional Analysis of Resistance Gene Clones with Inserted Restriction Sites

1. Sow seeds of *N. benthamiana* onto two sheets of filter paper wet with tap water and place in a plastic dish. Culture the plate in a plant growth room or growth chamber at 25 °C with 16 h light (50–100 μmol m^{-2} s^{-1}) and 8 h dark cycle.
2. Transfer the tiny seedlings to a potting soil within a week. Grow the plants under the same condition until the four to six true leaves fully expand, which may take a few weeks. Water the plants as necessary with liquid fertilizer once a week.
3. Prepare electrocompetent cells of Agrobacterium. Inoculate a single Agrobacterium colony formed on L broth R20G20 plate into 100 mL of L broth R20G20 in a 500 mL flask, culture overnight at 30 °C with vigorous shaking. Wash the Agrobacterium cells three times by spinning at 4000 × *g* for 5 min at 4 °C, and resuspending the pellet in 50 mL of ice-cold 10% glycerol. Finally, resuspend the cells in ice-cold 1 mL of 10% glycerol, aliquot into 40 μL and store at −80 °C.
4. Transform Agrobacterium with pBAL3, its derivatives with the insertion sites, and viral agroinfection clones. Add about 50 ng purified plasmids to electrocompetent Agrobacterium cells, transfer to a 2-mm gap electroporation cuvette on ice, and apply a pulse with the following setting: 2.5 kV, 25 μF and 400 Ω. Plate the Agrobacterium onto L broth R20G20K50 plates and culture for 2 days at 28 °C.
5. Pick up colonies and culture in 2 mL L broth R20G20K50 overnight at 28 °C.
6. Inoculate 2 mL fresh L broth R20G20K50 containing 4 mg/L acetosyringone with 20 μL of the overnight culture and culture again overnight at 28 °C. (Optional) Mix 1.5 mL of the culture with an equal volume of 45% glycerol to prepare glycerol stocks.
7. Transfer the overnight culture to a 2 mL microfuge tube, spin down the bacteria at 2300 × *g* for 5 min at room temperature. Resuspend the bacteria in 1 mL infiltration buffer, and measure the optical density (OD) at 600 nm.
8. Adjust the Agrobacterium harboring the resistance gene or its derivatives to OD = 1 with the infiltration buffer containing 30 mg/L acetosyringone. Adjust the Agrobacterium containing

viral clone to OD = 0.1 with the infiltration buffer, and add 10 μL of this bacterial suspension to 1 mL of the suspension of Agrobacterium harboring the resistance gene or its derivatives. Let stand at room temperature for at least 2 h.

9. Infiltrate the mixed bacterial suspensions into the intercellular space of fourth to sixth true leaves of *N. benthamiana* plants. In all the infiltrated leaves, a control experiment with wild-type L^3 gene and pBTPIW should be included. Grow the plants for 4–5 days and observe for the induction of hypersensitive cell death (*see* **Note 5**).

3.3 Insertion of Epitope-Tag to Resistance Gene Coding Sequence and Its Functional Analysis

1. Anneal the oligonucleotides (HA-U and -N, Myc-U and -N, FLAG-U and -N) in the following mixture: 5 μL of annealing buffer, 5 μL ultrapure water, 20 μL each of 250 μM oligonucleotides complementary to each other. Heat in boiling water for 5 min, transfer to a Styrofoam box containing about 3 L of hot water at 75 °C, place a lid and let cool under room temperature overnight.
2. Digest the clone(s) of resistance gene expression vector with an insertion site from Subheading 3.2, **step 9** first with SmaI. Purify the plasmid DNA using PCR purification kit. Digest the SmaI-digested plasmid with BlnI. Add alkaline phosphatase to BlnI reaction after 2 h of digestion, incubate for 30 min, heat-inactivate alkaline phosphatase, and purify the plasmid.
3. Ligate the double-digested plasmid with the annealed oligonucleotides. Transform *E. coli*, spread on LB-K50 plate, and culture overnight at 37 °C.
4. Pick up some colonies, culture overnight at 37 °C in 2 mL LB medium containing 50 mg/L kanamycin and extract the plasmid DNA. Examine the insertion of epitope tags by BglII digestion and confirm the sequence.
5. Transform Agrobacterium with the pBAL3 derivatives inserted with the epitope tags and examine their function as described in Subheading 3.2, **steps 4–9** (*see* **Note 6**).

4 Notes

1. We initially used an old version of software, GENETYX-MAC version 12 (GENETYX CORP, Tokyo, Japan) for predicting the secondary structure L^3 protein with Chou–Fasman method under default settings. We selected the insertion site candidates based on the criteria above but include those do not fit to the criteria. In this protocol, we present additional prediction by another method, Robson method in the same software and by web services, sopma (https://npsa-prabi.ibcp.fr/cgi-bin/npsa_automat.pl?page=npsa_sopma.html) and GOR4 (https://

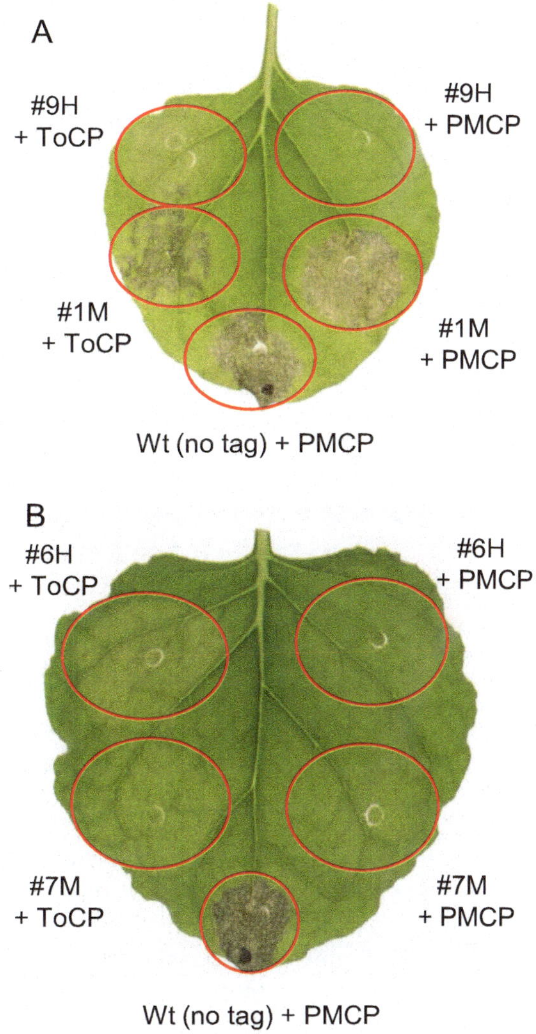

Fig. 4 Representative results of the agroinfiltration-mediated functional assay of L[3] protein with inserted epitope tags. Wild-type [Wt (no tag)] and epitope-tagged L[3] proteins (#9H, HA-tag inserted at the site 9; #1 M, c-Myc-tag at site 1; #6H, HA-tag at site 6; #7 M, c-Myc-tag at site 7) were coexpressed with PMMoV genome harboring its CP (PMCP) or ToMV CP (ToCP). Red circles indicate the approximate position of the Agroinfiltration. The #1 M conditions the response to both ToCP and PMCP like the wild type, but #9H, #6H and #7 M were barely responsive to the both CP

npsa-prabi.ibcp.fr/cgi-bin/npsa_automat.pl?page=npsa_gor4.html) for reference (Fig. 2).

2. pBAL3 and PMMoV agroinfection clones can be obtained from K.T. Sekine (k-sekine@agr.u-ryukyu.ac.jp) or K. Kobayashi (kappei@ehime-u.ac.jp).
3. Different brands of commercial soil mix can be used. We use Super Mix A (Sakata seed; Yokohama, Japan).

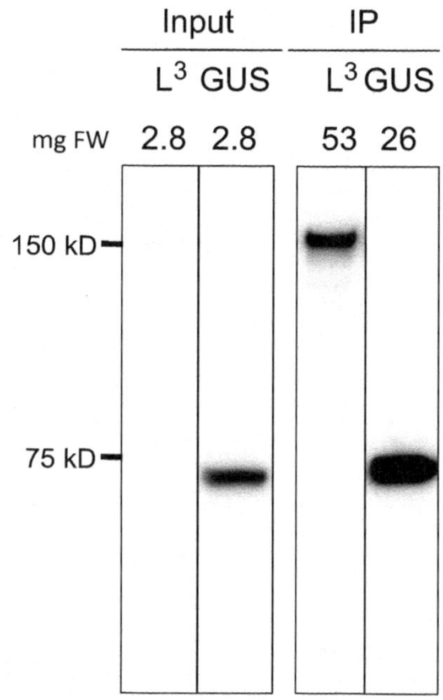

Fig. 5 Western blot detection of epitope-tagged L^3 protein. IP, immunoprecipitation; L^3, epitope-tagged L^3 protein #3H (HA tagged at site 3); GUS, HA tagged beta-glucuronidase as a control; FW, fresh weight equivalent. Preparation of protein samples are described in Chapter 1

4. The oligonucleotides were designed to have BglII sites, which partially overlap with the XbaI protruding ends, which are compatible with the BlnI-digested ends, and add single serine residues between the inserted restriction sites and epitope tags. The downstream of the epitope tag sequences was also added with single serine residues, and the proline codons were modified from CCC to CCG to eliminate SmaI restriction site (Fig. 3c).
5. L^3 protein induces the cell death response when coexpressed with cognate tobamovirus coat proteins, which can be observed after 4–5 days post-infiltration (Fig. 4). To detect the epitope-tagged protein, however, protein samples should be prepared 2 days postinfiltration as described in Chapter 1.
6. The detection of the epitope-tagged L^3 protein can be done as described in Chapter 1 (Fig. 5, Input). The epitope-tagged L^3 protein can be immunoprecipitated using the anit-epitope tag antibody (Fig. 5, IP). The detail of the immunoprecipitation will be described elsewhere.

Acknowledgements

This work was supported in part by JSPS KAKENHI Grant Numbers 21380032, 24658044, 26292026 and 15K14664.

References

1. Whitham S, Dinesh-Kumar SP, Choi D, Hehl R, Corr C, Baker B (1994) The product of the tobacco mosaic virus resistance gene N: similarity to toll and the interleukin-1 receptor. Cell 78(6):1101–1115. https://doi.org/10.1016/0092-8674(94)90283-6
2. Les EF, Holzberg S, Calderon-Urrea A, Handley V, Axtell M, Corr C, Baker B (1999) The helicase domain of the TMV replicase proteins induces the N-mediated defence response in tobacco. Plant J 18(1):67–75. https://doi.org/10.1046/j.1365-313X.1999.00426.x
3. Liu Y, Schiff M, Marathe R, Dinesh-Kumar SP (2002) Tobacco Rar1, EDS1 and NPR1/NIM1 like genes are required for N-mediated resistance to tobacco mosaic virus. Plant J 30(4):415–429. https://doi.org/10.1046/j.1365-313X.2002.01297.x
4. Liu Y, Schiff M, Dinesh-Kumar SP (2004) Involvement of MEK1 MAPKK, NTF6 MAPK, WRKY/MYB transcription factors, COI1 and CTR1 in N-mediated resistance to tobacco mosaic virus. Plant J 38(5):800–809. https://doi.org/10.1111/j.1365-313X.2004.02085.x
5. Liu Y, Schiff M, Serino G, Deng X-W, Dinesh-Kumar SP, Martin GB, Chai J (2002) Role of SCF ubiquitin-ligase and the COP9 signalosome in the N gene-mediated resistance response to tobacco mosaic virus. Plant Cell 14(7):1483–1496. https://doi.org/10.1105/tpc.002493.pathogen
6. Jin H, Liu Y, Yang KY, Kim CY, Baker B, Zhang S (2003) Function of a mitogen-activated protein kinase pathway in N gene-mediated resistance in tobacco. Plant J 33(4):719–731. https://doi.org/10.1046/j.1365-313X.2003.01664.x
7. Hoser R, Zurczak M, Lichocka M et al (2013) Nucleocytoplasmic partitioning of tobacco N receptor is modulated by SGT1. New Phytol 200(1):158–171. https://doi.org/10.1111/nph.12347
8. Caplan JL, Mamillapalli P, Burch-Smith TM, Czymmek K, Dinesh-Kumar SP (2008) Chloroplastic protein NRIP1 mediates innate immune receptor recognition of a viral effector. Cell 132:449–462
9. Dinesh-Kumar SP, Baker BJ (2000) Alternatively spliced N resistance gene transcripts: their possible role in tobacco mosaic virus resistance. Proc Natl Acad Sci 97(4):1908–1913. https://doi.org/10.1073/pnas.020367497
10. Mestre P, Baulcombe DC (2006) Elicitor-mediated oligomerization of the tobacco N disease resistance protein. Plant Cell 18:491–501
11. Rairdan GJ, Moffett P (2006) Distinct domains in the ARC region of the potato resistance protein Rx mediate LRR binding and inhibition of activation. Plant Cell 18:2082–2093
12. Rairdan GJ, Collier SM, Sacco MA, Baldwin TT, Boettrich T, Moffett P (2008) The coiled-coil and nucleotide binding domains of the potato Rx disease resistance protein function in pathogen recognition and signaling. Plant Cell 20:739–751
13. Moffett P, Farnham G, Peart J, Baulcombe DC (2002) Interaction between domains of a plant NBS-LRR protein in disease resistance-related cell death. EMBO J 21:4511–4519
14. Tameling WI, Baulcombe DC (2007) Physical association of the NB-LRR resistance protein Rx with a Ran GTPase-activating protein is required for extreme resistance to Potato virus X. Plant Cell 19:1682–1694
15. Sacco MA, Mansoor S, Moffett P (2007) A RanGAP protein physically interacts with the NB-LRR protein Rx, and is required for Rx-mediated viral resistance. Plant J 52:82–93
16. Takahashi H, Shoji H, Ando S, Kanayama Y, Kusano T, Takeshita M, Suzuki M, Masuta C (2012) *RCY1* -mediated resistance to *Cucumber mosaic virus* is regulated by LRR domain-mediated interaction with CMV(Y) following degradation of RCY1. Mol Plant-Microbe Interact 25:1171–1185
17. Jarvik JW, Telmer CA (1998) Epitope tagging. Annu Rev Genet 32:601–618
18. Sekine K-T, Tomita R, Takeuchi S et al (2012) Functional differentiation in the leucine-rich repeat domains of closely related plant virus-resistance proteins that recognize common avr proteins. Mol Plant-Microbe Interact 25:1219–1229
19. Nelson BD, Manoil C, Traxler B (1997) Insertion mutagenesis of the lac repressor and its implications for structure-function analysis. J Bacteriol 179(11):3721–3728
20. Tomita R, Sekine K-T, Mizumoto H, Sakamoto M, Murai J, Kiba A, Hikichi Y, Suzuki K, Kobayashi K (2011) Genetic basis for the hierarchical interaction between Tobamovirus spp. and L resistance gene alleles from different pepper species. Mol Plant-Microbe Interact 24:108–117

Chapter 3

Reverse Genetic Analysis of Antiviral Resistance Signaling and the Resistance Mechanism in *Arabidopsis thaliana*

Yukiyo Sato and Hideki Takahashi

Abstract

Antiviral RNA silencing and the resistance gene-conferred defense response are major antiviral immune systems in plants. Several of the components involved have been genetically or biochemically identified in *Arabidopsis thaliana*. One powerful tool to dissect antiviral immune systems involves a reverse genetic approach that analyzes *Arabidopsis* mutant lines with impaired antiviral defense responses. In particular, to better understand the signaling networks involved in the resistance gene-conferred antiviral response in host plants, establishment of mutant lines carrying the homozygous mutant allele and antiviral resistance gene is required. The information on well-characterized defense-related signaling mutant alleles and the PCR-based genotyping method provided in this chapter allows the efficient selection of *Arabidopsis* mutant lines that can be used to study antiviral resistance signaling networks and resistance mechanisms.

Key words *Arabidopsis thaliana*, Cucumber mosaic virus, *R* gene, Nucleotide-binding and leucine-rich repeat (NB-LRR) protein, Reverse genetics

1 Introduction

Plants harbor two major antiviral immune systems: RNA silencing as a primary broad antiviral defense response accompanied by small interfering RNA biogenesis and a dominant or recessive resistance (*R* or *r*) gene-conferred defense response that is more specific to virus species or strains [1–4]. Antiviral RNA silencing pathways and nucleotide-binding and leucine-rich repeat (NB-LRR) class *R* gene-conferred resistance to viruses have been well-characterized through genetic and biochemical approaches [1, 5, 6]. One powerful method to identify the components required for antiviral RNA silencing pathways and to dissect the defense signaling pathways underlying the molecular mechanism of NB-LRR class *R* gene-mediated resistance to viruses involves reverse genetic analysis of a series of mutant plants with impaired antiviral resistance to viruses. Furthermore, the role of endogenous small RNAs in the regulation of plant defense responses to viruses has been demonstrated [5].

Kappei Kobayashi and Masamichi Nishiguchi (eds.), *Antiviral Resistance in Plants: Methods and Protocols*,
Methods in Molecular Biology, vol. 2028, https://doi.org/10.1007/978-1-4939-9635-3_3,

Arabidopsis thaliana is a host plant to many viruses and has several geographical ecotypes that can be used as genetic resources. Accordingly, *Arabidopsis* mutants can be used for the reverse genetic analysis of antiviral resistance. Indeed, in *A. thaliana*, several signaling mutants impairing tolerance to environmental biotic or abiotic stresses or exhibiting unusual responses to plant hormones have been isolated by forward genetic screens using chemical mutagens such as ethyl methanesulfonate, physical mutagens such as fast-neutron, gamma ray, or X-ray irradiation, and insertional mutagenesis involving integration of T-DNA [7–11] or transposons [12, 13].

As an example, the scheme for the reverse genetic analysis of defense signaling pathways downstream of R protein is represented in Fig. 1. The defense signaling network activated by direct or indirect recognition of a viral avirulent determinant by R protein is coordinately regulated by positive and negative regulators (Fig. 1a). Loss-of-function mutations in a positive regulator of the defense signaling pathway compromise resistance to avirulent viruses (Fig. 1b). In contrast, loss-of-function mutations in a negative regulator of the defense signaling pathway enhance resistance to such avirulent viruses (Fig. 1c). Thus, changes in the phenotype of the antiviral responses of mutant lines to virus infection suggest that the wild-type version of the loss-of-function mutation gene is associated with antiviral defense signaling downstream of R protein.

In this chapter, the procedure for the establishment of *Arabidopsis* materials carrying a homozygous *R* gene and a homozygous characterized mutant allele is described for reverse genetic analysis of signaling networks in the defense system to Cucumber mosaic virus (CMV). We also provide a list of mutant alleles and information on genotyping markers that can be used to identify these mutants.

2 Materials

2.1 Plants and Cultivation

1. Seeds of *A. thaliana* wild ecotypes and mutant lines, which can be obtained from *Arabidopsis* stock centers (*see* **Note 1**) (Table 1).
2. Pro-Mix BX Mycorrhizae (Premier Tech Horticulture Ltd., Quakertown, PA, USA).
3. Autoclave bags.
4. Open-bottom plastic pots.
5. Plastic trays.
6. Liquid compound fertilizer such as HYPONeX (HYPONeX Japan Co., Ltd., Osaka, Japan).
7. Colorless and transparent plastic wrap.

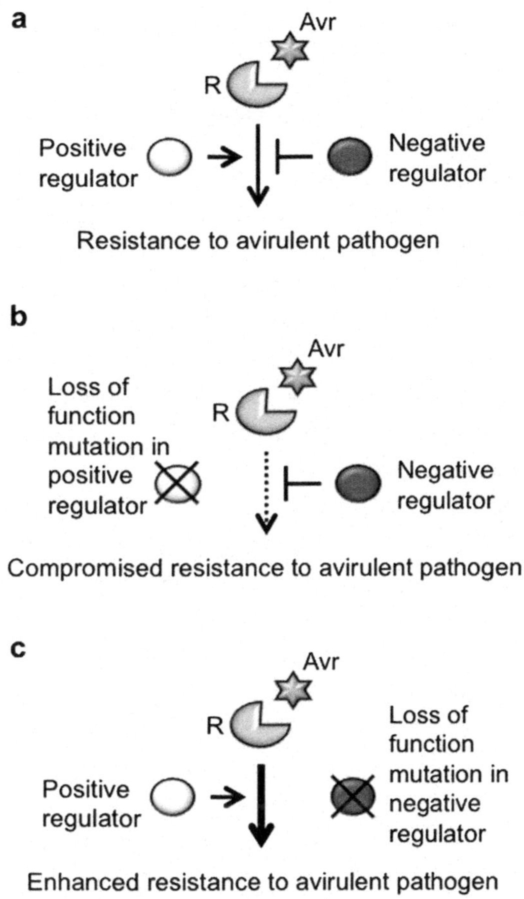

Fig. 1 Scheme for the genetic analysis of R protein-mediated resistance signaling. (**a**) In wild-type plants, direct or indirect interaction between a resistance (R) protein and a pathogen-derived avirulence (Avr) factor activates defense signaling. Resistance to avirulent pathogen is modulated by both positive and negative regulators of defense signaling. (**b**) Loss-of-function mutations in positive regulators of defense signaling compromise resistance to avirulent pathogens. (**c**) Loss-of-function mutations in negative regulators of defense signaling enhance resistance to avirulent pathogens

2.2 Plant Crossing

1. *Arabidopsis thaliana* plants.
2. Stereoscopic microscope.
3. Precision stainless steel tweezers.
4. Stainless steel scissors.

2.3 General Laboratory Equipment and Stock Reagents

1. 200-μL and 1000-μL pipette tips and appropriate micropipettes.
2. 1.5 mL microcentrifuge tubes, sterilized.
3. 0.2 mL PCR tubes, sterilized.

Table 1
Information on *Arabidopsis* stock centers

Stock center	Organization	Stock catalog URL
Arabidopsis Biological Resource Center (ABRC)	Ohio State University, USA	http://www.arabidopsis.org/servlets/Order?state=catalog
Nottingham Arabidopsis Stock Center (NASC)	Nottingham University, UK	http://arabidopsis.info/BrowsePage
RIKEN Bioresource Center (BRC)	RIKEN, Japan	http://epd.brc.riken.jp/en/seed
SENDAI Arabidopsis Seed Stock Center (SASSC)	Miyagi University of Education, Japan	http://sassc.epd.brc.riken.jp/top.php?mode=general
Versailles Arabidopsis Stock Center	INRA, France	http://publiclines.versailles.inra.fr/catalog/index
Lehle Seeds	Lehle Seeds, Inc., USA	http://www.arabidopsis.com/main/cat/!ct_seat.html

4. Microcentrifuge.
5. Nuclease-free water.
6. 1 M tris-hydroxymethyl aminomethane (Tris), adjusted to pH 8.0 with hydrochloric acid (HCl), sterilized by autoclave.
7. 1 M Tris, adjusted to pH 7.5 with HCl, sterilized by autoclave.
8. 0.5 M ethylenediaminetetraacetic acid (EDTA), adjusted to pH 8.0 with 5 N sodium hydroxide (NaOH), sterilized by autoclave.
9. 5 M sodium chloride (NaCl), sterilized by autoclave.
10. % (w/v) sodium dodecyl sulfate (SDS), dissolved in sterile distilled water.
11. TE: 10 mM Tris–HCl (pH 8.0), 1 mM EDTA (pH 8.0), sterilized by autoclave.
12. 3 M sodium acetate, adjusted to pH 5.2 with acetic acid, sterilized by autoclave.
13. Ethanol, molecular biology grade.
14. 70% (v/v) ethanol, molecular biology grade.

2.4 Genomic DNA Extraction

2.4.1 Simplified Method

1. Fresh leaf tissues of *A. thaliana* plants.
2. An electric drill and plastic pestles.
3. Extraction buffer: 200 mM Tris–HCl (pH 7.5), 250 mM NaCl, 25 mM EDTA (pH 8.0), 0.5% SDS, sterilized by autoclave before the addition of SDS.
4. 2-propanol, molecular biology grade.

2.4.2 CTAB Method

1. Fresh leaf tissues of *A. thaliana* plants.
2. An electric drill and plastic pestles.
3. 2% cetyltrimethylammonium bromide (CTAB) solution: 100 mM Tris–HCl (pH 8.0), 20 mM EDTA (pH 8.0), 1.4 M NaCl, 2% (w/v) CTAB, sterilized by autoclave before the addition of CTAB powder.
4. Chloroform–isoamyl alcohol (24:1), molecular biology grade.
5. 1% CTAB solution: 50 mM Tris–HCl (pH 8.0), 10 mM EDTA (pH 8.0), 1% (w/v) CTAB, sterilized by autoclave before the addition of CTAB powder.
6. TE containing 10 μg/mL RNaseA.
7. Phenol–chloroform–isoamyl alcohol (25:24:1), molecular biology grade.

2.5 PCR-Based Genotyping

2.5.1 PCR

1. GoTaq® DNA polymerase (Promega, Madison, WI, USA), containing 5 × Green or Colorless GoTaq® Reaction Buffer and 5 U/μL GoTaq® DNA polymerase (*see* **Note 2**).
2. 2 mM dNTP Mix: Aqueous solution containing dATP, dTTP, dGTP, and dCTP at 2 mM each.
3. 20 μM primers: Synthetic oligonucleotides ordered from suppliers (*see* Subheading 3.4, and Tables 2 and 3).
4. Template DNA extracted from plants.
5. Thermal cycler.

2.5.2 Restriction Enzyme Digestion

1. PCR products.
2. Spin columns: NucleoSpin® Gel and PCR Clean-up (Macherey-Nagel GmbH, Düren, Germany).
3. Restriction enzymes (New England Biolabs Inc., Ipswich, MA, USA), supplied with 10× reaction buffer and 6× Gel Loading Dye, Purple (*see* **Note 3**).
4. Incubator.

2.6 Gel Electrophoresis

2.6.1 Agarose Gel Electrophoresis

1. 50× TAE: 2 M Tris, 1 M acetic acid, 50 mM EDTA (pH 8.0).
2. Agarose, electrophoresis grade.
3. 6× Gel Loading Dye, Purple (New England Biolabs Inc) (*see* **Note 4**).
4. Size marker.
5. Gel casting set (Mupid, Co., Ltd., Tokyo, Japan).
6. Submarine electrophoresis apparatus (Mupid, Co., Ltd.).
7. 10 mg/mL ethidium bromide solution, stored in the shade at 4 °C.
8. UV transilluminator with the camera.

Table 2
Primer sequences used for detection of T-DNA and transposon insertion

Collections	Reference	Primer	Primer sequences (5′ → 3′)[a]	Detailed information on primers
SALK lines	[7]	LBa1	TGGTTCACGTAGTGGGCCATCG	[7], http://signal.salk.edu/tdnaprimers.2.html
		LBb1	GCGTGGACCGCTTGCTGCAACT	
		LBb1.3	ATTTTGCCGATTTCGGAAC	
		SALK_LB	GGCAATCAGCTGTTGCCCGTCTCACTGGTG	
SAIL lines	[8]	SAIL_LB1	GCCTTTTCAGAAATGGATAAATAGCCTTGCTTCC	[8], http://signal.salk.edu/tdnaprimers.2.html
		SAIL_LB2	GCTTCCTATTATATCTTCCCAAATTACCAATACA	
		SAIL_LB3	TAGCATCTGAATTTCATAACCAATCTCGATACAC	
		SAIL_LB	TTCATAACCAATCTCGATACAC	
GABI-Kat lines	[9, 10]	o8409	ATATTGACCATCATACTCATTGC	https://www.gabi-kat.de/faq/vector-a-primer-info.html
		o8474	ATAATAACGCTGCGGACATCTACATTTT	
Wisconsin DsLox lines	[11]	p745	AACGTCCGCAATGTGTTATTAAGTTGTC	[11]
SLAT insertion lines	[12]	dSpm1	CTTATTTCAGTAAGAGTGTGGGGTTTTGG	[12]
		dSpm11	GGTGCAGCAAAACCCACACTTTTACTTC	

[a]For SALK lines, SAIL lines, GABI-Kat lines, and Wisconsin DsLox lines, primer sequences on the left border of T-DNA are indicated

Table 3
The characterized mutant lines lacking signaling pathways for disease resistance in *A. thaliana* ecotype Col-0 background

Gene	AGI code	Function[a]	Polymorphism[b]	Reference 1[c]	Mutagen[d]	Marker[e]	Primer sequences (5′ → 3′)[f]	Reference 2[g]
CBP60g	AT5G26920	SA signaling	*cbp60g-1* (SALK_023199)	[14]	T-DNA	T-DNA insertion	F: TGTTTCGGTGGACTTGTGAC R: TAAATCCCTCAACGGTCCAG	[15]
EDS1	AT3G48090	SA signaling	*eds1-22* (SALK_071051)	[16]	T-DNA	T-DNA insertion	LP: CCATCATATAGTCTCGCAG RP: CTCGGATTACGCTTGCTCGAA LB: LBa1	[17]
EDS5, *SID1*	AT4G39030	SA signaling	*eds5-1*	[18, 19]	EMS	dCAPS/*Bgl*II	F: CCTGTTTCCTGGTGTCTACAC R: AAGAAAGGTATAAGCAGTCTATAGAT	[20]†
			eds5-3 (*sid1*)	[21]	EMS	dCAPS/*Hph*I	F: ACAGGTCCGGCGATGGGGAGGTG R: CAGCTTAGCTATTGGGTAG	[22]†‡
NDR1	AT3G20600	SA signaling	*ndr1-1*	[23]	Fast-neutron	INDEL	F: AATCTACTACGACGATGTCCAC R: GGGACGGTTTCAATTCTGTGATAG	[20]†

(continued)

Table 3
(continued)

Gene	AGI code	Function[a]	Polymorphism[b]	Reference 1[c]	Mutagen[d]	Marker[e]	Primer sequences (5′ → 3′)[f]	Reference 2[g]
NPR1	AT1G64280	SA signaling	*npr1-1*	[24, 25]	EMS	dCAPS/*Nla*III	F: AGGCACTTGACTCGGATGAT R: ATGCACTTGCACCTTTTTCC	[26]
						dCAPS/*Bsp*HI	F: AATGTGAAGACCGCAACAGAT R: CTTCCGCATCGCAGCATCAT	[20]†
NPR3	AT5G45110	SA signaling	*npr3-2* (SALK_043055)	[27]	T-DNA	T-DNA insertion	LP: AGTGATCCGAAAGTTGTTGCC RP: CCCTTCTCAAAATGGAAATCG LB: LBa1	[28]
			npr3-3 (SALK_009990)	[28]	T-DNA	T-DNA insertion	F: AGTGATCCGAAAGTTGTTGCC R: CCCTTCTCAAAATGGAAATCG RP: TGTGTTCCATTTCTATGCCTG LB: LBa1	[28]
NPR4	AT4G19660	SA signaling	*npr4-1* (SALK_027020) and *npr4-2* (SALK_098460)	[29]	T-DNA	T-DNA insertion	F: ATGGCTGCAACTGCAATAGAG R: TCATGTTGGATTCTCTAAGGC LB: GGCAATCAGCTGTTGCCCGTCTCACTGGT RB: GCTCATGATCAGATTGTCGTTTCCCGCCTT	[29]

PAD4	AT3G52430	SA signaling	*pad4-1*	[18, 30]	EMS	CAPS/*Bsm*F1	F: GCGATGCATCAGAAGAG R: TTAGCCCAAAAGCAAGTATC	[22]†‡
						dCAPS/*Dde*I	F: TCGCATAAGACTAGGTAAGTCTT R: GCGTAAATCCATTTCTTTCCTA	[20]†
						dCAPS/*Kpn*I	F: ACTCGAGATTCAATGGTACAAAGAT R: GTCTCCACCATTTTAATCACTGGGTAC	[31]
PAL1	AT2G37040	SA signaling	*pal1-2* (SALK_022804) and *pal1-3* (SALK_096474)	[32]	T-DNA	T-DNA insertion	F: CTGCAGCGGAGCAAATGA R: CACTCATCACCTCTGCGAAA	[32]
PAL2	AT3G53260	SA signaling	*pal2-2* (SALK_092252)	[32]	T-DNA	T-DNA insertion	LP: TATTCCGGCGTTCAAAAATC RP: CAATGGATCAAATCGAAGCA LB: TTGATTTGGGTGATGGTTCA	[32]
			pal2-3 (GABI_692H09)	[32]	T-DNA	T-DNA insertion	LP: CAATGGATCAAATCGAAGCA RP: TATTCCGGCGTTCAAAAATC LB: CCCATTTGGACGTGAATGTAGACAC	[32]
PAL3	AT5G04230	SA signaling	*pal3-1* (SM_3.39574) and *pal3-2* (SM_3.19684)	[32]	dSpm	Transposon insertion	F: AACCCCATTGTTTTCCCTCT R: GCTCTTTTTGAAGAGCAGCA	[32]

(continued)

Table 3
(continued)

Gene	AGI code	Function[a]	Polymorphism[b]	Reference 1[c]	Mutagen[d]	Marker[e]	Primer sequences (5′ → 3′)[f]	Reference 2[g]
PAL4	AT3G10340	SA signaling	*pal4-1* (SALK_070702) and *pal4-2* (SALK_022730)	[32]	T-DNA	T-DNA insertion	F: TCAAATACCGAATCGAAGCA R: TATTCCGGCATTCAAGAACC	[32]
SAG101	AT5G14930	SA signaling	*sag101-1* and *sag101-2*	[33]	dSpm	Transposon insertion	F: CACGCGTCCGAAGATCTTGGAGATACATA R: ACTTCCGGGTGTTCATAAACTCGGTCAAG RB: dSpm1 and dSpm11	[33]
			sag101-3 (SALK_022911)	[17, 34]	T-DNA	T-DNA insertion	LP: GGAGCAACTGCAAGAGACAT RP GGTCCTTCTCTTGTACAC LB: LBa1	[17]
			sag101-4 (GABI_476E10)	[17]	T-DNA	T-DNA insertion	LP: GGTCTGTTGCATCTCTCTATAC RP: CCTCTAAGAATATCTCCGGCG LB: o8409	[17]
SARD1	AT1G73805	SA signaling	*sard1-1* (SALK_138476)	[35]	T-DNA	T-DNA insertion	F: CTTCAGTGTCGGAGTAGTCG R: CAAGACCTCTCTAACCTAAC	[15]

SID2, ICS1	AT1G74710	SA signaling	*sid2-1*	[21]	EMS	CAPS/*Mfe*I	F: TGTCTGCAGTGAAGCTTTGG R: CACAAACAGCTGGAGTTGGA	[36]
						CAPS/*Mse*I	F: AATCAAAAGCCTTCTTC R: CATTTCTTGGATAATAGTTTGG	[22]†‡
			sid2-2 (*eds16-1*)	[37]	Fast neutron	INDEL	F1: TTCTTCATGCAGGGGAGGAG F2: CAACCACCTGGTGCACCAGC R: AAGCAAAATGTTTGAGTCAGCA	[38]†
						INDEL	F: TACGAGAGAATATAAGAGAAAAGTTA R: CAGATAGAAAGAAAAAGGGTTAAAGC	[20]†
			sid2-3 (SALK_042603)	[39]	T-DNA	T-DNA insertion	LP: ACCCTAATTTGGATTTGGTGC RP: AGCTCTAGGCCTAGTTGCAGC	[20]†
SIZ1	AT5G60410	SA signaling	*siz1-2* (SALK_065397) and *siz1-3* (SALK_034008)	[40]	T-DNA	T-DNA insertion	F: CTGATGGTAGCCTTGCCCCT R: CAACTAAACCTCCTGAAACGTCAG	[41]
ALD1	AT2G13810	SAR signaling	*ald1-T2* (SALK_007673)	[42]	T-DNA	T-DNA insertion	F: CACCAGTCCCAAGGCTTATCAA R: TCCTAGTCCCACTAAAGGTTGC	[43]
DIR1	AT5G48485	SAR signaling	*dir1-2* (GK_403C01)	[44]	T-DNA	T-DNA insertion	RP: GATCGTGATAATGGCTATGTTGGTCGATACATC LB: GTAAGGTAATGGGCTACACTGAATTGGTAGCTC	[44]

(continued)

Table 3
(continued)

Gene	AGI code	Function[a]	Polymorphism[b]	Reference 1[c]	Mutagen[d]	Marker[e]	Primer sequences (5′ → 3′)[f]	Reference 2[g]
FMO1	AT1G19250	SAR signaling	*fmo1-1* (SALK_026163)	[45]	T-DNA	T-DNA insertion	F: CTTTTCGGTTG GACTTGGAAC R: CGTAGGATACG TCCAAAGCA	[43]
CTR1	AT5G03730	ET signaling	*ctr1-1*	[46, 47]	DEB	dCAPS/ *Eco*RI	F: ATTCCTCTTGTG AGACAAGC R: CCTTCAATCGCGA GAGACCGAA	[48]
EBF1	AT2G25490	ET signaling	*ebf1-1* (SALK line)	[49]	T-DNA	T-DNA insertion	F: AGCAATTGCTG TTGGAACTGCTGGT R: TTCAAGCTGAA GACTCTCAAGTGA LB: SALK_LB	[48]
EBF2	AT5G25350	ET signaling	*ebf2-1* (SALK line)	[49]	T-DNA	T-DNA insertion	F: TTCTCATGAA CTGTGGCTCAA R: AGAAGAGTAT CCACAGTGCT LB: SALK_LB	[48]
EIL1	AT2G27050	ET signaling	*eil1-1*	[50]	En1/Spm	Transposon insertion	F: GGGAATGGTG GAAAGATAAG R: CTTTCGCCGTC ATCTTATCC LB: SALK_LB	[48]
			eil1-3 (SALK_042113)	[48]	T-DNA	T-DNA insertion	F: GTGGCTCGAGA GCTTTATCCCGAGTCA R: AGAATGGATCCT ACAGAGATTGTGTAATA LB: SALK_LB	[48]

EIN1, ETR1	AT1G66340	ET signaling	*etr1-1*	[51, 52]	EMS	dCAPS/*Apa*LI	F: AAGTTAATAAGATGAGTTGGTGCA R: AGAAATCAGCCGTGTTTCCG	[52]†‡
						dCAPS/*Apa*LI	F: CTTTGTGAAGAAATCAGCCGTGT R: CCATAAGTTAATAAGATGAGTTGGTGCA	[53]†‡
EIN2	AT5G03280	ET signaling	*ein2-1*	[51, 54]	EMS	CAPS/*Afl*II	F: TGCGAGTAACAGAGCGGAAG R: TACATCAGAGTCTTCTTTAAGACTAC	[38]†
						CAPS/*Bsr*BI	F: CGCCATCTTTGTTTCAACAATCAGATCC R: CCAGAGGAAAGAGAGTTGGATGTAAAGTACTCTACCGCT	[20, 55]†
			ein2-5	[54, 56]	EMS	CAPS/*Nla*III	F: GCTCTTGTTCTTCTCTAGTC R: GAAGCATCATTGCCACCAAG	[48]
EIN3	AT3G20770	ET signaling	*ein3-1*	[57]	EMS	CAPS/*Hae*III	F: TACCAAGTATCAAGCGGAG R: AGGCCACCAATCCTCTTTC	[58]
EIN4	AT3G04580	ET signaling	*ein4-1*	[47]	DEB	dCAPS/*Mse*I	F: GAGTCATTCCACATAGGACAT R: GTGATCTCTTAATAGCCATTG	[52]†‡

(continued)

Table 3
(continued)

Gene	AGI code	Function[a]	Polymorphism[b]	Reference 1[c]	Mutagen[d]	Marker[e]	Primer sequences (5′ → 3′)[f]	Reference 2[g]
EIN5, XRN4	AT1G54490	ET signaling	*ein5-1*	[47]	X-ray	dCAPS/ *Taq*I	F: GTTGATGACT GATCCCTCATCCT R: GAGTGTCAACT ATCCAGCATGAA	[59]
			ein5-7 (*ein7*)	[47]	X-ray	CAPS/ *Bcl*I	F: TTCAAATGTTCC GGGAGAAG R: GACGAAGCACC AACACCTTA	[59]
AOS, DDE2	AT5G42650	JA signaling	*dde2-2*	[60]	En1/Spm	CAPS/ *Bst*UI	F: GACACGAACCG GATCCAAAG R: GCCGAAATCCG CTTTCCCTTTA	[38]†
COI1	AT2G39940	JA signaling	*coi1-1*	[61]	EMS	CAPS/ *Xcm*I	F: GGTTCTCTTTA GTCTTTAC R: CAGACAACTAT TTCGTTACC	[61]†
			coi1-21 (SALK_035548)	[62]	T-DNA	T-DNA insertion	F: CTGCAGTGTG TAACGATGCTC R: TGGACCATATAA ATTCATGCAGTC LB: LBb1	[31]
JAR1	AT2G46370	JA signaling	*jar1-1*	[63]	EMS	CAPS/ *Hin*dIII	F: CAGTGTGTGTG TTTTTGATCATAAGCT R: CAAATTTAAACT ATACCTGTTTCTGAAGG	[36]

NTRA	AT2G17420	ROS signaling	*ntra* (SALK_539152)	[64]	T-DNA	T-DNA insertion	LP: CAAATCCGCCGTCTCTAGCC RP: CGCCCTAAACGTATCCCTCCT LB: LBa1 or LBb1	[65]
NTRB	AT4G35460	ROS signaling	*ntrb* (SALK_545978)	[64]	T-DNA	T-DNA insertion	LP: GACAAGCCATAGGGTCACAGAGC RP: TCGGAGCGATTCGGTACTACG LB: LBa1 or LBb1	[65]
RBOHD	AT5G47910	ROS signaling	*atrbohD*	[66]	dSpm	Transposon insertion	LP: ATGAAAATGAGACGAGGCAATTC RP: GGATACTGATCATAGGCGTGGCTCCA RB: dSpm1	[66]
RBOHF	AT1G64060	ROS signaling	*atrbohF*	[66]	dSpm	Transposon insertion	LP: CGAAGAAGATCTGGAGACGAGA RP: CTTCCGATATCCTTCAACCAACTC RB: dSpm1	[66], §
TRX5	AT1G45145	ROS signaling	*attrx5-4* (SALK_144259)	[65]	T-DNA	T-DNA insertion	LP: TTTTCGTGTTCGTGGTTGAA RP: TCTTGTTATGTCCAGGGCTTTT LB: LBa1 or LBb1	[65]

(continued)

Table 3
(continued)

Gene	AGI code	Function[a]	Polymorphism[b]	Reference 1[c]	Mutagen[d]	Marker[e]	Primer sequences (5′ → 3′)[f]	Reference 2[g]
EDR1	AT1G08720	MAPK signaling	*edr1-1*	[67]	Gamma ray	dCAPS/ *Kpn*I	F: CAGAGGCTGAAA GGACAGATTCTTGGTA R: CCTCACTGTTC TGATTGTAAGG	[31]
			edr1-2 (SALK_053889)	[31]	T-DNA	T-DNA insertion	F: GTTTCCATAACGA GGATCACG R: GTATCTCAAGGC ATCGTGCTC LB: LBb1	[31]
			edr1-3 (SALK_127158)	[31]	T-DNA	T-DNA insertion	F: CTCGATAACCTC TGGGCTACC R: AAAACTCCTTGGT TCTTTGGC LB: LBb1	[31]
MEKK1	AT4G08500	MAPK signaling	*mekk1-1* (SALK_052557)	[68]	T-DNA	T-DNA insertion	LP: GATTATTCCAC GAAACACCGCG RP: AGAAATAGCCA AATCATCAGGACC LB: LBa1	[68]
			mekk1-3 (WiscDsLox339H07)	[68]	T-DNA	T-DNA insertion	LP: GATCATCGGGG ATCGTCTTGGG RP: GAAACTTCCAT TTCCATACCCCC LB: p745	[68]
						T-DNA insertion	F: GAAGAGATGGGAGC TAGGTTTATCCAGTT R: ACTTTTGACTGCT CCATGAGATAATGACA LB: CAATGTGTTATT AAGTTGTC	[69]

MEKK2	AT4G08480	MAPK signaling	*mekk2* (SALK_150039)	[70]	T-DNA	T-DNA insertion	F1: GCTGGCTGTT CTACAGGATCA R1: CAGAGGGTTG GAGATCTTGTG F2: CATGAAGAAGT CGTCGGATAAGTC R2: AGAAGTCCCCA TCTCTACATTCAG F3: GTCAAACTTGC AGATTTTGGATTG R3: AATTGGAGATT AAATCCATCGAGA	[69]
MEKK3	AT4G08470	MAPK signaling	*mekk3* (WiscDsLox472D11)	[69]	T-DNA	T-DNA insertion	F: ATATGGTACATCG CTTATCCCACTGACAT R: GCGTTTTCACAAA TTCTGACCTTTAACTC LB: CAATGTGTTATTA AGTTGTC	[69]
MKK1	AT4G26070	MAPK signaling	*mkk1–2* (SALK_027645)	[71]	T-DNA	T-DNA insertion	LP: GACAAGTCTCTTA AGTCATAACATCTCG RP: AACATGCTATCTG CCATCTGC LB: LBa1	[72]
MKK2	AT4G29810	MAPK signaling	*mkk2-1* (SAIL_511_H01)	[71]	T-DNA	T-DNA insertion	LP: TCACTGGAAGGT AAAACAAGAAATC RP: GTTAAAGCCAT CCCTGACTCC LB: GCTTCCTATTATTCTTCC CAAATTACCAATACA	[72]

(continued)

Table 3
(continued)

Gene	AGI code	Function[a]	Polymorphism[b]	Reference 1[c]	Mutagen[d]	Marker[e]	Primer sequences (5′ → 3′)[f]	Reference 2[g]
MKK3	AT5G40440	MAPK signaling	*mkk3-1* (SALK_051970)	[73]	T-DNA	T-DNA insertion	LP: GAACAAACGTTTTCTCATGTGTG RP: AGAAGGATCCAGATGCTCGAC LB: LBa1	[72]
MKK4	AT1G51660	MAPK signaling	*mkk4*(SAIL_565_A12)	[72]	T-DNA	T-DNA insertion	LP: CATCAACCTGATTAGGTTTGAG RP: CAGAGAGAACCGGGGAAAAG LB: GCTTCCTATTATTCTTCCCAAATTACCAATACA	[72]
MKK5	AT3G21220	MAPK signaling	*mkk5* (SALK_067321)	[72]	T-DNA	T-DNA insertion	LP: TAACCAGGCAACCATCTCAAG RP: TGGAAAGAGCGTGGAATACAC LB: LBa1	[72]
MKK6	AT5G56580	MAPK signaling	*mkk6* (SALK_084332)	[72]	T-DNA	T-DNA insertion	LP: CGCAGTCCTGTTTTCAAATTC RP: CAAAAGCTTCGTTAAAGCTCTCTC LB: LBa1	[72]
MKK7	AT1G18350	MAPK signaling	*mkk7* (SM_3_17177)	[72]	T-DNA	T-DNA insertion	LP: AAGCACACGACGCACATTAAC RP: CAGGCTCCTCACTTAAATCCC LB: TACGAATAAGAGCGTCCATTTTAGAGTGA	[72]

MKK9	AT1G73500	MAPK signaling	*mkk9-1* (SAIL_60_H06)	[72]	T-DNA	T-DNA insertion	LP: TCCCCTAACATTCTGGAGTATA RP: TCAAACCGGCGAATCTTCTTC LB: GCTTCCTATTATTCTTCCCAAATTACCAATACA	[72]
			mkk9-2 (SALK_146400)	[72]	T-DNA	T-DNA insertion	LP: GAAACTCAACGTTCTCGGATG RP: CCCAAAACTTATGTACACGATTG LB: LBa1	[72]
MPK3	AT3G45640	MAPK signaling	*mpk3* (SALK_151594)	[74]	T-DNA	T-DNA insertion	F: CCGAGCAATCTTCTGTTGAACGCGAATTG R: TGCTGCACTTCTAACCGTATGTTGGATTG LB: p745	[75]
MPK4	AT4G01370	MAPK signaling	*mpk4-2* (SALK_056245)	[68]	T-DNA	T-DNA insertion	F: CGGTGAAACAATGACACGAGA R: CCGCTTCAACAGATGGTTACG	[69]
MPK6	AT2G43790	MAPK signaling	*mpk6-2* (SALK_073907)	[76]	T-DNA	T-DNA insertion	F: GCCTCAGATGCCTGGGATTGAGAATATTC R: AGAGTGGCTTACGGTCCATTAACTCCATG LB: p745	[75]
PP5	AT2G42810	MAPK signaling	*papp5-1* (SALK_021153)	[77]	T-DNA	T-DNA insertion	LP: TTCAAGCTCGCGCTATATCAC RP: CAGCCACTGACTTTAGTCATGC LB: LBa1	[77]

(continued)

Table 3
(continued)

Gene	AGI code	Function[a]	Polymorphism[b]	Reference 1[c]	Mutagen[d]	Marker[e]	Primer sequences (5′ → 3′)[f]	Reference 2[g]
ACD1, *PAO*	AT3G44880	HR cell death	*pao* (SALK_111333)	[78]	T-DNA	T-DNA insertion	LP: CGACGGTGAC AATTCAAAGGG RP: GGCTCACCTGA CGCTTGGTTA LB: LBb1	[78]
ACD2	AT4G37000	HR cell death	*acd2-2*	[79]	EMS	CAPS/*Tas*I	F: GTAGAAACAGAC TTCGATTTCGATGCAT R: CGGACAAAGAGAG AAGGAGACAC	[80]†‡
AtbZIP10, *BZO2H1*	AT4G02640	HR cell death	*atbzip10* (SALK_014867)	[81]	T-DNA	T-DNA insertion	F: CCATTGACGATTTCT CCGATCCTTTCTG R: GGGGAGACAGGAATCTG ATATGTATAAACAGGG LB: CCACCATCAAACAGGATT TTCGCCTGCTGGGGC	§
AtMC1	AT1G02170	HR cell death	*atmc1* (GK-096A10)	[82]	T-DNA	T-DNA insertion	LP: GCGTCACCTTCT CATCAACA RP: ACGGTACCACTAT GGCAAGC	§
AtMC2	AT4G25110	HR cell death	*atmc2* (SALK_009045)	[82]	T-DNA	T-DNA insertion	LP: CCTTTCCCCAAC TCATCTCC RP: GGCCGTACGAT TGTAGCATT LB: CCACCATCAAACAGGAT TTTCGCCTGCTGGGGC	§

DND1, CNGC2	AT5G15410	HR cell death	*dnd1-1*	[83, 84]	EMS	dCAPS/ *Mbo*I	F: TGCAGGCAGTGTTTTGGTTA R: ATGAGATTAAGAGCAAAACCCGA	[85]
DND2, CNGC4	AT5G54250	HR cell death	*dnd2-1*	[86, 87]	EMS	dCAPS/ *Nla*III	F: TCCAAATGGGTTCGAGCAT R: GCAATCTTGAACTGAATCC	[85]
LSD1	AT4G20380	HR cell death	*lsd1-2* (SALK_042687)	[81]	T-DNA	T-DNA insertion	LP: CACCAACTTTCCGCTTTCATCA RP: CTCCAATCAGGTTGCCCATGCT LB: LBa1	§
PEN1	AT3G11820	HR cell death	*pen1-1*	[88]	EMS	dCAPS/ *Mlu*I	F: GAAACACTCTCTTCATGTCACGCG R: GAGGACAGAGGTCCTGGTTCG	[36]
			SALK_004484C	[89]	T-DNA	T-DNA insertion	F: TTGCGAGCAGCTATCTTTAGC R: GGCGGTTTTATTGAAAAGTCC	[89]
PEN2	AT2G44490	HR cell death	*pen2-3*	[90]	EMS	CAPS/ *Bsm*AI	F: TGGCAAAAGGGATAGAGGAG R: CTGTAGACCCACGGCTCATT	[36]
			GABI_134C04	[89]	T-DNA	T-DNA insertion	F: AGGCTTTCTCTTTGGAACTGC R: TCCTTCGACATCATCTGGATC	[89]

(continued)

Table 3
(continued)

Gene	AGI code	Function[a]	Polymorphism[b]	Reference 1[c]	Mutagen[d]	Marker[e]	Primer sequences (5′ → 3′)[f]	Reference 2[g]
PEN3, PDR8	AT1G59870	HR cell death	*pen3-1*	[36]	EMS	CAPS/ *Hph*I	F: TGAAAGCTTCTG CTGCTCAA R: TGAGGTGAACGA TTTGTTGC	[36]
			atpdr8-1 (SALK_000578)	[91]	T-DNA	T-DNA insertion	RP: TGAAGATATCTTC TCATCTGGTTC LB: LBb1	[91]
			atpdr8-2 (SALK_142256)	[91]	T-DNA	T-DNA insertion	F: TGATTCCGATTCGTC ATCATC R: GTTCCTTCTTTTCC ATGGCTC	[89]
						T-DNA insertion	RP: TGATTGGTACA GTCTTCTGGC LB: LBb1	[91]
CPR1, CPR30	AT4G12560	NLR turnover	*cpr1-2 (cpr30-1)*	[20, 92]	Spontaneous	CAPS/ *Eco*RI	F: GAGGATGGTTCAGTTT AAGATGAATT R: CCCAATGAAGACT ATTACCAGCAA	[20]†
			cpr1-3 (*cpr30-2*, SALK_045148)	[92]	T-DNA	T-DNA insertion	LP: TTTCGTAAATTTTTA CACAAAATCG RP: TGTGAGTAGCCT TGTCTTGGG	[93]
HSP90.1	AT5G52640	NLR turnover	*hsp90.1-1* (SALK_007614) and *hsp90.1-2* (SALK_075596)	[94]	T-DNA	T-DNA insertion	RP: CTTAGCTTGT GCTCGATCTTC LB: LBb1	[94]

HSP90.2, MUSE12	AT5G56030	NLR turnover	*hsp90.2-3*	[95]	EMS	dCAPS/ *Ase*I	F: CTTCGATTTTTTCCGATCTACGACAATGGC R: CAAGGTTGTTGACCAAATCTAATAGAAGAAGCACG	§
			hsp90.2-7	[96]	EMS	dCAPS /*Hin*dIII	F: GATGATGAGGGAAAGCAACTGGTTGATCTAAG R: GGGCCTTAAAATGGCCCCCATCA	§
IMPA-3, MOS6	AT4G02150	NLR turnover	*mos6-2*	[97]	Fast neutron	INDEL	F: CATTAATCGAACTTGGCG R: CACATCGCAAGTTTTTGC	[98]
NUP88, MOS7	AT5G05680	NLR turnover	*mos7-1*	[99]	Fast neutron	INDEL	F: CGTGGCACCCTGATAGTG R: GAATAATATAAAGACCTGCACC	[98]
NUP96, MOS3	AT1G80680	NLR turnover	*mos3-1*	[100]	Fast neutron	INDEL	F: CTTGTATGATCCTTCTTCATTC R: CTGACAGAATAATAATGTCTG	[98]
RAR1, PBS2	AT5G51700	NLR turnover	*rar1-21*	[101]	EMS	CAPS/ *Cac*8I	F: AACTTTTGCCACCGGTTATG R: GGCCAGAACTGGTTTCTCAG	[102]
SGT1A	AT4G23570	NLR turnover	*sgt1a-3* (*sgt1aKO*, SALK_122139)	[103]	T-DNA	T-DNA insertion	LP: CTTGAAAAGGGTGCCTCTATCACGC RP: CATCAGATGACACTGAAGAAGGAAAAGG LB: CCACCATCAAACAGGATTTTCGCCTGCTGGGGC	§

(continued)

Table 3
(continued)

Gene	AGI code	Function[a]	Polymorphism[b]	Reference 1[c]	Mutagen[d]	Marker[e]	Primer sequences (5′ → 3′)[f]	Reference 2[g]
SGT1B, EDM1	AT4G11260	NLR turnover	*eta3*	[104]	EMS	CAPS/ *Alu*I	F: AGGATGAGAAGCTTGATGGAGATGCACC R: CGTCCCATTCGACTCTGCCTGTCAAAGC	[102]
SRFR1	AT4G37460	NLR turnover	*srfr1-3*	[103]	EMS	CAPS/ *Taq*I	F: CAGGGGAAGTAATCTTATCGGATATCAC R: CAATTTTCCTGTCTTGACCAGGGTTCG	[103]
			srfr1-4 (*snc5-2*, SAIL_412_E08)	[103, 105]	T-DNA	T-DNA insertion	F: TCATCACTAATTCCGCAACG R: CGACTTATGTAACGGATCAG	[103]
			srfr1-5 (*snc5-3*, SAIL_216_F11)	[103]	T-DNA	T-DNA insertion	F: CTATGGTTCTACTGAGCTCG R: TGCTCATGGTTTAGTTAGCC	[103]
AGO1	AT1G48410	Gene silencing	*ago1-3*	[106]	EMS	CAPS/ *Bse*GI	F: CACAGGAATCATCATGGTGAG R: CTCGGGAACTGATTGTCTCTG	[107]†
			ago1-27	[108]	EMS	CAPS/ *Bsp*1286I	F: ACCACGTTCTTTGGGATGAG R: TCTACCCATTCCACCTC	[109]‡
			ago1-36 (SALK_087076)	[110]	T-DNA	T-DNA insertion	LP: AATCAGGTATATTCCGGTGGG RP: CAATGAGGCTTTATCACCAGC LB: LBb1.3	[111]

AGO2	AT1G31280	Gene silencing	*ago2–1* (SALK_003380)	[112]	T-DNA	T-DNA insertion	LP: TAAGATGATAAAGGCTCGTA	[107]
							RP: ACAACAGCAACAATGGAC	
							LB: LBb1.3	
						T-DNA insertion	LP: TCAATCATCGGAAGAACAAGC	[113]
							RP: CATAAGCTCGACATTGCTTCC	
							LB: LBa1	
						T-DNA insertion	LP: CCATTGTAGGGCTGAGTATGC	[114]
							RP CGTTTCCCTGTGGCCTGAACA	
							LB: LBb1.3	
			SALK_037548	[115]	T-DNA	T-DNA insertion	LP: CTGCTTAAAGTTCTTCTCACGAG	[116]
							RP: TTGTGCGTGAGGAACCCAA	
							LB: LBb1.3	
AGO3	AT1G31290	Gene silencing	*ago3-2* (SALK_005335)	[113]	T-DNA	T-DNA insertion	LP: GTCTCCGACGCCTACTTGAC	[113]
							RP AAACAGAGAGACAGTGGACGC	
							LB: LBa1	
						T-DNA insertion	LP: CGATAGTCCCGACTGACTCTG	[114]
							RP AAACAGAGAGACAGTGGACGC	
							LB: LBb1.3	
AGO4	AT2G27040	Gene silencing	*ago4-2*	[117]	EMS	dCAPS/*Xba*I	F: ATGCCATCGTCTTCAGTTCCA	[117]
							R: GCCTGAACTCAATGTTAAGTCTA	
			ago4-6 (SALK_071772)	[118]	T-DNA	T-DNA insertion	LP: TTCTCCAGCTGGCTAGCTATG	[119]
							RP CCCAGAAAGGTGACATCTTTG	

(continued)

Table 3
(continued)

Gene	AGI code	Function[a]	Polymorphism[b]	Reference 1[c]	Mutagen[d]	Marker[e]	Primer sequences (5′ → 3′)[f]	Reference 2[g]
AGO5	AT2G27880	Gene silencing	*ago5-1* (*ago5-3*, SALK_063806)	[120]	T-DNA	T-DNA insertion	LP: ATCCACAACG TGGGCTAGTCC RP AGCATGGCTG TTCAAATAGAAGTC LB: LBa1	[121]†
			ago5-2 (SALK_118422)	[113]	T-DNA	T-DNA insertion	LP: CCTCGATTG AAGCTCGTGTAC RP: TTCAAGGCAAC ATTTTCCATG LB: LBa1	[113]
						T-DNA insertion	LP: TTGTGGTTTAAT GTGTGTTACC RP: ACTGAGAGACAA TCCCCAATTCTG LB: LBa1	[122]
			ago5-4 (SALK_050483)	[123]	T-DNA	T-DNA insertion	LP: CTACCCATCAGG GAGCTAAGG RP TTTCTGGACCA TATCACAGCC LB: LBb1.3	[114]
			ago5-5 (SALK_123506)	[122]	T-DNA	T-DNA insertion	LP: TTGTGGTTTAATG TGTGTTACC RP: AGGTATCTGGC CTGTAGCTCTC LB: LBa1	[122]
			ago5-6 (SALK_062912)	[122]	T-DNA	T-DNA insertion	LP: TGACCTGTTTTG TCTCAGCAGAAT RP: AGGTATCTGGCC TGTAGCTCTC LB: LBa1	[122]

AGO6	AT2G32940	Gene silencing	*ago6-2* (SALK_031553)	[124]	T-DNA	T-DNA insertion	LP: TCTTAGAACG ACAATGGTGG RP: ACTCTAAGTG CATCCTGAGC	[119, 124]
						T-DNA insertion	LP: TGAAAATGAC CCTTCCACGATG RP: TTTCGGAGAA TTTGCATGAAGC	[125]
			ago6-3 (SALK_106607)	[113]	T-DNA	T-DNA insertion	LP: CAGATGCATT GTGTGTGGATC RP: TATCCCACTT CTGGAGGTGTG LB: LBa1	[113]
AGO7, *ZYPPY*	AT1G69440	Gene silencing	*zip-1*	[126]	EMS	CAPS/ *Bsa*I	F: CTGTACTTTGAC AGCGGAAACC R: ACTGGCTTGGACT TTCTACTAGGTTC	[127]
			ago7-1 (SALK_037458)	[128]	T-DNA	T-DNA insertion	LP: CTCCCTGAAAC TCCACAGGG RP: AGCTCCTTTTCC TCTCAGCAG LB: LBb1	[128]
						T-DNA insertion	LP: CCATCAAACTT GGACACAAGC RP: TCCTCCTCCT CATCTTCTTCC LB: LBa1	[113]

(continued)

Table 3
(continued)

Gene	AGI code	Function[a]	Polymorphism[b]	Reference 1[c]	Mutagen[d]	Marker[e]	Primer sequences (5′ → 3′)[f]	Reference 2[g]
AGO8	AT5G21030	Gene silencing	*ago8-1* (SALK_139894)	[113]	T-DNA	T-DNA insertion	LP: GCAGTGTAGTCCACTTAGTCCG RP: TGTTCCTGTTTCCATGACTTG LB: LBa1	[113]
						T-DNA insertion	LP: CAAAAGACTAAAGCTTCC RP: GTCATCATTATAATTACTAGTC LB: LBa1	[122]
						T-DNA insertion	LP: TCCCTGTTTTGGTTCCTTTTC RP: TCCTGTTCCTGTTTCCATGAC LB: LBb1.3	[114]
						T-DNA insertion	LP: ATCGTTCACACCTTGATTTGC RP: AACATTTTGCTTATGATGGCG	[119]
			ago8-2 (SALK_151983)	[121]	T-DNA	T-DNA insertion	LP: CACTTACAATCTTTCCAG RP: CTTGGTGGATTGAATTCAGTTTTGG LB: LBa1	[121]†
			ago8-3 (SALK_060402)	[122]	T-DNA	T-DNA insertion	LP: GAATCTCAGTTTAACCAAG RP: CACTTACAATCTTTCCAG LB: LBa1	[122]

AGO9	AT5G21150	Gene silencing	*ago9-1* (SALK_127358)	[120]	T-DNA	T-DNA insertion	LP: AGCTGAAAGTGGGCAAGGGAG RP: CCAGGGGAACCATGGGATACA	[125]
						T-DNA insertion	LP: GGGATACATCCATCCCAACA RP: AGGCCGTATCTTACCACACC LB: LBb1.3	[114]
			ago9-2 (SALK_112059)	[113]	T-DNA	T-DNA insertion	LP: CAAGTTCTTGGAGGTCGTCTG RP: ATGTGTACGCACACAATCACG LB: LBa1	[113]
			ago9-3 (SAIL_34_G10)	[129]	T-DNA	T-DNA insertion	LP: TGCAGGAACAATCATTGACAG RP: TCACGAAAAGAGCGAAATTTG	[119]
AGO10	AT5G43810	Gene silencing	*ago10-1* (SALK_000457)	[113]	T-DNA	T-DNA insertion	LP: AGGTGGCAATCAAGTTTGTTG RP: AATTTTGCATGCCTACATTGG LB: LBa1	[113]
						T-DNA insertion	LP: ACTCGGTGAAGGGTTAGAGTC RP: AACTTGGAGACATGGCAAATGC LB: LBb1.3	[130]
			ago10-2 (SALK_047336)	[113]	T-DNA	T-DNA insertion	LP: CGGGTTTAGTTTGTGCTCAAG RP: CCGCGCATAGGTATAACAGAG LB: LBa1	[113]

(continued)

Table 3
(continued)

Gene	AGI code	Function[a]	Polymorphism[b]	Reference 1[c]	Mutagen[d]	Marker[e]	Primer sequences (5′ → 3′)[f]	Reference 2[g]
BRU1	AT3G18730	Gene silencing	*bru1-4* (SALK_034207)	[131]	T-DNA	T-DNA insertion	RP TAGCGGATTCCTTGAGGAG LB: GCTGAGAGTCTGCTACTAC	[131]
CMT3	AT1G69770	Gene silencing	*cmt3-11t* (SALK_148381)	[132]	T-DNA	T-DNA insertion	LP: AGATCGCTTCCAGAGTTAGCC RP: ATAAGAGAAGGAGCTGCTGCC LB: LBb1.3	[133]
						T-DNA insertion	LP: TAACGGAAGGATGCCAGATT RP: CAAGAAATGGGCTGTTGACAT LB: LBa1	[134]
DCL1	AT1G01040	Gene silencing	*dcl1-9* (*caf-1*)[h]	[135]	T-DNA	T-DNA insertion	RP AGTAGAAGAACCGCAGCTGAATC LB: CACACACACATCATCTCATTGATGCTTGG	[136]

DCL2	AT3G03300	Gene silencing	*dcl2-1* (SALK_064627)	[137]	T-DNA	T-DNA insertion	LP: CTCAGAAATAAAG ATAACAGTAAGCAAAT RP: TTGGATTGCAT GCACACATT LB: LBa1	[138]†
						T-DNA insertion	LP: TGTTAACAGA TCTCCGCTGCCA RP: TTGGCTGAGA TACCTCAAGGTGG	[125]
						T-DNA insertion	LP: TGAATCATCT GGAAGAGGTGG RP: CTTCACAGG AGTTTTTGGCTG LB: LBb1.3	[116]
			dcl2-3 (SALK_095069)	[128]	T-DNA	T-DNA insertion	LP: CATGCAGGG GGAAAGCACCC RP: CCCAAAAAA GACGCCACTCGC LB: LBb1	[128]

(continued)

Table 3
(continued)

Gene	AGI code	Function[a]	Polymorphism[b]	Reference 1[c]	Mutagen[d]	Marker[e]	Primer sequences (5′ → 3′)[f]	Reference 2[g]
DCL3	AT3G43920	Gene silencing	*dcl3-1* (SALK_005512)	[137]	T-DNA	T-DNA insertion	LP: TGAAGAACAG GTAACCTTGCC RP: CTGAAGAGCG TGAAGGAGTGG LB: LBb1	[128]
						T-DNA insertion	LP: TGATCCAGAACA TGACTCTTTCC RP: CCGATACATTG GTGGAGGGGT	[125]
						T-DNA insertion	LP: ACAGGTAACCT TGCCATGTTG RP: TGGAAAAGTTT GCTACAACGG LB: LBb1.3	[116]
						T-DNA insertion	LP: GGACTCAATGCA ATATAGAGCTTTG RP: CTGAATATGGATA ATAAGTTTGA GACATATC LB: LBa1	[139]†
DCL4	AT5G20320	Gene silencing	*dcl4-2*	[140]	EMS	dCAPS/*Mbo*II	F: TGAAGAGCATG TCAAGAAGGAG R: GAGCACGACCT CTGGACTGT	[140]
DML2	AT3G10010	Gene silencing	*dml2-2* (SALK_113573)	[141]	T-DNA	T-DNA insertion	F: GATTTGGTGG TCAACAAGGAA R: TCGAGCTTTGA CGAGAATGTT	[141]

DML3	AT4G34060	Gene silencing	*dml3-2* (SALK_056440)	[142]	T-DNA	T-DNA insertion	F: CAGAACGACCC ATCAAGAAAA R: TCAACAACTGA GCCTTTCCAC	[141]
DRB4	AT3G62800	Gene silencing	*drb4-1* (SALK_000736)	[143]	T-DNA	T-DNA insertion	F: CCCTCGAAGTG TCTTCACC R: GTGGCGTCAAA GAAGCAACTG	[143]
DRM1	AT5G15380	Gene silencing	*drm1-2* (SALK_031705)	[144]	T-DNA	T-DNA insertion	LP: CCTGTGTTGAT TGGGATTCAG RP: GTCGATGGAGTG CAACTTCTC LB: LBb1.3	[133]
DRM2	AT5G14620	Gene silencing	*drm2-2* (SALK_150863)	[144]	T-DNA	T-DNA insertion	LP: AGATCGCTTCCA GAGTTAGCC RP: TTGTCGCAAAA AGCAAAAGAG LB: LBb1.3	[133]
HEN1	AT4G20910	Gene silencing	*hen1-5* (SALK_049197)	[145]	T-DNA	T-DNA insertion	LP: CCAGATCCGC ATCCAAAGTC RP: GCCTGTTCTGC CACCAGCA	[125]
HYL1	AT1G09700	Gene silencing	*hyl1-2* (SALK_064863)	[145]	T-DNA	T-DNA insertion	F: AGTGTATCTGT CCCTTGTGG R: GCCTAAGAAT CTTCCCTCTA	[146]
KYP, *SUVH4*	AT5G13960	Gene silencing	*kyp-4* (*suvh4*, SALK_044606)	[131]	T-DNA	T-DNA insertion	F: AGCTGCCTCG GCAGTTAC R: CCAGGTACGT CTGTCTCTC	[131]

(continued)

Table 3
(continued)

Gene	AGI code	Function[a]	Polymorphism[b]	Reference 1[c]	Mutagen[d]	Marker[e]	Primer sequences (5′ → 3′)[f]	Reference 2[g]
MET1	AT5G49160	Gene silencing	*met1-3*	[147]	T-DNA	T-DNA insertion	LP: TTCGCAAACCA TTCTTCACAGAGC RP: TAGCCAACAAG TTATCGCTTACTC LB: TAATTGCGTCG AATCTCAGCATCG	§
MOM1	AT1G08060	Gene silencing	*mom1-2* (SAIL_610_G01)	[148]	T-DNA	T-DNA insertion	LP: GATTGTCCACCA CCTGCAGATGCAGG RP: GCAACTGTAGCA CATGCATCCAGCTC LB: SAIL_LB3	[139]†
MORC1, *CRT1*	AT4G36290	Gene silencing	*atmorc1-2* (*crt1-2*, SAIL_893_B06)	[149]	T-DNA	T-DNA insertion	LP: TGAGTTTTGAC GACGATGATG RP: TTGCAGTTTG GAACCAAAATC	[134]
			atmorc1-4 (SAIL_1239_C08) and *atmorc1-5* (*crt1-5*, SAIL_131_H11)	[134]	T-DNA	T-DNA insertion	LP: AAGCAGCTGC AGTGGATTATG RP: CGTATCTCAG CCGCTAACTTG	[134]
MORC2, *CRH1*	AT4G36280	Gene silencing	*atmorc2-2* (*crh1-1*, SALK_072774)	[149]	T-DNA	T-DNA insertion	LP: GTTGTAGCTG TATGGGGCTTG RP: CTACTCAGAGC GTTGGCATTC	[134]
			atmorc2-4 (*crh1-4*, SALK_021267)	[134]	T-DNA	T-DNA insertion	LP: GGTTGACTCT TCCACTGCTTG RP: TTTCGTCATCA TTGCTTTTCC	[134]

NRPD1a	AT1G63020	Gene silencing	*nrpd1a-3* (SALK_128428)	[150]	T-DNA	T-DNA insertion	LP: TTCACCTTTTTGATCCCTTGATC RP: TGTTCATTCAGTTACAAGTCGTG LB: LBa1	[150]†‡
						T-DNA insertion	F: TGAGAAAGCATTGACTTGGGC R: AGCCTTCTTGAAACTTCGCCG	[131]
						T-DNA insertion	LP TTTTGATCCCTTGATCACCTG RP CTGATGGTCCCGTTGAAGATA LB: LBb1.3	[116]
			nrpd1a-4 (SALK_083051)	[150]	T-DNA	T-DNA insertion	F: GGGTTCGAATACGGGTCACTTGA R: TGTTACATACTGAGAAGCATGCT LB: LBa1	[150]†‡*
						T-DNA insertion	LP: TGGGTTTGCCATTTTCATATC RP: CCACAAACTCTCCCGTAACAG	[151]
NRPD1b	AT2G40030	Gene silencing	*nrpd1b-11* (SALK_029919)	[152]	T-DNA	T-DNA insertion	LP: CCCAGTTCTGCTGGAAATAGTG RP TCTGTGAATGGAGCCTTGATG LB LBb1.3	[116]
			nrpd1b-12 (SALK_033852)	[152]	T-DNA	T-DNA insertion	LP: AGGCACCAAGAAAATGTTTTG RP: TTATTTTTGTCCCTGGAACCC	[151]

(continued)

Table 3
(continued)

Gene	AGI code	Function[a]	Polymorphism[b]	Reference 1[c]	Mutagen[d]	Marker[e]	Primer sequences (5′ → 3′)[f]	Reference 2[g]
NRPD2a	AT3G23780	Gene silencing	*nrpd2a-1* (SALK_095689)	[150]	T-DNA	CAPS/*Dra*I	F: TTGGCTTTTACATCTTGCAGGTTC R: TCAGCAGAGCTTGGTGTCGAAGTTGAGAG	[150]†‡
						T-DNA insertion	F: CATTGTCTCTTGGTTTTAGCTCG R: TCATCAGTGGCTCGGTTTTAC	[151]
NRPD2b	AT3G18090	Gene silencing	*nrpd2b-1* (SALK_008535)	[150]	T-DNA	T-DNA insertion	F: CACCATGGCATGCAGGCTTCAGAACATGACATAC R: AAATCCACAATCTCTTTGTGCACA LB: LBa1	[150]†‡*
RDR1	AT1G14790	Gene silencing	*rdr1-1* (SAIL_672F11)	[137]	T-DNA	T-DNA insertion	LP: CCAAGATTTGAGTTGGTTCAGG RP: GGGAAGTTGTGGATCGTCTCGACGC LB: SAIL_LB2	[107]

RDR2	AT4G11130	Gene silencing	*rdr2-1* (SAIL_1277H08)	[137]	T-DNA	T-DNA insertion	LP: ATGGTGTCAGAGACGACGACGAACCGATCAAC RP: ACACATTAGGACTAACAAATTTACC LB: SAIL_LB2	[107]
			rdr2-2 (SALK_059661)	[128]	T-DNA	T-DNA insertion	LP: GGTAAGGTTCTGAACTCG RP: CATTTTCTCGGGCTTACC LB: LBa1	[150]†‡
						T-DNA insertion	LP: ATCACACCTTTGTAGCCACCG RP: GAAGGGATTACCGTTGGGCC LB: LBb1	[128]
RDR6, *SDE1*, *SGS2*	AT3G49500	Gene silencing	*sgs2-18*	[143, 153]	EMS	CAPS/*Stu*I	F: CTTAAGTTGGATGTGCACTAAGGCC R: AGAACTTCTTCATACTGTCTCGAAG	[107]
			rdr6-11	[127]	T-DNA	CAPS/*Taq*I	F: TACTGTCCCTGGCGATCTCT R: CCACCTCACACGTTCCTCTT	[127]
			rdr6-12	[127]	Fast neutron	CAPS/*Bfa*I	F: TGCAAGAGGAACGTGTGAGGTG R: GCTTCAACCTCTTGTACGCATC	[107]

(continued)

Table 3
(continued)

Gene	AGI code	Function[a]	Polymorphism[b]	Reference 1[c]	Mutagen[d]	Marker[e]	Primer sequences (5′ → 3′)[f]	Reference 2[g]
ROS1	AT2G36490	Gene silencing	*ros1-4* (SALK_045303)	[154]	T-DNA	T-DNA insertion	LP: CGTGGATTACATAACATGTTATTTG RP: TTGTTCCCAACAAATCTCCTG LB: LBb1.3	[133]
						T-DNA insertion	F: ACATGTCCAGCGCTTTAGTTG R: ATTGATTGGGTTTAGGCTGGATG	[141]
SDC	AT2G17690	Gene silencing	*sdc* (SALK_017593)	[151]	T-DNA	T-DNA insertion	F: TCATAAACTGCGGTTCCAGAC R: CGTTCCATATTCCTCCTTTCC	[151]
SDE3	AT1G05460	Gene silencing	*sde3-4* (SALK_092019)	[128]	T-DNA	T-DNA insertion	LP: AGGCCAGGAACTTCGAGGGA RP: TTATGTGAAGGGATCACGCCC LB: LBb1	[128]

SDE5	AT3G15390	Gene silencing	*sde5-2* (SALK_020726)	[155]	T-DNA	T-DNA insertion	LP: TCTACATCCTGCAACCAAATGG RP: CCTTCTCCACGAGTCTGTGTG LB: LBa1	[122]
			sde5-5 (SALK_026595)	[122]	T-DNA	T-DNA insertion	LP: CTTTCAAGTACCTCAGAGTCC RP: GAATCAAGGGAATGCTCAGCACG LB: LBa1	[122]
			sde5-6 (SALK_115496)	[122]	T-DNA	T-DNA insertion	LP: CCTTCTCCACGAGTCTGTGTG RP: TCTACATCCTGCAACCAAATGG LB: LBa1	[122]
			sde5-7 (SALK_026511)	[122]	T-DNA	T-DNA insertion	LP: CAGATGAAAGAATATTATGGAG RP: CCAGAGAAGTAAATAAGTTGG LB: LBa1	[122]

(continued)

Table 3
(continued)

Gene	AGI code	Function[a]	Polymorphism[b]	Reference 1[c]	Mutagen[d]	Marker[e]	Primer sequences (5′ → 3′)[f]	Reference 2[g]
SGS3, *SDE2*	AT5G23570	Gene silencing	*sgs3-1*	[153]	EMS	CAPS/*Bfa*I	F: TCACAAGATGA TGGAGGAGACT R: CTAGGAGGATT CGAATTCTGAG	[107]
			sgs3-11	[127]	EMS	dCAPS/*Pst*I	F: CAAAAAACCTG TGGTGGTCTGCA R: ACAACCTTGGC ACGTTCCTGC	[127]
			sgs3-14 (SALK_001394)	[127]	T-DNA	T-DNA insertion	LP: GGTAACGATAC CTGCGTCTTACTG RP: TTATCATACTTT CTGCTCAACGCTC LB: LBb1.3	[107]

[a]All genes are classified into nine groups depending on the function of encoding proteins [salicylic acid (SA) signaling; systemic acquired resistance (SAR) signaling; ethylene (ET) signaling; jasmonic acid (JA) signaling; reactive oxygen species (ROS) signaling; mitogen-activated protein kinase (MAPK) signaling; hypersensitive response (HR) cell death; nucleotide-binding site-leucine-rich repeat protein (NLR) turnover; or gene silencing]

[b]Only polymorphisms in ecotype Col-0 background are listed

[c]Reference to each mutant line

[d]EMS: ethyl methanesulfonate, DEB: Diepoxybutane; T-DNA: T-DNA insertion [7–11]; dSpm: dSpm transposon insertion [12]; En1/Spm: En1/Spm transposon insertion [13]

[e]For CAPS and dCAPS markers, restriction enzymes are indicated

[f]A primer pair of "F (Forward)" and "R (Reverse)" or a primer pair of "RP" and "LP" corresponds to *A. thaliana* genomic sequence. Primer "LB (left border)" or "RB (right border)" corresponds to the sequence of exogeneous insert. In the case of T-DNA or transposon insertion, wild-type alleles are specifically amplified by a pair of LP and RP primers or a pair of F and R primers. By contrast, mutant alleles are specifically amplified by a pair of RP and LB primers or a pair of LP and RB primers

[g]Reference to primer sequence for PCR genotyping. References with a dagger (†) describes product size of molecular markers. Reference with a double dagger (‡) describes condition of PCR genotyping. A section mark (§) indicates that primer information was referred from seed catalog. It was experimentally confirmed that the multiplex PCR can be performed with the primer sets marked with an asterisk (∗)

[h]Mutant line in Ws/Ler background. Partial loss-of-function mutation. Null dcl1 mutant alleles are embryo-lethal

2.6.2 Polyacrylamide Gel Electrophoresis

1. 30% acrylamide solution: 29% (w/v) acrylamide plus 1% (w/v) *N,N'*-methylenebisacrylamide.
2. 5× TBE: 445 mM Tris, 445 mM boric acid, 10 mM EDTA (pH 8.0).
3. 10% (w/v) ammonium persulfate (APS), molecular biology grade.
4. *N,N,N',N'*-tetramethylethylenediamine (TEMED), molecular biology grade.
5. 6× Gel Loading Dye, Purple (New England Biolabs Inc.) (*see* **Note 4**).
6. Size marker.
7. 1-mm Dual Mini Gel Cast (gel size of 90 mm × 80 mm × 1 mm; ATTO Co., Ltd., Tokyo, Japan).
8. Mini Slab (ATTO Co., Ltd.).
9. Power supply.
10. 10 mg/mL ethidium bromide solution, stored in the shade at 4 °C.
11. UV transilluminator with the camera.

2.7 Analysis of Resistance to Virus Infection

1. Virus inoculum: 10–100 μg/mL CMV.
2. Carborundum: 600 mesh. Put carborundum in a conical flask, cover the mouth of the flask with quadruple folded-cotton gauze, fasten the gauze with kite strings, and then sterilize the flask by dry heat.
3. Cotton swabs, sterilized by autoclave.

3 Methods

The procedure for the reverse genetic analysis of defense signaling downstream of antiviral resistance genes is summarized in Fig. 2.

3.1 Plant Cultivation

1. Put "Pro-Mix BX Mycorrhizae" soil into the autoclave bag. Wet the soil with water and mix by hand in the bag. Sterilize by autoclave for 20 min at 120 °C. Keep at room temperature.
2. Put the sterilized soil into the open-bottom plastic pots placed on the tray.
3. Pour an appropriate volume of liquid fertilizer (e.g., 1:1000 HYPONeX–water) into the tray.
4. Sow *A. thaliana* seeds on the soil. Overlay the entire surface of the tray with the plastic wrap. Place the tray with the seed-sown pots overnight at 4 °C (*see* **Note 5**).

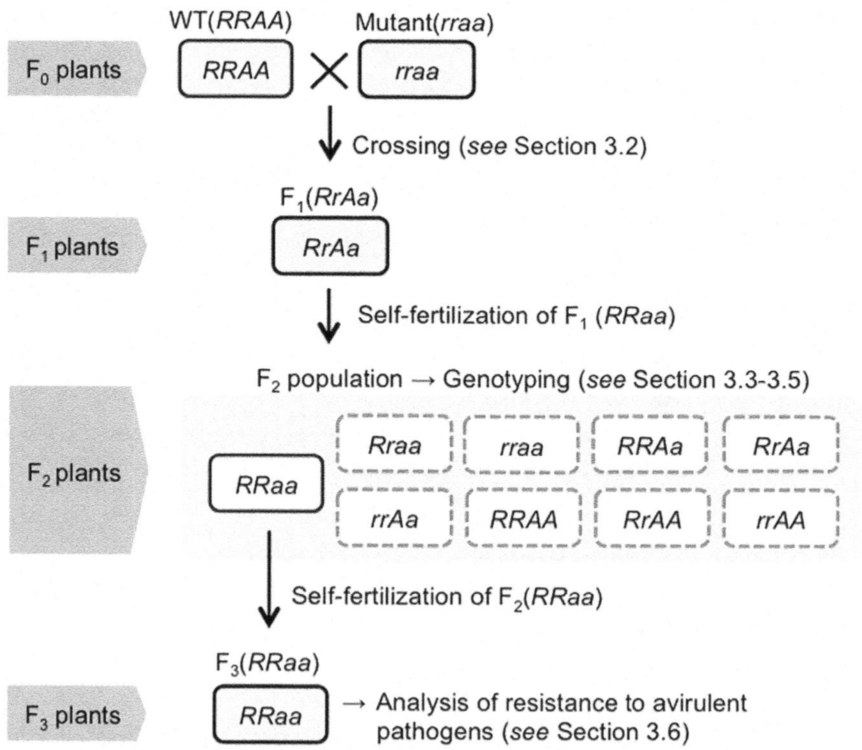

Fig. 2 Summarized procedure for the reverse genetic analysis of R protein-conferred resistance signaling. Functional haplotypes of the ***R*** gene and a gene encoding a defense signaling component are indicated by upper case "***R***" and "***A***," respectively. Lower case "***r***" and "***a***" indicate the corresponding recessive alleles. Plant genotypes are shown in rounded rectangles. A wild-type plant "WT(***RRAA***)" carrying a homozygous functional ***R*** gene (***R***) and homozygous defense signaling gene (***A***) was reciprocally crossed with a mutant line "mutant(***rraa***)" carrying a non-functional ***R*** gene allele (***r***) and a mutant ***A*** gene allele "***a***." The F_1 plant with a heterozygous genotype of ***R/r*** allele and ***A/a*** allele (***RrAa***) was self-pollinated to obtain F_2 progenies. F_2 lines carrying the "***RRaa***" genotype can be isolated from the F_2 population by PCR-based genotyping. Resistance to avirulent pathogens was analyzed in each F_3 line (***RRaa***) obtained from the individual self-fertilized F_2 line (***RRaa***)

5. Transfer the set into a growth chamber. Remove the plastic wrap after seed germination. Grow plants at 23 °C under short-day conditions (10:14-h light–dark cycle) or at 25 °C under constant illumination with 8000–10,000 lux lights (*see* **Note 6**).
6. Pour water into the tray frequently and irrigate with a liquid fertilizer every 3 weeks.

3.2 Plant Crossing

1. Grow both *A. thaliana* mutant lines [mutant(*rraa*)] and *A. thaliana* containing the antiviral resistance gene of interest [WT(*RRAA*)] until they flower. Define the maternal and paternal F_0 plant lines (*see* **Note 7**).
2. Pick all of the reproductive tissues off from the stem of the maternal plants, except for the mature bud.

3. Peel all of the sepals, petals, and stamens off with the precision stainless steel tweezers and expose the pistil of the buds in the maternal plants under the stereoscopic microscope.
4. Pick the mature stamens from the fresh flowers in the paternal plants and pollinate the stigma in the maternal plants under the stereoscopic microscope.
5. Grow the pollinated plants in the growth chamber. Harvest the fully ripened and dried F_1(*RrAa*) seeds at about 2 weeks after the pollination.
6. Grow the F_1(*RrAa*) plants, self-fertilize them, and harvest the F_2 seeds. Use the F_2 plants for PCR-based genotyping.

3.3 Genomic DNA Extraction

Genomic DNA, which can be used as a template in PCR, can be extracted from the leaf tissues of *A. thaliana* by a quick and simplified method (Subheading 3.3.1). When PCR products are not amplified using genomic DNA extracted by the simplified method, use genomic DNA extracted by the CTAB method (Subheading 3.3.2) [156].

3.3.1 Simplified Method

1. Put the leaf samples in a 1.5-mL microcentrifuge tube. Add 10 volumes (w/v) of extraction buffer and grind the leaf samples using the electric drill with the plastic pestle. Centrifuge for 1 min at 12,000 × *g*. Transfer the supernatant to a new microcentrifuge tube.
2. Add one volume of 2-propanol to the supernatant, mix well, and incubate for 2 min at room temperature. Centrifuge for 5 min at 12,000 × *g*. Discard the supernatant and dry the pellet.
3. Dissolve the pellet in 100 μL TE.

3.3.2 CTAB Method

1. Put the leaf samples in a 1.5-mL microcentrifuge tube. Add 10 volumes (w/v) of 2% CTAB solution and grind the leaf samples thoroughly using the electric drill with the plastic pestle.
2. Add an equal volume of chloroform–isoamyl alcohol (24:1) and mix well by inversion for 5 min. Centrifuge for 5 min at 12,000 × *g*. Transfer the aqueous phase to a new microcentrifuge tube.
3. Repeat **step 2** once.
4. Add 1.5 volumes of 1% CTAB solution, mix well, and leave for 1 h at room temperature. Centrifuge for 10 min at 5500 × *g*. Discard the supernatant. Dissolve the pellet in 100 μL TE.
5. Add 0.1 volumes of 3 M sodium acetate (pH 5.2) and 2.5 volumes of ethanol. Mix well and place at −20 °C for 20 min.
6. Centrifuge for 15 min at 12,000 × *g*. Discard the supernatant. Wash the DNA pellet with 70% ethanol and centrifuge for 5 min at 12,000 × *g*. Discard the supernatant and dry the pellet.
7. Dissolve the pellet in 50 μL TE containing 10 μg/mL RNaseA. Incubate for 30 min at 37 °C.

8. Add an equal volume of phenol–chloroform–isoamyl alcohol (25:24:1) and mix well by inversion for 5 min. Centrifuge for 5 min at 12,000 × *g*. Transfer the aqueous phase to a new microcentrifuge tube.
9. Repeat **step 2** once.
10. Repeat **steps 5** and **6** once.
11. Dissolve the pellet in 50 μL TE. The DNA solution can be stored at −20 °C.

3.4 PCR-Based Genotyping

Point mutations that overlap with the recognition sites of restriction enzymes can be detected by cleaved amplified polymorphic sequence (CAPS) markers (Fig. 3a) [157]. Point mutations not overlapping with recognition sites can be alternatively detected by derived CAPS (dCAPS) markers (Fig. 3a) [158]. Insertion/deletion (INDEL) mutations containing T-DNA and transposon insertions can be detected by the size differences in the PCR products amplified using primers corresponding to sequences on the inside and near-outside of the INDELs (Fig. 3b). Fragments containing these molecular markers are first amplified by PCR (Subheading 3.4.1). To detect CAPS and dCAPS markers, PCR products are further digested with restriction enzymes (Subheading 3.4.2). Primers corresponding to the border sequences of T-DNA and transposons are useful to detect insertional mutants (Table 2). The mutant lines characterized are listed in Table 3 with information on the genotyping markers and primer sequences.

3.4.1 PCR

1. Prepare the reaction mix in 0.2-mL PCR tubes as follows (*see* **Note 2**). Genomic DNA extracted from WT(*RRAA*), mutant(*rraa*), F_1(*RrAa*) (optional), and F_2 plants (genotype uncharacterized) should be used as templates (Fig. 2) (*see* **Note 8**).

	Volume	Final conc.
5× Green or Colorless GoTaq Reaction Buffer	2 μL	1×
2 mM dNTP Mix	1 μL	0.2 mM each
20 μM forward primer	0.25 μL	0.5 μM
20 μM reverse primer	0.25 μL	0.5 μM
Genomic DNA solution	*x* μL	<100 ng/10 μL
5 μ/μL GoTaq DNA polymerase	0.05 μL	1.25 U
Nuclease-free water	(6.45-*x*) μL	
In total	10 μL	

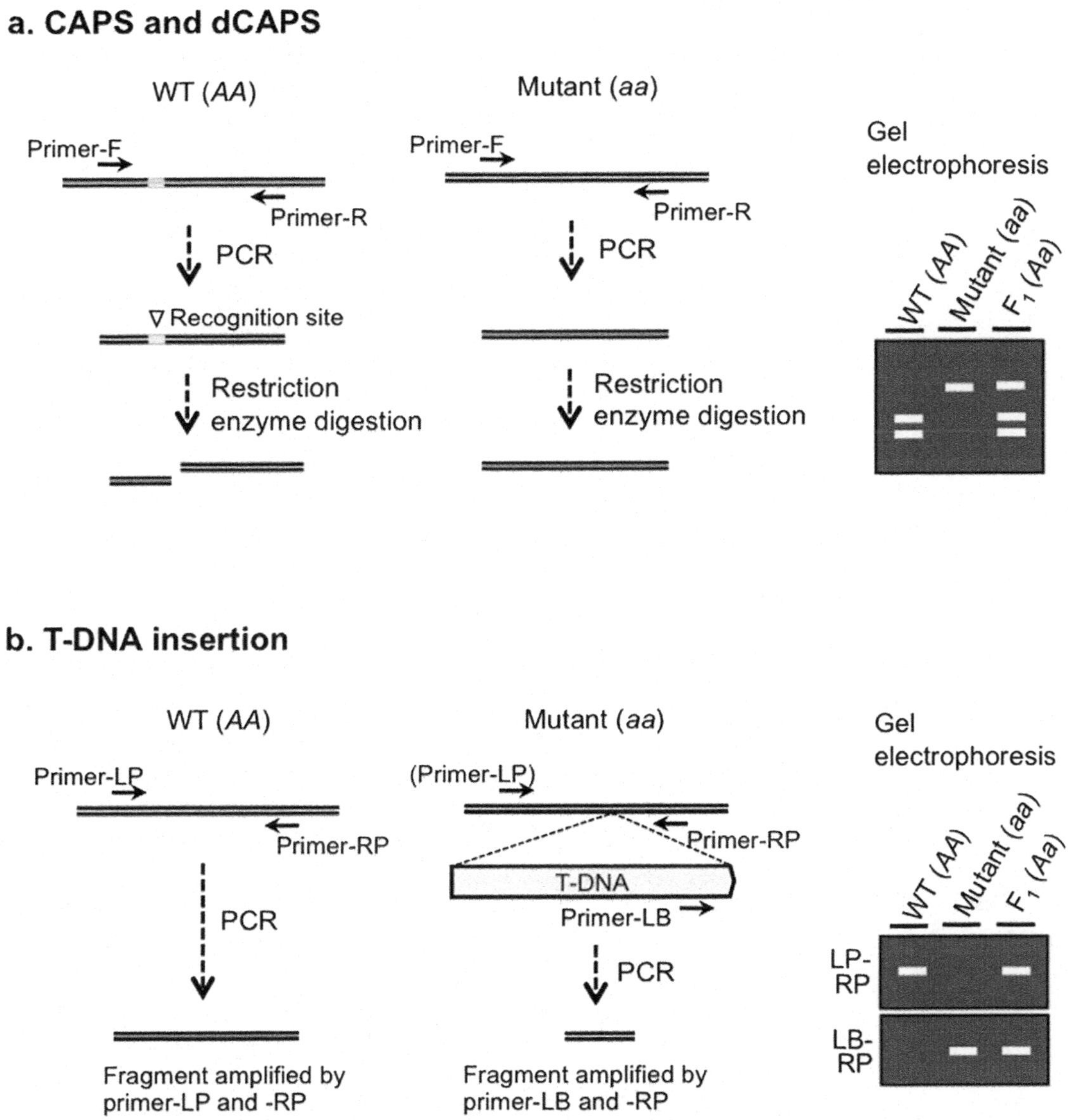

Fig. 3 Diagram of PCR-based genotyping. (**a**) Genotyping by CAPS and dCAPS markers. Fragments containing recognition site polymorphisms are amplified by PCR using a pair of primers (represented as primer-F and -R) and are then digested with restriction enzymes. Because CAPS and dCAPS are codominant markers, homozygous (***AA*** or ***aa***) and heterozygous (***Aa***) genotypes can be easily distinguished. (**b**) Genotyping of T-DNA insertion mutants. The genotyping schemes for the transposon insertion and INDEL markers are almost the same. Wild-type allele (represented as "***A***") can be detected by PCR using a pair of primers corresponding to sequences outside the insert (represented as primer-RP and -LP). Mutant allele (represented as "***a***") can be detected by PCR using a primer corresponding to the sequence inside of the insert (represented as primer-LB). Homozygous and heterozygous genotypes can be distinguished by PCR using each set of primers or by multiplex PCR using all three primers

2. Prepare the PCR reaction for the detection of each marker on both loci containing the *R* and "*A*" genes. The genotyping marker for the "*A*" gene locus can be obtained from the information in Table 3. Design markers for the genotyping of the *R* gene on your own. Codominant CAPS and dCAPS markers require a pair of primers (Table 3) corresponding to the regions between which the polymorphisms are located (Fig. 3a).
3. Three primers are generally used to detect T-DNA or transposon insertions, as described in Fig. 3b. Perform the PCR reaction using either a pair of primers specific for the wild-type gene or a pair of primers specific for the mutant gene containing an exogenous insert. Alternatively, multiplex PCR can be performed using three primers at once according to primer design.
4. Place the tubes in the thermal cycler and run the following three steps: 95 °C for 2 min, 30 cycles of 95 °C for 30 s/50–60 °C for 45 s (annealing)/72 °C for 1 min/kb, followed by 72 °C for 5 min. The optimum annealing temperature depends on the Tm value of the primers. In general, the annealing temperature is set to be from −5 °C to ±0 °C of the Tm value of the primers.
5. To detect CAPS and dCAPS markers, check the concentration of the PCR amplicons by agarose gel electrophoresis using 2 μL of the PCR product solution. To detect INDEL markers, use the PCR product solution directly for gel electrophoresis.

3.4.2 Restriction Enzyme Digestion

1. To detect CAPS and dCAPS markers, purify the PCR products by ethanol precipitation or using spin columns.
2. Prepare the reaction mix as follows.

	Volume	Final conc.
10× Optimized Reaction Buffer	1 μL	1×
PCR product solution	*y* μL	200 ng/10 μL
Restriction enzyme	*z* (≤1) μL	4 U
Nuclease-free water	(9-*y*-*z*) μL	
In total	10 μL	

3. Incubate the solution for 2 h at the optimum temperature for the restriction enzymes suggested by the suppliers (*see* **Note 9**).

3.5 Gel Electrophoresis

PCR products are detected by gel electrophoresis for genotyping. Agarose gel electrophoresis (Subheading 3.5.1) is simple and is suited to the separation of larger molecular markers greater than 200 bp. Polyacrylamide gel electrophoresis (Subheading 3.5.2) offers higher separation resolution and can be applied for the separation of molecular markers smaller than 200 bp. Thus, dCAPS markers should be detected by polyacrylamide gel electrophoresis.

3.5.1 Agarose Gel Electrophoresis

1. Set up the gel plate according to the instructions from the supplier.
2. Add 1% (w/v) or 2% (w/v) agarose into 1× TAE (*see* **Note 10**). Dissolve agarose completely by heating and pour it into the gel cast.
3. Pour 1× TAE (*see* **Note 11**) into the submarine electrophoresis tank and immerse the gel in the buffer.
4. If you are using the GoTaq Green Master Mix for PCR, load the PCR product solution directly into each well of the gel. If the DNA sample does not contain loading dyes, mix the DNA sample with a 1:6 volume of the 6× Gel Loading Dye, Purple (*see* **Note 4**). The appropriate size marker should also be applied.
5. Run the gel for 30 min at a constant volume of 100 V at room temperature. Stop the gel running once the yellow dye (contained in the GoTaq Green Master Mix) or red dye (contained in the 6× Gel Loading Dye, Purple) reaches the end of the gel or two-thirds the distance of the gel, respectively.
6. Immerse the gel in the 10 μg/mL ethidium bromide solution for 20 min.
7. Detect the DNA bands on the stained gel using the UV transilluminator with the camera.

3.5.2 Polyacrylamide Gel Electrophoresis

1. Set up the gel plate according to the instructions from the supplier.
2. Mix 5.27 mL of distilled water, 2.66 mL of 30% acrylamide solution, 2.00 mL of 5× TBE, and 70 μL of 10% APS (*see* **Note 12**).
3. Add 3.5 μL of TEMED, mix well, and quickly pour the running gel into the plate (*see* **Note 13**). Then, quickly put the comb into the gel. It should take about 60 min to polymerize at room temperature.
4. Pour 1× TBE into the lower part of the electrophoretic tank. Then, carefully slant the gel plate to release any air bubbles. Pour 1× TBE into the upper part of the electrophoretic tank. Wash each well with running buffer by pipetting up and down using a P-1000 micropipette.
5. If the sample does not contain any loading dyes, mix the DNA sample with a 1:6 volume of the 6× Gel Loading Dye, Purple (*see* **Note 4**). Then, load the mixture into each well of the gel.
6. Run the gel for about 1 h at a constant volume of 120 V per gel (set the current to the maximum) at room temperature.

Stop the gel running once the yellow dye (contained in the GoTaq Green Master Mix) or red dye (contained in the 6× Gel Loading Dye, Purple) reaches the end of the gel or two-thirds the distance of the gel, respectively.

7. Immerse the gel in the 10 μg/mL ethidium bromide solution for 20 min.
8. Detect the DNA bands on the stained gel using the UV transilluminator with the camera.

3.6 Analysis of Resistance to Virus Infection

3.6.1 Preparation of Plants

1. After the genotyping of each F_2 plant, select F_2(*RRaa*) lines containing both the homozygous *R* gene and homozygous mutant allele in the defense signaling component (*see* **Note 14**). If the parent plant carrying the antiviral gene does not have an ecotype Col-0 background, at least three independent F_2 lines should be selected to analyze resistance to virus infection (*see* **Note 15**).
2. Obtain F_3(*RRaa*) lines by self-fertilization of F_2(*RRaa*).

3.6.2 Virus Inoculation and Analysis of Resistance

1. Prepare WT(*RRAA*), mutant(*rraa*), and F_3(*RRaa*) plants of the same age (*see* **Note 14**).
2. Sprinkle carborundum thinly on fully expanded leaves of *A. thaliana*. Rub-inoculate fully expanded leaves of *A. thaliana* with a virus using cotton swabs. In the case of CMV, 10–100 μg/mL of the virus is optimal for inoculation.
3. Rinse the carborundum from the surface of the leaves with distilled water.
4. Overlay the whole plants with the plastic wrap. Grow the inoculated plants in the growth chamber.
5. Resistance phenotypes to avirulent viruses can be estimated by local or systemic responses in plants. The hypersensitive response (HR) is observed in inoculated leaves 2 or 3 days after the inoculation. As an example, Fig. 4 shows the genotyping of the *cpl3-1* mutation in an *A. thaliana* ecotype C24 carrying the NB-LRR class *R* gene, *RCY1* conferring resistance to CMV. The phenotypic analysis indicates that the *cpl3-1* mutation confers enhanced *RCY1*-mediated CMV resistance (*see* **Note 16**).

4 Notes

1. Arabidopsis Biological Resource Center (ABRC) and Nottingham Arabidopsis Stock Center (NASC) (Table 1) supply the various characterized mutant lines. Several mutant lines listed in Table 3 can be supplied by ABRC and NASC. Other organizations that can supply *A. thaliana* ecotypes and mutagenized bulk seed stocks are shown in Table 1.

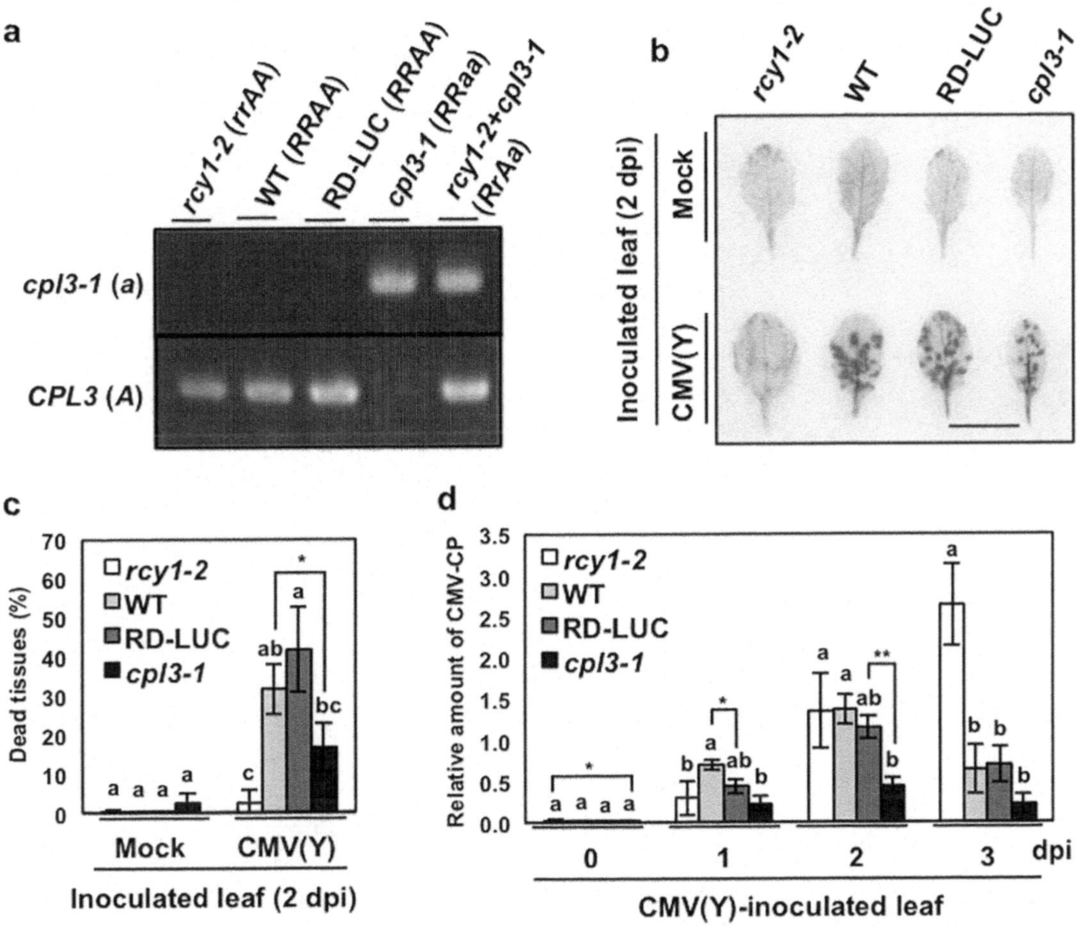

Fig. 4 Genotyping and phenotypic analysis of ***cpl3-1*** mutant. WT (wild-type C24), ***rcy1-2*** mutant (in a C24 background), and RD-LUC (in a C24 background) were used as controls (***see*** **Note 16**). The mutant line ***cpl3-1*** has an RD-LUC background. (**a**) The genotype of the ***CPL3*** locus was analyzed. WT, RD-LUC, and ***rcy1-2*** plants have homozygous ***CPL3*** alleles, whereas ***cpl3-1*** plant has homozygous ***cpl3-1*** alleles. (**b**) Necrotic local lesions (NLLs) in mock- or CMV (Y)-inoculated leaf (2 dpi) were visualized by trypan blue staining. (**c**) The ratio of the NLL area to the total leaf area was quantitated by ImageJ. (**d**) The relative accumulated amount of CP in CMV(Y)-inoculated leaf was immunologically detected over time by enzyme-linked immunosorbent assay with anti-CMV-CP antibody. In (**b**), (**c**), and (**d**), fully expanded leaves of 5-week-old plants, which were grown under short-day conditions, were inoculated with mock and CMV(Y). In (**c**) and (**d**), the average values of four plants are indicated with standard deviation error bars. The data were statistically analyzed by two-way analysis of variance, followed by Tukey's honest significant difference test. Different letters indicate significant differences between plant lines ($n = 4$, $P < 0.01$). Asterisks also indicate significant differences between plant lines ($n = 4$, $^{**}P < 0.05$, $^{*}P < 0.1$)

2. Other PCR reaction systems are available. The Green GoTaq® Reaction Buffer, which is premixed with loading dye, shortens the time required for the gel electrophoresis procedure. Use Colorless GoTaq® Reaction Buffer to detect INDEL and CAPS/dCAPS markers.

3. Restriction enzymes are also available from multiple suppliers. Restriction enzyme digestion can be performed easily with the system offered by New England Biolabs. New England Biolabs provides the reaction buffers premixed with BSA stabilizer and the convenient CutSmart Buffer that is applicable to various restriction enzymes.
4. Other loading dyes are also available. The 6× Gel Loading Dye, Purple (New England Biolabs Inc.) is provided with the restriction enzymes purchased from New England Biolabs. This dye is useful because it does not cast any UV shadows over the underlying bands, as noted by the supplier.
5. The harvested seeds require vernalization for 1 week at 4 °C for seed germination.
6. Flower formation in *A. thaliana* can be promoted under continuous illumination. It takes about 4 weeks for the major wild-type of *A. thaliana* ecotypes to develop seeds under continuous illumination, whereas 7 weeks are required under short-day conditions. In contrast, the larger plant, which is useful for pathogen infection assays, can be obtained under short-day conditions promoting the vegetative growth of *A. thaliana*.
7. If somatic mutations are involved in the phenotype, a reciprocal cross is necessary.
8. Genomic DNA of WT(*RRAA*), mutant(*rraa*), and F_1(*RrAa*) can be used as control markers for homozygous intact gene pairs, homozygous mutant gene pairs, and a heterozygous genotype, respectively (Fig. 3). Control markers of the heterozygous genotype can also be obtained from the mixture of genomic DNA extracted from both WT(*RRAA*) and mutant(*rraa*).
9. If digestion efficiency is low, longer reaction time may improve efficacy.
10. The effective range of separation in 1.0% and 2.0% agarose is 250–12,000 and 100–3000 bp, respectively, according to the information from a supplier (Nippon Gene Co., Ltd., Toyama, Japan) and the general protocol.
11. Close markers for sizes less than 1000 bp can be separated more clearly using an agarose gel consisting of 0.5× TBE. The composition of 5× TBE is described in Subheading 2.6.2.
12. The effective range of separation in an 8% acrylamide gel is 60–400 bp. The blue dye found in Green GoTaq® Reaction Buffer moves with 160 bp fragments in 8% acrylamide gels. The red dye contained in 6× Gel Loading Dye, Purple moves with 45 bp fragments in 8% acrylamide gels.
13. Note that once APS and TEMED are added, the gel will begin to polymerize.

14. F_3 lines with the "*RRAA*" [the same as WT(*RRAA*) in F_0], "*rraa*" [the same as Col-0 mutant(*rraa*) in F_0], and "*rrAA*" (not containing functional *R* gene alleles but containing homozygous wild-type gene "*A*" alleles) genotypes can be used as controls for F_3(*RRaa*) in phenotypic analysis. If required, please screen and self-fertilize F_2 lines with the "*RRAA*," "*rraa*," and "*rrAA*" genotypes.
15. *RCY1*, which was isolated from *A. thaliana* ecotype C24, encodes CC-NB-LRR class R protein in a yellow strain of CMV (Y) [159]. *RCY1*-mediated resistance signaling pathways have been dissected by the reverse genetic approach in *A. thaliana* [159–162]. Many defense signaling mutant lines have a CMV(Y)-susceptible Col-0 background carrying the corresponding allele *rcy1*. For reverse genetic analysis of defense signaling downstream of *RCY1*, the mutant lines need to be crossed with wild ecotype C24. More than three independent F_2 and F_3 lines obtained from crosses between ecotype C24 and Col-0 mutants are required for phenotypic analysis [160]. Otherwise, a defense signaling mutant of Col-0 that was crossed with *RCY1*-transformed Col-0 [163] is also available for reverse genetic analysis [161, 162].
16. The *cpl3-1* mutant line carries a T-DNA insertion in the coding sequence of *C-TERMINAL DOMAIN PHOSPHATASE 3* (*CPL3*) in an RD-LUC background [164]. RD-LUC is the transgenic line of *A. thaliana* ecotype C24 and contains the exogenous *luciferase* gene driven by the *RD29A* promoter [164]. To investigate the role of CPL3 in *RCY1*-mediated resistance to CMV(Y), the response to CMV was compared among *cpl3-1* mutant, RD-LUC, wild-type C24, and *rcy1-2* carrying a point mutation at the CC domain of RCY1 in a C24 background and susceptible to CMV(Y) [165] (Fig. 4). The *cpl3-1* allele was specifically detected by PCR using SK-LB3 (5′-TTGACCATCATACTCATTGCTG-3′) and CPL3-RP (5′-GGACCAAGAAGCCCAAATT-3′) primers (Fig. 4a). *CPL3* allele was specifically detected by PCR using CPL3-LP (5′-GAGGATGTGTCAGTTTCAGCAGC-3′) and CPL3-RP primers (Fig. 4a). WT and RD-LUC exhibited HR resistance to CMV(Y), suggesting that the *RD29A-luciferase* transgene does not affect the response to CMV in a C24 background (Fig. 4b, c). In CMV(Y)-inoculated leaves of *cpl3-1*, the ratio of the necrotic local lesion area to the total leaf area was significantly lower than that in WT and RD-LUC (Fig. 4b, c). Furthermore, in CMV(Y)-inoculated leaves of *cpl3-1*, the accumulated amount of CMV(Y)-coat protein (CP) was clearly lower than in *rcy1-2*, WT, and RD-LUC (Fig. 4d). These results suggest that the *cpl3-1* mutation enhances *RCY1*-mediated resistance to CMV(Y). Thus, CPL3 may act as a negative regulator of *RCY1*-mediated defense signaling pathways.

Acknowledgments

This work was supported by grants for JSPS Research Fellows (15J01964), for "Scientific Research on Innovative Areas" from the Ministry of Education, Culture, Sports, Science and Technology (MEXT) of Japan (Grant numbers: 16H06429, 16K21723, and 16H06435), and by the Japan Society for the Promotion of Science (JSPS) through the JSPS Core-to-Core Program (Advanced Research Networks) entitled "Establishment of International Agricultural Immunology Research-Core for Quantum Improvement in Food Safety."

References

1. Voinnet O (2005) Induction and suppression of RNA silencing: insights from viral infections. Nat Rev Genet 6:206–220
2. Csorba T, Kontra L, Burgyán J (2015) Viral silencing suppressors: tools forged to fine-tune host-pathogen coexistence. Virology 479-480:85–103
3. de Ronde D, Butterbach P, Kormelink R (2014) Dominant resistance against plant viruses. Front Plant Sci 5:1–17
4. Hashimoto M, Neriya Y, Yamaji Y et al (2016) Recessive resistance to plant viruses: potential resistance genes beyond translation initiation factors. Front Microbiol 7:1695
5. Peláez P, Sanchez F (2013) Small RNAs in plant defense responses during viral and bacterial interactions: similarities and differences. Front Plant Sci 4:343
6. Cournoyer P, Dinesh-kumar SP (2011) NB-LRR immune receptors in plant virus defence. In: Caranta C, Aranda MA, Tepfer M, Lopez-Moya JJ (eds) Recent advances in plant virology, 1st edn. Caister Academic Press, Norfolk
7. Alonso JM, Stepanova AN, Leisse TJ et al (2003) Genome-wide insertional mutagenesis of *Arabidopsis thaliana*. Science 301:653–657
8. Sessions A, Burke E, Presting G et al (2002) A high-throughput Arabidopsis reverse genetics system. Plant Cell 14:2985–2994
9. Rosso MG, Li Y, Strizhov N et al (2003) An *Arabidopsis thaliana* T-DNA mutagenized population (GABI-Kat) for flanking sequence tag-based reverse genetics. Plant Mol Biol 53:247–259
10. Kleinboelting N, Huep G, Kloetgen A et al (2012) GABI-Kat SimpleSearch: new features of the *Arabidopsis thaliana* T-DNA mutant database. Nucleic Acids Res 40:D1211–D1215
11. Woody ST, Austin-Phillips S, Amasino RM et al (2007) The *WiscDsLox* T-DNA collection: an arabidopsis community resource generated by using an improved high-throughput T-DNA sequencing pipeline. J Plant Res 120:157–165
12. Tissier AF, Marillonnet S, Klimyuk V et al (1999) Multiple independent defective *suppressor-mutator* transposon insertions in Arabidopsis: a tool for functional genomics. Plant Cell 11:1841–1852
13. Wisman E, Hartmann U, Sagasser M et al (1998) Knock-out mutants from an *En-1* mutagenized *Arabidopsis thaliana* population generate phenylpropanoid biosynthesis phenotypes. Proc Natl Acad Sci U S A 95:12432–12437
14. Wang L, Tsuda K, Sato M et al (2009) Arabidopsis CaM binding protein CBP60g contributes to MAMP-induced SA accumulation and is involved in disease resistance against *Pseudomonas syringae*. PLoS Pathog 5:e1000301
15. Sun T, Zhang Y, Li Y et al (2015) ChIP-seq reveals broad roles of SARD1 and CBP60g in regulating plant immunity. Nat Commun 6:10159
16. Yang S, Hua J (2004) A haplotype-specific *resistance* gene regulated by *BONZAI1* in Arabidopsis. Plant Cell 16:1060–1071
17. Chen QF, Xu L, Tan WJ et al (2015) Disruption of the *Arabidopsis* defense regulator genes *SAG101*, *EDS1*, and *PAD4* confers enhanced freezing tolerance. Mol Plant 8:1536–1549
18. Glazebrook J, Rogers EE, Ausubel FM (1996) Isolation of Arabidopsis mutants with

enhanced disease susceptibility by direct screening. Genetics 143:973–982

19. Reuber TL, Plotnikova JM, Dewdney J et al (1998) Correlation of defense gene induction defects with powdery mildew susceptibility in *Arabidopsis* enhanced disease susceptibility mutants. Plant J 16:473–485
20. Gou M, Su N, Zheng J et al (2009) An F-box gene, *CPR30*, functions as a negative regulator of the defense response in Arabidopsis. Plant J 60:757–770
21. Nawrath C, Métraux JP (1999) Salicylic acid induction-deficient mutants of Arabidopsis express *PR-2* and *PR-5* and accumulate high levels of camalexin after pathogen inoculation. Plant Cell 11:1393–1404
22. Heck S, Grau T, Buchala A et al (2003) Genetic evidence that expression of NahG modifies defence pathways independent of salicylic acid biosynthesis in the *Arabidopsis-Pseudomonas syringae* pv. *tomato* interaction. Plant J 36:342–352
23. Century KS, Holubt EB, Staskawiczt BJ (1995) NDR1, a locus of *Arabidopsis thaliana* that is required for disease resistance to both a bacterial and a fungal pathogen. Proc Natl Acad Sci U S A 92:6597–6601
24. Cao H, Bowling SA, Gordon AS et al (1994) Characterization of an Arabidopsis mutant that is nonresponsive to inducers of systemic acquired resistance. Plant Cell 6:1583–1592
25. Cao H, Glazebrook J, Clarke JD et al (1997) The Arabidopsis *NPR1* gene that controls systemic acquired resistance encodes a novel protein containing ankyrin repeats. Cell 88:57–63
26. Nishimura MT, Stein M, Hou B-H et al (2003) Loss of callose synthase results in salicylic acid-dependent disease resistance. Science 301:969–972
27. Zhang Y, Cheng YT, Qu N et al (2006) Negative regulation of defense responses in Arabidopsis by two *NPR1* paralogs. Plant J 48:647–656
28. Shi Z, Maximova S, Liu Y et al (2013) The salicylic acid receptor NPR3 is a negative regulator of the transcriptional defense response during early flower development in *Arabidopsis*. Mol Plant 6:802–816
29. Liu G, Holub EB, Alonso JM et al (2005) An Arabidopsis *NPR1*-like gene, *NPR4*, is required for disease resistance. Plant J 41:304–318
30. Jirage D, Tootle TL, Reuber TL et al (1999) *Arabidopsis thaliana PAD4* encodes a lipase-like gene that is important for salicylic acid signaling. Proc Natl Acad Sci U S A 96:13583–13588
31. Hiruma K, Nishiuchi T, Kato T et al (2011) *Arabidopsis ENHANCED DISEASE RESISTANCE 1* is required for pathogen-induced expression of plant defensins in nonhost resistance, and acts through interference of *MYC2*-mediated repressor function. Plant J 67:980–992
32. Huang J, Gu M, Lai Z et al (2010) Functional analysis of the Arabidopsis *PAL* gene family in plant growth, development, and response to environmental stress. Plant Physiol 153:1526–1538
33. Feys BJ, Wiermer M, Bhat RA et al (2005) *Arabidopsis* SENESCENCE-ASSOCIATED GENE101 stabilizes and signals within an ENHANCEDDISEASESUSCEPTIBILITY1 complex in plant innate immunity. Plant Cell 17:2601–2613
34. Langenbach C, Campe R, Schaffrath U et al (2013) UDP-glucosyltransferase UGT84A2/BRT1 is required for Arabidopsis nonhost resistance to the Asian soybean rust pathogen *Phakopsora pachyrhizi*. New Phytol 198:536–545
35. Zhang Y, Xu S, Ding P et al (2010) Control of salicylic acid synthesis and systemic acquired resistance by two members of a plant-specific family of transcription factors. Proc Natl Acad Sci U S A 107:18220–18225
36. Stein M, Dittgen J, Sánchez-Rodríguez C et al (2006) *Arabidopsis* PEN3/PDR8, an ATP binding cassette transporter, contributes to nonhost resistance to inappropriate pathogens that enter by direct penetration. Plant Cell 18:731–746
37. Dewdney J, Reuber TL, Wildermuth MC et al (2000) Three unique mutants of *Arabidopsis* identify eds loci required for limiting growth of a biotrophic fungal pathogen. Plant J 24:205–218
38. Tsuda K, Sato M, Stoddard T et al (2009) Network properties of robust immunity in plants. PLoS Genet 5:e1000772
39. Gross J, Won KC, Lezhneva L et al (2006) A plant locus essential for phylloquinone (vitamin K1) biosynthesis originated from a fusion of four eubacterial genes. J Biol Chem 281:17189–17196
40. Miura K, Rus A, Sharkhuu A et al (2005) The *Arabidopsis* SUMO E3 ligase SIZ1 controls phosphate deficiency responses. Proc Natl Acad Sci U S A 102:7760–7765
41. Lee J, Nam J, Park HC et al (2007) Salicylic acid-mediated innate immunity in Arabidopsis is regulated by SIZ1 SUMO E3 ligase. Plant J 49:79–90
42. Song JT, Lu H, McDowell JM et al (2004) A key role for *ALD1* in activation of local and

systemic defenses in *Arabidopsis*. Plant J 40:200–212
43. Xu E, Brosché M (2014) Salicylic acid signaling inhibits apoplastic reactive oxygen species signaling. BMC Plant Biol 14:155
44. Carella P, Kempthorne CJ, Wilson DC et al (2017) Exploring the role of DIR1, DIR1-like and other lipid transfer proteins during systemic immunity in Arabidopsis. Physiol Mol Plant Pathol 97:49–57
45. Bartsch M, Gobbato E, Bednarek P et al (2006) Salicylic acid-independent ENHANCEDDISEASESUSCEPTIBILITY1 signaling in *Arabidopsis* immunity and cell death is regulated by the monooxygenase *FMO1* and the Nudix hydrolase *NUDT7*. Plant Cell 18:1038–1051
46. Kieber JJ, Rothenberg M, Roman G et al (1993) *CTR1*, a negative regulator of the ethylene pathway in Arabidopsis, encodes a member of the Raf family of protein kinases. Cell 72:427–441
47. Roman G, Lubarsky B, Kieber JJ et al (1995) Genetic analysis of ethylene signal transduction in *Arabidopsis thaliana*: five novel mutant loci integrated into a stress response pathway. Genetics 139:1393–1409
48. An F, Zhao Q, Ji Y et al (2010) Ethylene-induced stabilization of ETHYLENE INSENSITIVE3 and EIN3-LIKE1 is mediated by proteasomal degradation of EIN3 binding F-Box 1 and 2 that requires EIN2 in *Arabidopsis*. Plant Cell 22:2384–2401
49. Guo H, Ecker JR (2003) Plant responses to ethylene gas are mediated by $SCF^{EBF1/EBF2}$-dependent proteolysis of EIN3 transcription factor. Cell 115:667–677
50. Alonso JM, Stepanova AN, Solano R et al (2003) Five components of the ethylene-response pathway identified in a screen for weak *ethylene-insensitive* mutants in *Arabidopsis*. Proc Natl Acad Sci U S A 100:2992–2997
51. Chang C, Kwok S, Bleecker A et al (1993) *Arabidopsis* ethylene-response gene *ETR1*: similarity of product to two-component regulators. Science 262:539–544
52. Hua J, Meyerowitz EM (1998) Ethylene responses are negatively regulated by a receptor gene family in *Arabidopsis thaliana*. Cell 94:261–271
53. Kachroo P, Yoshioka K, Shah J et al (2000) Resistance to turnip crinkle virus in Arabidopsis is regulated by two host genes and is salicylic acid dependent but *NPR1*, ethylene, and jasmonate independent. Plant Cell 12:677–690
54. Guzmán P, Ecker JR (1990) Exploiting the triple response of *Arabidopsis* to identify ethylene-related mutants. Plant Cell 2:513–523
55. Resnick JS, Wen C-K, Shockey JA et al (2006) *REVERSION-TO-ETHYLENE SENSITIVITY1*, a conserved gene that regulates ethylene receptor function in *Arabidopsis*. Proc Natl Acad Sci U S A 103:7917–7922
56. Alonso JM, Hirayama T, Roman G et al (1999) EIN2, a bifunctional transducer of ethylene and stress responses in *Arabidopsis*. Science 284:2148–2152
57. Chao Q, Rothenberg M, Solano R et al (1997) Activation of the ethylene gas response pathway in Arabidopsis by the nuclear protein ETHYLENE-INSENSITIVE3 and related proteins. Cell 89:1133–1144
58. Chen H, Xue L, Chintamanani S et al (2009) ETHYLENE INSENSITIVE3 and ETHYLENE INSENSITIVE3-LIKE1 repress *SALICYLIC ACID INDUCTION DEFICIENT2* expression to negatively regulate plant innate immunity in *Arabidopsis*. Plant Cell 21:2527–2540
59. Potuschak T, Vansiri A, Binder BM et al (2006) The exoribonuclease XRN4 is a component of the ethylene response pathway in *Arabidopsis*. Plant Cell 18:3047–3057
60. Von Malek B, Van Der Graaff E, Schneitz K et al (2002) The *Arabidopsis* male-sterile mutant *dde2-2* is defective in the *ALLENE OXIDE SYNTHASE* gene encoding one of the key enzymes of the jasmonic acid biosynthesis pathway. Planta 216:187–192
61. Xie D-X, Feys BF, James S et al (1998) COI1: an *Arabidopsis* gene required for jasmonate-regulated defense and fertility. Science 280:1091–1094
62. Trusov Y, Sewelam N, Rookes JE et al (2009) Heterotrimeric G proteins-mediated resistance to necrotrophic pathogens includes mechanisms independent of salicylic acid-, jasmonic acid/ethylene- and abscisic acid-mediated defense signaling. Plant J 58:69–81
63. Staswick PE, Su W, Howell SH (1992) Methyl jasmonate inhibition of root growth and induction of a leaf protein are decreased in an *Arabidopsis thaliana* mutant. Proc Natl Acad Sci U S A 89:6837–6840
64. Reichheld J-P, Meyer E, Khafif M et al (2005) AtNTRB is the major mitochondrial thioredoxin reductase in *Arabidopsis thaliana*. FEBS Lett 579:337–342
65. Sweat TA, Wolpert TJ (2007) Thioredoxin *h5* is required for victorin sensitivity mediated by a CC-NBS-LRR gene in *Arabidopsis*. Plant Cell 19:673–687
66. Torres MA, Dangl JL, Jones JDG (2002) *Arabidopsis* $gp91^{phox}$ homologues *AtrbohD* and *AtrbohF* are required for accumulation of

reactive oxygen intermediates in the plant defense response. Proc Natl Acad Sci U S A 99:517–522

67. Frye CA, Innes RW (1998) An Arabidopsis mutant with enhanced resistance to powdery mildew. Plant Cell 10:947–956
68. Nakagami H, Soukupová H, Schikora A et al (2006) A mitogen-activated protein kinase kinase kinase mediates reactive oxygen species homeostasis in *Arabidopsis*. J Biol Chem 281:38697–38704
69. Su S-H, Bush SM, Zaman N et al (2013) Deletion of a tandem gene gamily in *Arabidopsis*: increased MEKK2 abundance triggers autoimmunity when the MEKK1-MKK1/2-MPK4 signaling cascade is disrupted. Plant Cell 25:1895–1910
70. Bush SM, Krysan PJ (2010) iTILLING: a personalized approach to the identification of induced mutations in Arabidopsis. Plant Physiol 154:25–35
71. Gao M, Liu J, Bi D et al (2008) MEKK1, MKK1/MKK2 and MPK4 function together in a mitogen-activated protein kinase cascade to regulate innate immunity in plants. Cell Res 18:1190–1198
72. Yoo S-D, Cho Y-H, Tena G et al (2008) Dual control of nuclear EIN3 by bifurcate MAPK cascades in C_2H_4 signalling. Nature 451:789–795
73. Dóczi R, Brader G, Pettkó-Szandtner A et al (2007) The *Arabidopsis* mitogen-activated protein kinase kinase MKK3 is upstream of group C mitogen-activated protein kinases and participates in pathogen signaling. Plant Cell 19:3266–3279
74. Wang H, Ngwenyama N, Liu Y et al (2007) Stomatal development and patterning are regulated by environmentally responsive mitogen-activated protein kinases in *Arabidopsis*. Plant Cell 19:63–73
75. Bush SM, Krysan PJ (2007) Mutational evidence that the *Arabidopsis* MAP kinase MPK6 is involved in anther, inflorescence, and embryo development. J Exp Bot 58:2181–2191
76. Liu Y, Zhang S (2004) Phosphorylation of 1-aminocyclopropane-1-carboxylic acid synthase by MPK6, a stress-responsive mitogen-activated protein kinase, induces ethylene biosynthesis in Arabidopsis. Plant Cell 16:3386–3399
77. Barajas-López Jde D, Kremnev D, Shaikhali J et al (2013) PAPP5 is involved in the tetrapyrrole mediated plastid signalling during chloroplast development. PLoS One 8:e60305
78. Pruzinská A, Tanner G, Aubry S et al (2005) Chlorophyll breakdown in senescent Arabidopsis leaves. Characterization of chlorophyll catabolites and of chlorophyll catabolic enzymes involved in the degreening reaction. Plant Physiol 139:52–63
79. Greenberg JT, Guo A, Klessig DF et al (1994) Programmed cell death in plants: a pathogen-triggered response activated coordinately with multiple defense functions. Cell 77:551–563
80. Mène-Saffrané L, Dubugnon L, Chételat A et al (2009) Nonenzymatic oxidation of trienoic fatty acids contributes to reactive oxygen species management in *Arabidopsis*. J Biol Chem 284:1702–1708
81. Kaminaka H, Näke C, Epple P et al (2006) bZIP10-LSD1 antagonism modulates basal defense and cell death in *Arabidopsis* following infection. EMBO J 25:4400–4411
82. Coll NS, Vercammen D, Smidler A et al (2010) *Arabidopsis* type I metacaspases control cell death. Science 330:1393–1397
83. Yu I-C, Parker J, Bent AF (1998) Gene-for-gene disease resistance without the hypersensitive response in *Arabidopsis dnd1* mutant. Proc Natl Acad Sci U S A 95:7819–7824
84. Clough SJ, Fengler KA, Yu IC et al (2000) The *Arabidopsis dnd1* "defense, no death" gene encodes a mutated cyclic nucleotide-gated ion channel. Proc Natl Acad Sci U S A 97:9323–9328
85. Genger RK, Jurkowski GI, McDowell JM et al (2008) Signaling pathways that regulate the enhanced disease resistance of *Arabidopsis* "*defense, no death*" mutants. Mol Plant-Microbe Interact 21:1285–1296
86. Yu I, Fengler KA, Clough SJ et al (2000) Identification of Arabidopsis mutants exhibiting an altered hypersensitive response in gene-for-gene disease resistance. Mol Plant-Microbe Interact 13:277–286
87. Jurkowski GI, Smith RK, Yu I et al (2004) *Arabidopsis DND2*, a second cyclic nucleotide-gated ion channel gene for which mutation causes the "*defense, no death*" phenotype. Mol Plant-Microbe Interact 17:511–520
88. Collins NC, Thordal-Christensen H, Lipka V et al (2003) SNARE-protein-mediated disease resistance at the plant cell wall. Nature 425:973–977
89. Johansson ON, Fantozzi E, Fahlberg P et al (2014) Role of the penetration-resistance genes *PEN1*, *PEN2* and *PEN3* in the hypersensitive response and race-specific resistance in *Arabidopsis thaliana*. Plant J 79:466–476
90. Lipka V, Dittgen J, Bednarek P et al (2005) Pre- and postinvasion defenses both contribute to nonhost resistance in *Arabidopsis*. Science 310:1180–1183

91. Kobae Y, Sekino T, Yoshioka H et al (2006) Loss of AtPDR8, a plasma membrane ABC transporter of *Arabidopsis thaliana*, causes hypersensitive cell death upon pathogen infection. Plant Cell Physiol 47:309–318
92. Cheng YT, Li Y, Huang S et al (2011) Stability of plant immune-receptor resistance proteins is controlled by SKP1-CULLIN1-F-box (SCF)-mediated protein degradation. Proc Natl Acad Sci U S A 108:14694–14699
93. Dong OX, Tong M, Bonardi V et al (2016) TNL-mediated immunity in *Arabidopsis* requires complex regulation of the redundant *ADR1* gene family. New Phytol 210: 960–973
94. Takahashi A, Casais C, Ichimura K et al (2003) HSP90 interacts with RAR1 and SGT1 and is essential for RPS2-mediated disease resistance in *Arabidopsis*. Proc Natl Acad Sci U S A 100:11777–11782
95. Hubert DA, Tornero P, Belkhadir Y et al (2003) Cytosolic HSP90 associates with and modulates the Arabidopsis RPM1 disease resistance protein. EMBO J 22:5679–5689
96. Hubert DA, He Y, McNulty BC et al (2009) Specific *Arabidopsis* HSP90.2 alleles recapitulate RAR1 cochaperone function in plant NB-LRR disease resistance protein regulation. Proc Natl Acad Sci 106:9556–9563
97. Palma K, Zhang Y, Li X (2005) An importin α homolog, MOS6, plays an important role in plant innate immunity. Curr Biol 15: 1129–1135
98. Gu Y, Zebell SG, Liang Z et al (2016) Nuclear pore permeabilization is a convergent signaling event in effector-triggered immunity. Cell 166:1–13
99. Cheng YT, Germain H, Wiermer M et al (2009) Nuclear pore complex component MOS7/Nup88 is required for innate immunity and nuclear accumulation of defense regulators in *Arabidopsis*. Plant Cell 21: 2503–2516
100. Zhang Y, Li X (2005) A putative nucleoporin 96 is required for both basal defense and constitutive resistance responses mediated by *suppressor of npr1-1, constitutive 1*. Plant Cell 17:1306–1316
101. Tornero P, Merritt P, Sadanandom A et al (2002) *RAR1* and *NDR1* contribute quantitatively to disease resistance in Arabidopsis, and their relative contributions are dependent on the *R* gene assayed. Plant Cell 14:1005–1015
102. Gloggnitzer J, Akimcheva S, Srinivasan A et al (2014) Nonsense-mediated mRNA decay modulates immune receptor levels to regulate plant antibacterial defense. Cell Host Microbe 16:376–390
103. Li Y, Li S, Bi D et al (2010) SRFR1 negatively regulates plant NB-LRR resistance protein accumulation to prevent autoimmunity. PLoS Pathog 6:e1001111
104. Gray WM, Muskett PR, Chuang H-W et al (2003) Arabidopsis SGT1b is required for SCF ^{TIR1}-mediated auxin response. Plant Cell 15:1310–1319
105. Kim SH, Gao F, Bhattacharjee S et al (2010) The Arabidopsis resistance-like gene *SNC1* is activated by mutations in *SRFR1* and contributes to resistance to the bacterial effector AvrRps4. PLoS Pathog 6:e1001172
106. Bohmert K, Camus I, Bellini C et al (1998) *AGO1* defines a novel locus of *Arabidopsis* controlling leaf development. EMBO J 17: 170–180
107. Arribas-Hernández L, Marchais A, Poulsen C et al (2016) The slicer activity of ARGONAUTE1 is required specifically for the phasing, not production, of *trans*-acting short interfering RNAs in Arabidopsis. Plant Cell 28:1563–1580
108. Morel J, Godon C, Mourrain P et al (2002) Fertile hypomorphic *ARGONAUTE* (*ago1*) mutants impaired in post-transcriptional gene silencing and virus resistance. Plant Cell 14:629–639
109. Mason GA, Vergara TL, Queitsch C (2016) The mechanistic underpinnings of an *ago1*-mediated, environmentally-dependent, and stochastic phenotype. Plant Physiol 170: 2420–2431
110. Baumberger N, Baulcombe DC (2005) *Arabidopsis* ARGONAUTE1 is an RNA slicer that selectively recruits microRNAs and short interfering RNAs. Proc Natl Acad Sci U S A 102:11928–11933
111. Lam P, Zhao L, Eveleigh N et al (2015) The exosome and trans-acting small interfering RNAs regulate cuticular wax biosynthesis during Arabidopsis inflorescence stem development. Plant Physiol 167:323–336
112. Lobbes D, Rallapalli G, Schmidt DD et al (2006) SERRATE: a new player on the plant microRNA scene. EMBO Rep 7:1052–1058
113. Takeda A, Iwasaki S, Watanabe T et al (2008) The mechanism selecting the guide strand from small RNA duplexes is different among *Argonaute* proteins. Plant Cell Physiol 49:493–500
114. Oliver C, Santos JL, Pradillo M (2014) On the role of some ARGONAUTE proteins in meiosis and DNA repair in *Arabidopsis thaliana*. Front Plant Sci 5:1–10
115. Mi S, Cai T, Hu Y et al (2008) Sorting of small RNAs into *Arabidopsis* Argonaute complexes is directed by the 5′ terminal nucleotide. Cell 133:116–127

116. Wei W, Ba Z, Gao M et al (2012) A role for small RNAs in DNA double-strand break repair. Cell 149:101–112
117. Agorio A, Vera P (2007) ARGONAUTE4 is required for resistance to *Pseudomonas syringae* in *Arabidopsis*. Plant Cell 19:3778–3790
118. Strickler SR, Tantikanjana T, Nasrallah JB (2013) Regulation of the S-locus receptor kinase and self-incompatibility in *Arabidopsis thaliana*. Genes Genomes Genet 3:315–322
119. Hernández-Lagana E, Rodríguez-Leal D, Lúa J et al (2016) A multigenic network of ARGONAUTE4 clade members controls early megaspore formation in *Arabidopsis*. Genetics 204:1045–1056
120. Katiyar-Agarwal S, Gao S, Vivian-Smith A et al (2007) A novel class of bacteria-induced small RNAs in *Arabidopsis*. Genes Dev 21:3123–3134
121. Harvey JJW, Lewsey MG, Patel K et al (2011) An antiviral defense role of AGO2 in plants. PLoS One 6:e14639
122. Havecker ER, Wallbridge LM, Fedito P et al (2012) Metastable differentially methylated regions within Arabidopsis inbred populations are associated with modified expression of non-coding transcripts. PLoS One 7: e45242
123. Tucker MR, Okada T, Hu Y et al (2012) Somatic small RNA pathways promote the mitotic events of megagametogenesis during female reproductive development in *Arabidopsis*. Development 139:1399–1404
124. Zheng X, Zhu J, Kapoor A et al (2007) Role of *Arabidopsis* AGO6 in siRNA accumulation, DNA methylation and transcriptional gene silencing. EMBO J 26:1691–1701
125. Smith LM, Pontes O, Searle I et al (2007) An SNF2 protein associated with nuclear RNA silencing and the spread of a silencing signal between cells in *Arabidopsis*. Plant Cell 19: 1507–1521
126. Hunter C, Sun H, Poethig RS et al (2003) The *Arabidopsis* heterochronic gene ZIPPY is an *ARGONAUTE* family member. Curr Biol 13:1734–1739
127. Peragine A, Yoshikawa M, Wu G et al (2004) *SGS3* and *SGS2/SDE1/RDR6* are required for juvenile development and the production of trans-acting siRNAs in *Arabidopsis*. Genes Dev 18:2368–2379
128. Vazquez F, Vaucheret H, Rajagopalan R et al (2004) Endogenous *trans*-acting siRNAs regulate the accumulation of *Arabidopsis* mRNAs. Mol Cell 16:69–79
129. Olmedo-Monfil V, Durán-Figueroa N, Arteaga-Vázquez M et al (2010) Control of female gamete formation by a small RNA pathway in *Arabidopsis*. Nature 464: 628–632
130. Arribas-Hernández L, Kielpinski LJ, Brodersen P (2016) mRNA decay of most Arabidopsis miRNA targets requires slicer activity of AGO1. Plant Physiol 171: 2620–2632
131. Elmayan T, Proux F, Vaucheret H (2005) *Arabidopsis RPA2*: a genetic link among transcriptional gene silencing, DNA repair, and DNA replication. Curr Biol 15:1919–1925
132. Zhang X, Yazaki J, Sundaresan A et al (2006) Genome-wide high-resolution mapping and functional analysis of DNA methylation in *Arabidopsis*. Cell 126:1189–1201
133. Xue W, Ruprecht C, Street N et al (2012) Paramutation-like interaction of T-DNA loci in Arabidopsis. PLoS One 7:e51651
134. Moissiard G, Bischof S, Husmann D et al (2014) Transcriptional gene silencing by *Arabidopsis* microrchidia homologues involves the formation of heteromers. Proc Natl Acad Sci U S A 111:7474–7479
135. Jacobsen SE, Running MP, Meyerowitz EM (1999) Disruption of an RNA helicase/RNAse III gene in *Arabidopsis* causes unregulated cell division in floral meristems. Development 126:5231–5243
136. Ye R, Chen Z, Lian B et al (2016) A dicer-independent route for biogenesis of siRNAs that direct DNA methylation in *Arabidopsis*. Mol Cell 61:222–235
137. Xie Z, Johansen LK, Gustafson AM et al (2004) Genetic and functional diversification of small RNA pathways in plants. PLoS Biol 2:642–652
138. Mlotshwa S, Pruss GJ, Peragine A et al (2008) *DICER-LIKE2* plays a primary role in transitive silencing of transgenes in *Arabidopsis*. PLoS One 3:e1755
139. Mlotshwa S, Pruss GJ, Gao Z et al (2010) Transcriptional silencing induced by Arabidopsis T-DNA mutants is associated with 35S promoter siRNAs and requires genes involved in siRNA-mediated chromatin silencing. Plant J 64:699–704
140. Yoshikawa M, Peragine A, Park MY et al (2005) A pathway for the biogenesis of *trans*-acting siRNAs in *Arabidopsis*. Genes Dev 19:2164–2175
141. Yamamuro C, Miki D, Zheng Z et al (2014) Overproduction of stomatal lineage cells in *Arabidopsis* mutants defective in active DNA demethylation. Nat Commun 5:4062
142. Schoft VK, Chumak N, Choi Y et al (2011) Function of the DEMETER DNA glycosylase in the *Arabidopsis thaliana* male gametophyte. Proc Natl Acad Sci U S A 108:8042–8047

143. Adenot X, Elmayan T, Lauressergues D et al (2006) DRB4-dependent *TAS3 trans*-acting siRNAs control leaf morphology through AGO7. Curr Biol 16:927–932
144. Chan SWL, Henderson IR, Zhang X et al (2006) RNAi, DRD1, and histone methylation actively target developmentally important non-CG DNA methylation in *Arabidopsis*. PLoS Genet 2:e83
145. Vazquez F, Gasciolli V, Crété P et al (2004) The nuclear dsRNA binding protein HYL1 is required for microRNA accumulation and plant development, but not posttranscriptional transgene silencing. Curr Biol 14: 346–351
146. Zhong S-H, Liu J-Z, Jin H et al (2013) Warm temperatures induce transgenerational epigenetic release of RNA silencing by inhibiting siRNA biogenesis in *Arabidopsis*. Proc Natl Acad Sci U S A 110:9171–9176
147. Saze H, Mittelsten Scheid O, Paszkowski J (2003) Maintenance of CpG methylation is essential for epigenetic inheritance during plant gametogenesis. Nat Genet 34:65–69
148. Habu Y, Mathieu O, Tariq M et al (2006) Epigenetic regulation of transcription in intermediate heterochromatin. EMBO Rep 7:1279–1284
149. Kang HG, Oh CS, Sato M et al (2010) Endosome-associated CRT1 functions early in *Resistance* gene-mediated defense signaling in *Arabidopsis* and tobacco. Plant Cell 22: 918–936
150. Herr AJ, Jensen MB, Dalmay T et al (2005) RNA polymerase IV directs silencing of endogenous DNA. Science 308:118–120
151. Vu TM, Nakamura M, Calarco JP et al (2013) RNA-directed DNA methylation regulates parental genomic imprinting at several loci in *Arabidopsis*. Development 140:2953–2960
152. Pontier D, Yahubyan G, Vega D et al (2005) Reinforcement of silencing at transposons and highly repeated sequences requires the concerted action of two distinct RNA polymerases IV in *Arabidopsis*. Genes Dev 19:2030–2040
153. Mourrain P, Béclin C, Elmayan T et al (2000) *Arabidopsis SGS2* and *SGS3* genes are required for posttranscriptional gene silencing and natural virus resistance. Cell 101:533–542
154. Penterman J, Zilberman D, Huh JH et al (2007) DNA demethylation in the *Arabidopsis* genome. Proc Natl Acad Sci U S A 104: 6752–6757
155. Hernandez-Pinzon I, Yelina NE, Schwach F et al (2007) SDE5, the putative homologue of a human mRNA export factor, is required for transgene silencing and accumulation of *trans*-acting endogenous siRNA. Plant J 50:140–148
156. Murray MG, Thompson WF (1980) Rapid isolation of high molecular weight plant DNA. Nucleic Acids Res 8:4321–4326
157. Konieczny A, Ausubel FM (1993) A procedure for mapping *Arabidopsis* mutations using co-dominant ecotype-specific PCR-based markers. Plant J 4:403–410
158. Neff MM, Neff JD, Chory J et al (1998) dCAPS, a simple technique for the genetic analysis of single nucleotide polymorphisms: experimental applications in *Arabidopsis thaliana* genetics. Plant J 14:387–392
159. Takahashi H, Miller J, Nozaki Y et al (2002) *RCY1*, an *Arabidopsis thaliana RPP8/HRT* family resistance gene, conferring resistance to cucumber mosaic virus requires salicylic acid, ethylene and a novel signal transduction mechanism. Plant J 32:655–667
160. Takahashi H, Kanayama Y, Zheng MS et al (2004) Antagonistic interactions between the SA and JA signaling pathways in *Arabidopsis* modulate expression of defense genes and gene-for-gene resistance to cucumber mosaic virus. Plant Cell Physiol 45:803–809
161. Takahashi H, Kai A, Yamashita M et al (2012) Cyclic nucleotide-gated ion channel-mediated cell death may not be critical for *R* gene-conferred resistance to *Cucumber mosaic virus* in *Arabidopsis thaliana*. Physiol Mol Plant Pathol 79:40–48
162. Ando S, Obinata A, Takahashi H (2014) WRKY70 interacting with RCY1 disease resistance protein is required for resistance to *Cucumber mosaic virus* in *Arabidopsis thaliana*. Physiol Mol Plant Pathol 85:8–14
163. Sekine K, Kawakami S, Hase S et al (2008) High level expression of a virus resistance gene, *RCY1*, confers extreme resistance to *Cucumber mosaic virus* in *Arabidopsis thaliana*. Mol Plant-Microbe Interact 21:1398–1407
164. Koiwa H, Barb AW, Xiong L et al (2002) C-terminal domain phosphatase-like family members (AtCPLs) differentially regulate *Arabidopsis thaliana* abiotic stress signaling, growth, and development. Proc Natl Acad Sci U S A 99:10893–13898
165. Sekine K-T, Ishihara T, Hase S et al (2006) Single amino acid alterations in *Arabidopsis thaliana* RCY1 compromise resistance to *Cucumber mosaic virus*, but differentially suppress hypersensitive response-like cell death. Plant Mol Biol 62:669–682

Chapter 4

Analysis of Antiviral Resistance Signaling Pathways by Virus-Induced Gene Silencing in *Nicotiana benthamiana*

Masayoshi Hashimoto, Yasuyuki Yamaji, and Ken Komatsu

Abstract

Owing to the absence of antiviral chemicals, strategies to deploy antiviral resistance in plants are desirable. Deciphering antiviral resistance mechanisms has been extensively promoted by virus-induced gene silencing (VIGS) technique, which decreases the transcriptional level of the gene of interest via RNA silencing machinery triggered by the partial gene fragment inserted into a viral vector. This technique has contributed to addressing the function of a number of host genes, which are involved in signaling pathways leading to Resistance (R) protein-mediated resistance and the viral disease symptoms such as systemic necrosis. Here we describe the general VIGS protocol and its tips to analyze antiviral resistance mechanism using *Tobacco rattle virus* (TRV)-based VIGS vector in tobacco plant *Nicotiana benthamiana*. In most cases, the knockdown of host genes by TRV vector is highly efficient and uniform at the whole plant level but is not associated with any severe symptoms. Using this method together with inoculation of a challenge virus and transient overexpression of a viral elicitor or another host signaling component by agroinfiltration, as well as measurement of hallmarks of antiviral responses, we can address the role of the host factors and the epistatic relationship between several host factors in the resistance signaling pathway against plant virus infection. The protocol described here using the highly susceptible host *N. benthamiana* to a variety of plant viruses provides the opportunity to study resistance mechanisms underlying many host–virus interactions.

Key words Virus-induced gene silencing, Tobacco rattle virus, Antiviral resistance, Signaling pathway

1 Introduction

Plant viruses are obligate parasites that completely depend on living plant cells for their propagation. Because of this property, direct and specific control strategies against plant viruses by chemicals have not been established yet, which poses difficult challenges for viral disease management. Therefore, one of the most promising ways to develop a novel, durable strategy for viral disease management is to exploit host resistance mechanisms against viruses, most of which remains elusive. However, deciphering resistance mechanisms against plant virus infection has long been technically

Kappei Kobayashi and Masamichi Nishiguchi (eds.), *Antiviral Resistance in Plants: Methods and Protocols*, Methods in Molecular Biology, vol. 2028, https://doi.org/10.1007/978-1-4939-9635-3_4,

challenging. The best way for us to overcome this technical limitation is to employ the model plant *Arabidopsis thaliana*, in which mutant screening and genetic analysis are easily accessible. In fact, previous studies using *A. thaliana* have successfully identified a number of critical host factors in plant antiviral machineries including R protein-mediated resistance, RNA silencing and recessive resistance [1–3]. However, in this cruciferous plant, it is not always true for the virus and the disease phenotype which scientists are interested in on another host species to be infectious and reproducible, respectively. Therefore, to analyze resistance mechanisms against plant viruses, it is highly desired to have an alternative host species other than *A. thaliana* in which genetic modification techniques and genome information is accessible.

In this decade, virus-induced gene silencing (VIGS) technique has been extensively used to overcome these limitations in non-model plants. VIGS is a technique exploiting the RNA silencing-based antiviral defense mechanism, in which a recombinant virus possessing a partial host gene fragment induces silencing of the endogenous host target gene depending on the sequence homology. The inherent nature of VIGS technique makes the method advantageous simultaneously knocking down multicopy genes and moderately compromising the function of the genes involved in fundamental plant growth processes. Since the antiviral RNA silencing machinery is common among a wide range of plant species, the VIGS technique is applicable for many plants including the solanaceous, leguminous, and monocotyledon plants including wheat and rice [4]. Thus far, this simple and robust gene-knockdown method enabled to identify many genes involving several antiviral resistance mechanisms such as R protein-mediated disease resistance and RNA silencing [5, 6]. The most widely used VIGS vector for functional analysis of plant genes is based on *Tobacco rattle virus* (TRV), which can be used for many plant species including *N. benthamiana* and *A. thaliana* [7]. Recently, apple spherical latent virus has been increasingly used for several researches that aim to address the function of plant genes [8].

In this chapter, we describe the general protocol to analyze resistance mechanisms using VIGS in *N. benthamiana*. This protocol includes several tips and update of the methods mentioned in the previously published papers [9–11] and can be applicable to address resistance mechanisms against plant viruses and the epistatic relationship among several plant factors in the resistance signaling pathway.

2 Materials

2.1 Biological Materials and Virus Vectors

1. *Nicotiana benthamiana* seeds.
2. *Agrobacterium tumefaciens* C58C1 for pBINTRA6, GV3101 with pSoup [12] for pTV00 and derivatives, and EHA105 for other constructs.

3. Plant growth chambers (e.g., Plant Growth Chamber MLR-352H-PA).
4. TRV vectors [7]. pBINTRA6 is a binary vector containing RNA1, and pTV00 is a vector containing RNA2 with multiple cloning sites to insert a partial cDNA fragment of target genes.

2.2 Culture and Transformation of Agrobacterium

1. SOB medium: 4 g Bacto tryptone, 1 g yeast extract, 500 μL of 1 M KCl, 400 μL of 5 M NaCl in 200 mL of distilled water. Autoclaved for 15 min and add 500 microliters of 2 M Mg^{2+}.
2. LB medium: 10 g Bacto tryptone, 5 g yeast extract, 5 g NaCl in 1 L of distilled water. Autoclaved for 15 min and add 1 mL of 50 mg/mL kanamycin. For plate medium, add 15 g agar before autoclaving.
3. SOC medium: 4 g Bacto tryptone, 1 g yeast extract, 500 μL of 1 M KCl, 400 μL of 5 M NaCl in 200 mL of distilled water. Autoclaved for 15 min and add 2 mL of 2 M $Mg^{2+}$and 3.6 mL of 2 M glucose.
4. Antibiotics stock solutions: 25 mg/mL rifampicin dissolved in methanol and 25 mg/mL chloramphenicol dissolved in ethanol. 50 mg/mL kanamycin is dissolved in water and filter-sterilized. Store at −25 °C.
5. 20 mM $CaCl_2$, autoclaved.
6. Liquid nitrogen.
7. 50% glycerol, autoclaved and used in a clean bench.
8. Agroinfiltration buffer: 10 mM $MgCl_2$ and 10 mM MES (pH 5.6 by KOH), sterilized by autoclave, and then add 1 mL of 200 mM filter-sterilized acetosyringone solubilized in DMSO (0.2 mM final concentration). Store at 4 °C protected from light.
9. 1-mL needleless syringe (e.g., 1-mL Terumo syringe without needle SS-01P, Terumo Medical Products).

2.3 RNA Extraction and Conventional RT-PCR

1. Sterile mortars and pestles.
2. Liquid nitrogen.
3. ISOGEN (NipponGene) or TRIZOL (Thermo Fisher Scientific).
4. Chloroform.
5. Isopropanol.
6. 70% and 100% ethanol.
7. RNase Inhibitor (TaKaRa Bio).
8. SuperScript III One-Step RT-PCR System with Platinum Taq (Thermo Fisher Scientific).

9. Primer sets for checking TRV vector infection (FR69 (5′-GCTGCTAGTTCATCTGCAC-3′) and FR70 (5′-GCACGGATCTACTTAAAGAAC-3′)).
10. 1% agarose gel in TAE buffer for electrophoresis.
11. 10 mg/mL ethidium bromide solution.
12. Virus-specific primers.

2.4 Quantitative RT-PCR (RT-qPCR)

1. RQ1 RNase-free DNase (Promega).
2. High capacity cDNA reverse transcription kit (ThermoFisher Scientific).
3. GoTaq qPCR master mix (Promega).
4. Host gene-specific primers (*see* Subheading 3.4 for detail).
5. Thermal cycler dice real-time system (TaKaRa Bio), or any other qPCR instruments.

2.5 General Laboratory Equipment

1. 1.5 mL microcentrifuge tubes.
2. Filtered pipette tips.
3. 200 μL PCR tubes, PCR tubes for RT-qPCR.
4. DNase/RNase-free distilled water.
5. A thermal cycler.
6. A high-speed centrifuge.
7. A benchtop centrifuge, a vortex mixer, and a heat block.
8. A spectrophotometer and its cuvettes.

3 Methods

cDNA synthesis and cloning into a binary vector are performed following conventional methods. Position and length of the partial fragment in the target gene for cloning into pTV00 influence the efficiency of VIGS [13]. If you find gene knockdown inefficient in your first trial, another cDNA fragment from a different position of the target gene should be tested.

3.1 Preparation of *Agrobacterium* Cultures and Transformation with a Binary Vector

For successful infection of TRV, two genomic segments, RNA1 and RNA2 are needed. In the experiments shown below, you should always include both wild-type RNA1 and RNA2 as a negative control, and RNA2 vector with a target host gene fragment, and control RNA2 vector with marker gene insert such as *phytoene desaturase* (*PDS*), or *magnesium chelatase* (*SU*), which provides us a visual positive control for VIGS.

1. Prepare competent *Agrobacterium* cells. Culture *Agrobacterium* from a glycerol stock in SOB medium (50 mL) supplemented

with 25 μg/mL each of chloramphenicol and rifampicin at 28 °C overnight (*see* **Note 1**), transfer the culture (about OD_{600} = 0.5 is the best) to 50 mL centrifuge tubes, and centrifuge at 3000 × *g*, 5 min. Discard the supernatant and suspend the pellet in 5 mL of 20 mM $CaCl_2$. Centrifuge again at 3000 × *g*, 5 min, and resuspend the pellet in 0.5 mL of 20 mM $CaCl_2$. After dividing the final solution into sterile 1.5 mL microcentrifuge tubes (around 50 μL each), freeze in liquid nitrogen and store at −80 °C until use.

2. After taking 200–300 ng of a binary vector into a 1.5 mL microcentrifuge tube, add 15–30 μL of the competent *Agrobacterium* cells described above, frozen in liquid nitrogen in 1 min for cold shock treatment, and add 1 mL of SOC medium. Incubate at 28 °C for 3 h for preculture (*see* **Note 2**).
3. Plate the *Agrobacterium* cells to solid LB medium with 50 μg/mL kanamycin. Incubate at 28 °C for 2 days for colony formation.
4. Pick 2–3 single, round shape colonies by a toothpick and dip it into about 4 mL of liquid LB medium in a 15 mL centrifuge tube with 50 μg/mL kanamycin. Incubate with rapid shaking overnight at 28 °C.
5. In the next day, prepare glycerol stock by mixing *Agrobacterium* cultures 1:1 with 50% glycerol and stored at −80 °C (*see* **Note 3**).

3.2 Agroinfiltration

1. Pick up the glycerol stock made at **step 5** of Subheading 3.1, and culture *Agrobacterium* in 3–4 mL of liquid LB medium with rapid shaking overnight at 28 °C.
2. Centrifuge at 5000 × *g*, 10 min to collect bacterial cells. Discard supernatant LB medium, and resuspend the cell with a similar amount of agroinfiltration buffer (about 3–4 mL).
3. Measure the optical density at 600 nm (OD600), and adjust the concentration at about OD600 = 1.0 by adding fresh agroinfiltration buffer.
4. Incubate for about 3 h in the dark at room temperature to induce the gene expression for T-DNA transfer by acetosyringone.
5. Mix *Agrobacterium* culture containing TRV RNA1 and that containing RNA2 and derivatives with a 1:1 ratio.
6. By a needleless syringe, inject the suspension of *Agrobacterium* culture into the abaxial part of the leaves of 3–4 weeks old *N. benthamiana* seedlings.
7. Incubate the inoculated plants in the dark overnight, and in the next day move the plants to a growth chamber and incubate at 25 °C.

3.3 Extraction of Plant Total RNA and Detection of TRV Infection

About 10 days after inoculation, positive control plants inoculated with TRV-SU or TRV-PDS will start to develop the beached phenotype. Use plants at about 20 days after inoculation, which have 2–3 bleached, upper leaves, for subsequent analysis.

1. Take a piece of leaf disks (at most 50 mg) from the uninoculated upper leaves showing the bleached phenotype in TRV-SU or TRV-PDS-inoculated plants. Crush the leaf disks into a fine powder using mortar and pestle in liquid nitrogen. Immediately add 1 mL of ISOGEN or TRIZOL and follow the manufacturers' instructions to extract total RNA.
2. Dissolve the total RNA into about 50 μL of fresh MilliQ water with 1% RNase Inhibitor. Measure the concentration of isolated RNA by spectrophotometer.
3. Perform one-step RT-PCR with the extracted total RNA (about 300–500 ng) as a template, using SuperScript III One-Step RT-PCR System in a 10 μL reaction mixture according to the manufacturer's instruction. Primers described in Subheading 2.2 should be used. The reaction is performed at 50 °C for 30 min, and 94 °C for 2 min, followed by 40 cycles of 94 °C for 15 s, 60 °C for 30 s and 68 °C for 1 min, and a final extension for 5 min at 68 °C. Time of the extension step (68 °C for 1 min) should be modified according to the insert length (*see* **Note 4**).
4. Carry out gel electrophoresis to check the size of the amplified product. Confirm whether the insert of the host gene cDNA was retained (*see* **Note 5**).

3.4 Quantification of Target Host mRNA by RT-qPCR

When the inoculated TRV vector successfully infects *N. benthamiana* plants and retains the host gene insert, check whether host gene expression is downregulated as expected using RT-qPCR.

1. Design host gene-specific RT-qPCR primers using the program such as Primer3 (http://bioinfo.ut.ee/primer3-0.4.0/). In the Primer3 website, input several parameters as follows: 80–150 for Product Size Range; 17 and 25 for Primer size; 60 and 65 for Primer Tm; 45 and 55 for Primer GC %. Note that, in order to discriminate host gene expression from virus infection, the primers to quantify the target gene transcript level should not be designed within the region cloned into the virus vector. Since it is difficult to predict the specificity and amplification efficiency of primers, the author would normally design three distinct sets of primers for the following steps.
2. To test the amplification efficiency and specificity of the designed primers and to establish a standard curve, total RNA extracted from healthy *N. benthamiana* plants by ISOGEN or TRIZOL is used as a template.

3. Treat the total RNA by DNaseI to remove contaminated genomic DNA, which would harm the final results. Dissolve 2 μg of total RNA in 8 μL RNase-free water and add 1 μL each of 10 × DNase I buffer and DNaseI enzyme. Incubate at 37 °C for 30 min, add 1 μL of RQ1 DNase stop solution, and incubate at 65 °C for 10 min using 0.2 mL PCR tubes.
4. Purify RNA by an equal volume of acid phenol, and acid phenol–chloroform–isoamyl alcohol (25:24:1). After these treatments, precipitate RNA by adding half volume of 7.5 M ammonium acetate and 3.5 volumes of 100% ethanol, and centrifuge at full-speed for at least 30 min.
5. Dissolve pellet with 15 μL of RNase-free water with 1% RNase inhibitor and measure the concentration of RNA by spectrophotometer.
6. Synthesis of cDNA from the DNase-treated RNA. 7.1 μL of the DNase-treated RNA was added with 1 μL each of 10 × reverse transcription buffer and 10 × random primers, 0.4 μL of 25 × dNTP, and 0.5 μL of reverse transcriptase.
7. Incubate at 25 °C for 10 min, 37 °C for 120 min, and 85 °C for 5 min.
8. A dilution series of the synthesized cDNA is used as template for qPCR reaction to determine the amplification efficiency and specificity of the designed primers and to establish a standard curve. Recommended number of dilution steps is six in order to have a wide detection range.
9. The reaction mix of qPCR is 2 μL of the cDNA template, 1 μL each of forward and reverse primers (10 μM each), 8.5 μL of water, and 12.5 μL of GoTaq qPCR Master Mix. Fresh MilliQ water is used as negative control. Three technical replicates are needed for each dilution step.
10. qPCR condition is as follows: Initial denaturation at 95 °C for 2 min, and 40 cycles at 95 °C for 5 s and 60 °C for 30 s, followed by a dissociation step at 95 °C for 15 s and 60 °C for 30 s, and 95 °C for 15 s.
11. After the reaction is done, check the dissociation curve in which a single peak indicates the specific amplification of the target using the primers.
12. Acceptable standard curves should have 5–6 data points, amplification efficiency ranging from around 90–110% with R^2 value over 0.99. If you do not gain an acceptable standard curve in terms of the number of data points, decrease the dilution ratio from 10 to 5 or 4 times. Primers for a reference gene (e.g., eEF1A, ubiquitin, and 18S rRNA genes) and the standard curve should be also prepared similarly.

13. Quantification of the target gene expression in TRV-inoculated plants using RT-qPCR. Total RNAs extracted in Subheading 3.3 are used as templates. The synthesized cDNA is normally diluted to 50 or 100 times as templates of qPCR for their amplification to fall within the range of the standard curve. Three technical replicates are recommended for each biological replicate.
14. After the qPCR reaction, the target gene expression level is calculated based on the crossing point method or second derivative maximum method.

3.5 Inoculation of Challenge Virus or Expression of a Virus Elicitor Protein and a Host Resistant Signaling Component

If the host target gene expression is downregulated properly, it is possible to test whether the target gene product is involved in the antiviral response by measuring the viral accumulation or the specific resistance phenotypes such as cell death. Moreover, this enables us to test the epistatic relationship of the VIGS target gene product with another signaling component in a resistant signaling pathway, such as mitogen-activated protein kinase (MAPK) pathway. If a downstream component were knocked down by VIGS, the signaling output such as cell death induced by transient expression of an upstream component would be compromised.

1. Select the expanded leaves where a target host gene was knocked down successfully. In every experiment, the authors recommend to include the positive control of VIGS using TRV-PDS to roughly estimate the knocked-down leaves by counting the level of the bleached leaves from the TRV-PDS-inoculated ones.
2. To analyze the involvement of the target host factor in antiviral resistance associated with virus infection, inoculate a virus to the expanded leaves where the target is downregulated and where only TRV vector control is infected. Viral inoculation is obtained by infiltration of *Agrobacterium* culture containing a viral genomic cDNA, mechanical inoculation of leaf sap from virus-infected plants or purified virion (*see* **Note 6**). It is critical to adjust the concentration of inoculum as lowest as possible wherein the antiviral resistant phenotype can be observed repeatedly. As a negative control, *Agrobacterium* culture expressing GFP or GUS, leaf sap from mock-inoculated plants, or the phosphate buffer for virion inoculation should be included (*see* **Note 7**). The authors recommend that infiltration of negative control *Agrobacterium* cells together with challenge inoculation is performed by the side-by-side manner in the same leaf.
3. To address the function of the target host factor in antiviral resistance uncoupled with a viral infection, infiltrate *Agrobacterium* culture containing a viral elicitor protein which

causes antiviral resistant phenotypes such as cell death and upregulation of defense-related genes. For inoculation, it is critical to employ the lowest concentration of *Agrobacterium* cells at which the antiviral resistant phenotype can be observed repeatedly.

3.6 Examination of Resistance Reactions by Analyzing Virus RNA Accumulation and Defense-Related Phenotypes

1. At the predetermined time interval after challenge inoculation when the intended defense phenotype can be observed, take photos of the infiltrated leaves or upper ones to see whether there are any phenotypic differences between control and tested tissues.
2. In order to see whether virus accumulation level was changed, extract total RNA from leaves and perform RT-qPCR to compare virus accumulation between VIGS and control plants.
3. For detection of reactive oxygen species, carry out 3,3′-diaminobenzidine (DAB) staining, which can histologically detect H_2O_2 production. Excise the infiltrated or inoculated leaves and immersed them in 1 mg/mL DAB solution with vacuum infiltration. Incubate at room temperature in the dark for about 3 h. The stained leaves can be decolorized in 70% (v/v) ethanol in a hot water bath before taking images.
4. For detection of dead cells, perform trypan blue staining. Excise the infiltrated or inoculated leaves and immersed them in trypan blue–lactophenol solution (mixture of 10 mL lactic acid, 10 mL glycerol, 10 g of phenol crystal, 10 mg trypan blue, and 10 mL water) with several times of boiling. The leaves are decolorized by incubation with chloral hydrate solution (2.5 g chloral hydrate in 1 mL water) under a fume hood.
5. For quantification of electrolyte leakage, which is correlated with an extent of cell death, punch out leaf disks (about 7 mm in diameter) from infiltrated or inoculated leaves. Immense three to four leaf disks in 4 mL of distilled water in 15 mL tube with vacuum infiltration. After an hour of incubation with shaking at room temperature, measure electrolyte leakage using a conductivity meter (B-173, HORIBA).

4 Notes

1. Different *A. tumefaciens* strains have different degrees of antibiotics resistance. It is necessary to modify antibiotic concentrations depending on *A. tumefaciens* strains used.
2. This transformation method is based on the freeze-thaw method described by Weigel and Glazebrook [14]. You can also use commercial electrocompetent *A. tumefaciens* cells. Either method produces enough amounts of transformants.

3. In most cases, the transformation of *A. tumefaciens* with a binary vector will be successful. If you would like to confirm successful transformation, take several microliters of culture, and use those for direct PCR detection and sequencing of the gene fragment in the binary vector. For that aim, GoTaq Green Master Mix (Promega) is useful.
4. The empty TRV vector produces an amplified product of about 350 bp. Do not forget to perform control PCR reaction using the empty virus vector, and apply the amplified PCR products on 1.2 or higher percentages of agarose gel, in order to discriminate between insert-deleted and insert-retained virus vectors in gel electrophoresis.
5. When the insert is deleted from the virus vector, redesign primers amplifying another fragment in the target host gene. The length and position of a target sequence are critical for the insert stability and efficiency of VIGS.
6. Leaf position used for virus inoculation considerably affects the outcome of experiments because the extent of gene knockdown, as well as cell physiology, differs depending on leaves.
7. You can isolate protoplasts from plant leaves whose targeted gene was knocked down by VIGS [15]. This experiment allows us to address whether the gene of interest plays a role in antiviral defense in a single cell level. One excellent example is the experiment analyzing the relationship between *Red clover necrotic mosaic virus* and its host factor (Kaido et al. [15]). Comparison between results obtained by leaf and protoplast provides us with useful information on mechanisms underlying resistance against viruses.

References

1. Ding SW, Voinnet O (2007) Antiviral immunity directed by small RNAs. Cell 130:413–426
2. Mandadi KK, Scholthof KBG (2013) Plant immune responses against viruses: how does a virus cause disease? Plant Cell 25:1489–1505
3. Hashimoto M, Neriya Y, Yamaji Y, Namba S (2016) Recessive resistance to plant viruses: potential resistance genes beyond translation initiation factors. Front Microbiol 26:1695
4. Lange M, Yellina AL, Orashakova S, Becker A (2013) Virus-induced gene silencing (VIGS) in plants: an overview of target species and the virus-derived vector systems. In: Becker A (ed) Virus-induced gene silencing. Methods in molecular biology, vol 975. Humana Press, Totowa, NJ, pp 1–14
5. Liu Y, Schiff M, Dinesh-Kumar SP (2004) Involvement of MEK1 MAPKK, NTF6 MAPK, WRKY/MYB transcription factors, *COI1* and *CTR1* in *N*-mediated resistance to tobacco mosaic virus. Plant J 38:800–809
6. Odokonyero D, Mendoza MR, Moffett P, Scholthof HB (2017) *Tobacco rattle virus* (TRV)-mediated silencing of *Nicotiana benthamiana ARGONAUTES* (*NbAGOs*) reveals new antiviral candidates and dominant effects of TRV-*NbAGO1*. Phytopathology 107:977–987
7. Ratcliff F, Martin-Hernandez AM, Baulcombe DC (2001) *Tobacco rattle virus* as a vector for analysis of gene function by silencing. Plant J 25:237–245
8. Kasajima I, Ito M, Yamagishi N, Yoshikawa N (2017) *Apple latent spherical virus* (ALSV) vector as a tool for reverse genetic studies and

non-transgenic breeding of a variety of crops. In: Rajewsky N, Jurga S, Barciszewski J (eds) Plant epigenetics. RNA Technologies. Springer, Cham, pp 513–536

9. Komatsu K, Hashimoto M, Ozeki J, Yamaji Y, Maejima K, Senshu H, Himeno M, Okano Y, Kagiwada S, Namba S (2010) Viral-induced systemic necrosis in plants involves both programmed cell death and the inhibition of viral multiplication, which are regulated by independent pathways. Mol Plant-Microbe Interact 23:283–293
10. Komatsu K, Hashimoto M, Maejima K, Shiraishi T, Neriya Y, Miura C, Minato N, Okano Y, Sugawara K, Yamaji Y, Namba S (2011) A necrosis-inducing elicitor domain encoded by both symptomatic and asymptomatic *Plantago asiatica mosaic virus* isolates, whose expression is modulated by virus replication. Mol Plant-Microbe Interact 24:408–420
11. Hashimoto M, Komatsu K, Maejima K, Okano Y, Shiraishi T, Ishikawa K, Takinami Y, Yamaji Y, Namba S (2012) Identification of three MAPKKKs forming a linear signaling pathway leading to programmed cell death in *Nicotiana benthamiana*. BMC Plant Biol 12:103
12. Hellens RP, Edwards EA, Leyland NR, Bean S, Mullineaux PM (2000) pGreen: a versatile and flexible binary Ti vector for *Agrobacterium*-mediated plant transformation. Plant Mol Biol 42:819–832
13. Liu E, Page JE (2008) Optimized cDNA libraries for virus-induced gene silencing (VIGS) using tobacco rattle virus. Plant Methods 4:5. https://doi.org/10.1186/1746-4811-4-5
14. Weigel D, Glazebrook J (2006) Transformation of *Agrobacterium* using the freeze-thaw method. CSH Protoc 2006. https://doi.org/10.1101/pdb.prot4666
15. Kaido M, Abe K, Mine A, Hyodo K, Taniguchi T, Taniguchi H, Mise K, Okuno T (2014) GAPDH-A recruits a plant virus movement protein to cortical virus replication complexes to facilitate viral cell-to-cell movement. PLoS Pathog 10:e1004505

Chapter 5

Random Mutagenesis of Virus Gene for the Experimental Evaluation of the Durability of NB-LRR Class Plant Virus Resistance Gene

Kengo Idehara, Reiko Tomita, Ken-Taro Sekine, Masamichi Nishiguchi, and Kappei Kobayashi

Abstract

NB-LRR class plant virus resistance gene is a one of the key players that shape the plant–virus interaction. Evolutionary arms race between plants and viruses often results in the breakdown of virus resistance in plants, which leads to a disastrous outcome in agricultural production. Although studies have analyzed the nature of plant virus resistance breakdown, it is still difficult to foresee the breakdown of a given virus resistance gene. In this chapter, we provide a protocol for evaluating the durability of plant virus resistance gene, which comprises the random mutagenesis of a virus gene, the introduction of the mutagenized gene into a virus context with highly efficient inoculation system, and the efficient screening of virus mutants that can overcome or escape a virus resistance.

Key words Agroinfection, Durability, Error-prone PCR, Infectious clone, Random mutagenesis, Resistance breakage

1 Introduction

In plant disease control, the use of resistant cultivars is one of the ideal strategies because it is economic, labor-saving, and, more importantly, effective unless resistance-breaking pathogens emerge [1]. Unfortunately, the emergence of resistance-breaking pathogen strains is often encountered, which may cause severe losses in the field. Once a resistance-breaking pathogen emerges, production of the crop will continuously face a threat by the pathogen until a new, effective resistance gene (*R*-gene) is introduced into cultivars of the crop. However, search for new *R*-genes in natural germplasm is laborious and time-consuming. A possible solution would be to use an *R*-gene from different crop species. For example, transgenic tomato plants expressing tobamovirus *R*-gene *N* from *Nicotiana glutinosa* were reportedly resistant to the cognate viruses [2].

Kappei Kobayashi and Masamichi Nishiguchi (eds.), *Antiviral Resistance in Plants: Methods and Protocols*, Methods in Molecular Biology, vol. 2028, https://doi.org/10.1007/978-1-4939-9635-3_5,

Although the transgenic approach would be promising to confer resistance to a crop species, its durability could be a different matter. In the case of *N* gene, which has been intensively studied in the original *N. glutinosa* and *Nicotiana tabacum* introgressed with the *R*-gene, it is known to be durable, that is, it is effective against different cognate viruses or virus mutants [3]. Moreover, tomato and tobacco plants largely share their tobamovirus pathogens. Therefore, one can expect durable resistance of *N* gene transgenic tomato against different tobamoviruses. In contrast to the *N* gene, different *R*-genes have been known to be broken by virus mutations and studies have analyzed the nature of plant virus resistance breakdown [1, 3]. Among them, a recent study has pointed out that some parameters largely influence the risk of resistance breakdown: number and nature of mutations involved in the breakdown, and fitness of mutant genotypes [4]. Although mathematical modeling highlighted the significance of these parameters, those for a particular virus could only be determined through an experimental approach.

Plant virus resistance is categorized into two groups based on the mode of action: one involves the hypersensitive reaction (HR) or the programmed cell death, and the other does not [1]. The virus resistance of the former type is genetically dominant, and the avirulent viruses, which elicit the resistance responses leading to the HR, are dominant within the virus population [5]. By contrast, the resistance of the latter type comprises both dominant and recessive resistances. The dominant *R*-genes of this class include tomato *Tm-1* and its cognate [6, 7] and *Arabidopsis thaliana JAX1* [8], and their resistance-breaking virus mutants were isolated through the repeated inoculation of a resistant host with a virus [7] or the grafting of a resistant scion onto a virus-infected susceptible rootstock [9], respectively. In case of the dominant *R*-genes involved in HR-mediated resistance, although several resistance-breaking virus isolates were reported, experimental attempts to isolate resistance-breaking mutants have not been successful. The failure would be because avirulent strains are genetically dominant over resistance-breaking virulent strains in the activation of resistance response, that is, when they coexist in a resistant host, the spread of resistance-breaking strains would be restricted by the resistance response elicited by avirulent strains. This fact makes it difficult to evaluate experimentally the resistance durability of *R*-genes that conditions virus resistance accompanying HR.

1.1 Pathosystem in Which the Protocol Was Established

In this chapter, we describe the protocol for introducing random mutations into a viral gene of interest, which is useful for evaluating resistance durability. As the example for such protocol, we introduced random mutations into coat protein (CP) gene of *Pepper mild mottle virus* (PMMoV) and tested the ability of the mutants to break *N′* gene from *Nicotiana sylvestris*, a wild tobacco.

In pepper production, *Tobamovirus* species have been among the most important pathogens. Pepper plants have a gene providing resistance to *Tobamovirus* species designated *L* gene has been used to protect the plants from virus infection [10–12]. Molecular cloning of *L* gene alleles has shown that they code for typical resistance protein (R-protein) with coiled-coil (CC), nucleotide binding (NB) and leucine-rich repeat (LRR) domains, and that they are seven alleles of a single copy *L* gene, namely, L^1, L^{1a}, L^{1c}, L^2, L^{2b}, L^3, and L^4 [13]. Their protein products recognize tobamovirus coat protein (CP) [14–17] and trigger resistance response accompanying HR with an increasing pathogen recognition spectrum [13]. However, even the L^4 gene with the broadest recognition spectrum has been broken down by virus mutations. As an alternative for broken *L* gene, we assumed the *N′* gene, which shares the viral avirulence factor with *L* gene [18–20] and codes for a CC-NB-LRR protein highly homologous to L protein [21]. The *N′* gene has not been broken by any *L* gene breaking PMMoV strains [21], although *N′* gene does not confer resistance to *Tobacco mosaic virus* (TMV). In contrast, although the *L* gene has easily been broken by several PMMoV strains, it has been durable to TMV [3]. Although the combined use of *N′* and *L* genes is expected to provide robust resistance to any *Tobamovirus* species, a possible pitfall of this strategy is that a "super strain" (an analogy to "super race" often mentioned for microbial pathogens; [22]) of PMMoV, which simultaneously breaks both *N′* and *L* genes, could emerge. The protocol presented in this chapter was designed to evaluate the durability of PMMoV resistance in the *N′*. Some preliminary results are also shown to prove the efficacy of the protocol.

1.2 Design of the Protocol

The protocol was designed to enable the random mutagenesis of PMMoV CP gene and the efficient inoculation of *N. sylvestris* containing *N′* gene based on the following criteria: firstly, it should be easy to introduce a population of the randomly mutagenized CP gene into the infectious virus construct, and secondly, it should be easy to infect plants efficiently with each clone from the virus construct population with the mutagenized gene. A protocol that fulfills these requirements is shown in Fig. 1a. PMMoV CP gene was randomly mutagenized (Fig. 1a(1)) and introduced into a PMMoV agroinfection construct (Fig. 1a(2 and 3)). Optimization of this process fulfilled the first requirement. A library of PMMoV agroinfection constructs with mutagenized CP genes was amplified in *Escherichia coli* (Figs. 1a(3)). The amplified library was introduced into *Agrobacterium tumefaciens* by transforming the bacteria using a plasmid DNA preparation from the bulk of *E. coli* colonies (Fig. 1a(4)). The *A. tumefaciens* colonies were clonally propagated (Fig. 1a(5)) and used to inoculate *N. sylvestris* plants that contain *N′* gene using toothpick inoculation method (Fig. 1a(6)), which fulfills the second requirement.

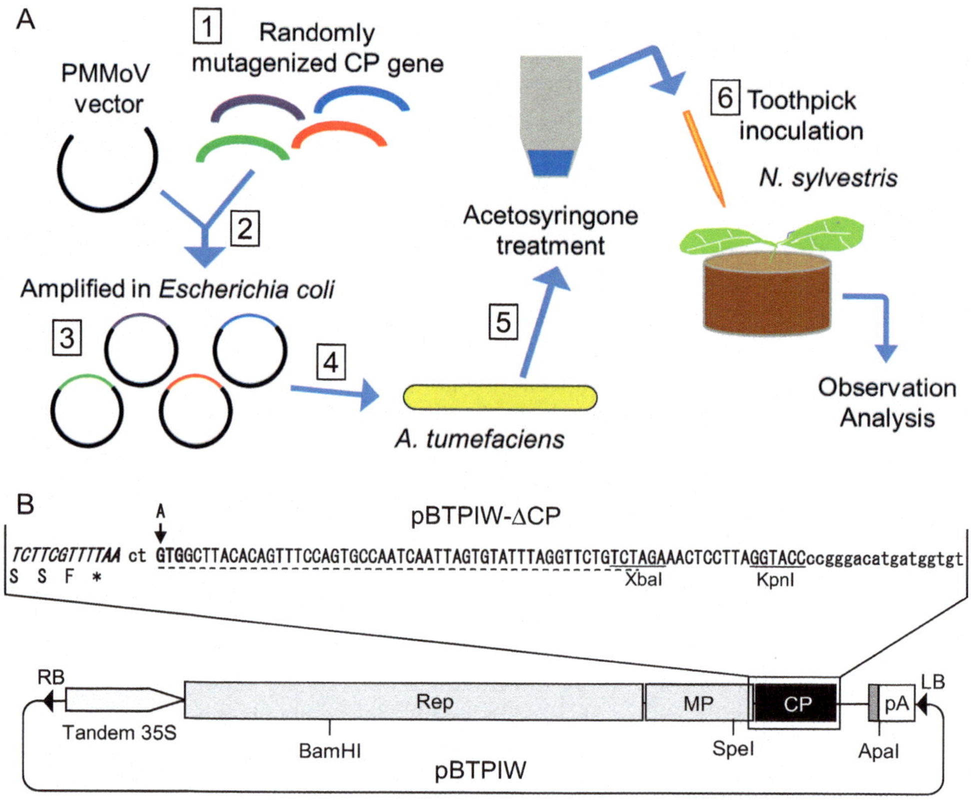

Fig. 1 Schemes of the protocol (**a**) and the construct (**b**). The outline of the protocol is described in the Introduction. The CP gene of agroinfection construct of PMMoV, pBTPIW, was replaced with the sequence indicated on the top of B, in which restriction sites are underlined and a partial movement protein gene sequences are italicized with the termination codon in boldface. RB and LB, right and left border sequences of T-DNA region; Tandem 35S, Cauliflower mosaic virus 35S promoter with duplicated enhancer regions; Rep, PMMoV replication proteins gene; MP, PMMoV movement protein gene; CP, PMMoV coat protein gene; pA, nopaline synthase gene polyadenylation signal sequence

Gene manipulation of plant RNA virus was first documented by the production of infectious in vitro transcripts [23], and this strategy has been adapted to many viruses. However, the production of in vitro transcripts is costly and laborious. Moreover, infection of plants with in vitro transcripts is not very efficient in some plants. In contrast, agroinfection method [24] is easier and cheaper to infect hosts with numbers of viral clones. An agroinfection construct of PMMoV, pBTPIW [13] (Fig. 1b), was efficient enough to achieve nearly 100% infection of inoculated plants. It was modified so that the CP gene can easily be replaced with randomly mutagenized PMMoV CP genes (*see* Subheadings 2.1 and 3.1). A modified PMMoV agroinfection construct pBTPIWΔCP (Fig. 1b) was constructed by replacing about 90% of the CP gene with restriction sites. The first 50 nucleotides in CP coding regions were shown to be involved in the CP subgenomic mRNA promoter

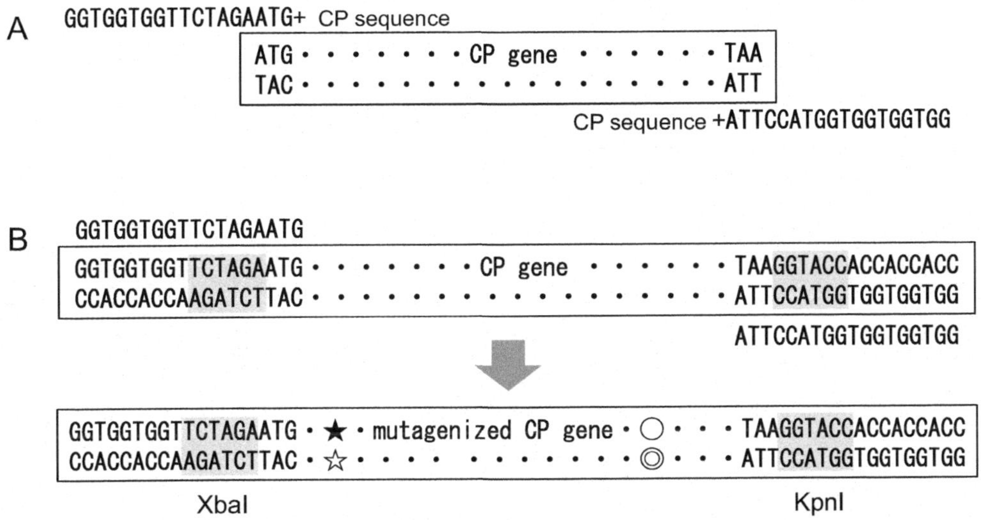

Fig. 2 Schemes of error-prone PCR. (**a**) Addition of restriction sites and scaffold sequence to CP gene to prepare template for error-prone PCR. The CP gene sequence is shown by the box, and the primers RM-PMMoV-F and RM-PMMoV-R are schematized in the upper left and lower right, respectively. (**b**) Error-prone PCR using primers comprised of scaffold sequence, restriction sites, and initiation or termination codons. The wild-type template sequence is shown by the box with the sequences of the primers ScXbATG-F and TAAKpSc-R are shown in the upper left and lower right, respectively. The error-prone PCR product is schematized in the lower box

activity in TMV [25], and the corresponding sequence in PMMoV is also likely to have a role in subgenomic CP mRNA transcription. Therefore, the first 50 nucleotides of PMMoV CP gene were kept in the upstream of restriction sites, with the initiation codon abolished by the substitution of adenine to guanine (Fig. 1b). Random mutagenesis of PMMoV CP gene was carried out by an error-prone PCR. A wild-type CP gene was first amplified to add the restriction sites and scaffold sequence (Fig. 2a). Error-prone PCR was carried out with primers containing the scaffold sequence, restriction site, and initiation or termination codon (Fig. 2b).

2 Materials

2.1 Modification of PMMoV Agroinfection Construct

1. pBTPIW: An agroinfection construct of PMMoV [13].
2. Primers.

PM-SnaF:	CAGTTCCAAT GGCTGACAGA
PMCP-A65R:	AACTTATTTACACCATCATGTCC CGGGGTACCTAAGGAGTTG TAG
PMCP-A65F:	TACAACTCCTTAGGTACCCCG GGACATGATGGTGTAAATAA GTTG

RzDn-Apa:	GCGGGCCCGTTTCGTCC TTTAGGGACTCGTCAGTGTACTGATA TAA
PMCP-InitMu-R:	AAACTGTGTAAGCCACA GTTAAAACGAAGAAGAC
PMCP-InitMu-F:	CGTTTTAACTGTGGCTTACACAGTTT CCAGTGCCAATCA ATTAGTGTATTTAG
PMCP5-XbA65R:	GGGGTACCTAAGGAGTTTCTAGACAGAA CCTAAATACA CTAATTGATTGG

3. High-fidelity PCR enzyme: PrimeSTAR GXL DNA polymerase (Takara Bio) or equivalent. Supplied with buffer and dNTP mixture.
4. PCR cloning kit: TOPO Zero blunt PCR cloning kit (Thermo) or equivalent.
5. Competent cells: Electrocompetent cells of *E. coli* HST08 strain (Takara Bio) or other strains, which are suitable for the transformation of large plasmids.
6. LB medium: 1% (w/v) tryptone, 0.5% (w/v) yeast extract, 1% (w/v) NaCl. Adjust pH to 7.0 with 1 N NaOH. Sterilize by autoclave.
7. Kanamycin: 50 mg/mL stock solution.
8. LB-K50 plates: Add 1.5% (w/v) agar to LB medium and sterilize by autoclave. After cooling down to 55 °C, add 50 mg/L (final concentration) kanamycin and pour into plastic petri dishes.
9. Incubator at 37 °C.
10. Shaking incubator at 37 °C.
11. Miniprep kit.
12. Restriction enzymes: ApaI, KpnI (or Acc65I, an isoschizomer producing 5′-protruding ends), SnaBI, XbaI.
13. Alkaline phosphatase: Molecular biology grade. Alkaline Phosphatase (Shrimp) (Takara Bio) or equivalent.
14. Ligation kit: Ligation kit ver. 2.1 (Takara Bio) or equivalent.
15. Suspension buffer: 50 mM glucose, 10 mM EDTA, 25 mM Tris–HCl, pH 8.0.
16. Lysis solution: 0.2 N NaOH, 1% (w/v) sodium dodecyl sulfate (SDS).
17. Neutralization solution: 3 M potassium acetate, pH 4.8.
18. Chloroform.
19. 99.5% ethanol.
20. 70% ethanol.
21. TE: 10 mM Tris–HCl, pH 8.0, 1 mM EDTA.
22. TER: TE containing 25 μg/mL RNase A (DNase-free).

23. 1% agarose gel in TAE buffer.
24. Agarose gel electrophoresis apparatus.
25. Gel-purification kit.
26. Coprecipitant: Ethachinmate (NIPPON GENE; Tokyo, Japan) or equivalent.
27. 3 M sodium acetate pH 5.2.
28. Electroporation cuvettes: 1 mm gap between electrodes.
29. Electroporation apparatus.
30. Midiprep kit: Genopure plasmid midi kit.

2.2 Random Mutagenesis of PMMoV CP Gene by an Error-Prone PCR

1. Primers.

RM-PMMoV-F:	GGTGGTGGTTCTAGAATGGCTTACACAGTT TCCAG
RM-PMMoV-R:	GGTGGTGGTGGTACCTTAAGGAGTTGTAGC CCAGG
ScXbATG-F:	GGTGGTGGTTCTAGAATG
TAAKpSc-R:	GGTGGTGGTGGTACCTTA
M3:	GTAAAACGACGGCCAGT
RV:	CAGGAAACAGCTATGAC

2. PCR cloning reagents as in Subheading 2.1, **items 3–11**.
3. Recombinant Taq DNA polymerase.
4. 10 × Taq buffer: 100 mM Tris–HCl pH 9.0, 25 mM $MgCl_2$, 500 mM KCl, 1% Triton X-100.
5. 25 mM $MgCl_2$: Sterilized by autoclave.
6. 50 mM $MnCl_2$: Sterilized by autoclave.
7. dNTPs: Prepare separately the four dNTP stocks, 10 mM dATP, 10 mM dGTP, 10 mM dCTP, and 10 mM dTTP.
8. TA cloning kit: TOPO TA cloning kit (Thermo) or equivalent.
9. 1% agarose gel in TAE buffer.
10. Agarose gel electrophoresis apparatus.
11. PCR purification kit.
12. Sequencing facility.

2.3 Construction of PMMoV Agroinfection Constructs Having Randomly Mutagenized CP Gene

1. pBTPIWΔCP: An agroinfection construct of PMMoV lacking CP coding sequence, constructed in Subheading 3.1.2.
2. Randomly mutagenized CP gene fragment: Prepared in Subheading 3.2.
3. Reagents and equipment for cloning as in Subheading 2.1, **items 5–27**.

4. 15 mL conical tube.
5. Miniprep kit.
6. Agrobacterium: *Agrobacterium tumefaciens* GV3101 strain.
7. L-broth: 1% (w/v) tryptone, 0.5% (w/v) yeast extract, 0.5% (w/v) NaCl, 0.1% (w/v) glucose. Sterilize by autoclave.
8. Antibiotics: 50 mg/mL kanamycin, 20 mg/mL rifampicin, 20 mg/mL gentamycin.
9. L broth R20G20K50 plates: Add 1.5% (w/v) agar to L-broth and sterilize by autoclave. After cooling down to 55 °C, add 50 mg/L (final concentration) kanamycin, 20 mg/L (final concentration) rifampicin and 20 mg/L (final concentration) gentamycin, and pour into plastic petri dishes.
10. L broth R20G20 plates: Add 1.5% (w/v) agar to L-broth and sterilize by autoclave. After cooling down to 55 °C, add 20 mg/L (final concentration) rifampicin and 20 mg/L (final concentration) gentamycin, and pour into plastic petri dishes.
11. L broth R20G20: Add 20 mg/L (final concentration) rifampicin and 20 mg/L (final concentration) gentamycin.
12. 500 mL Erlenmeyer flask, sterilized.
13. 10% glycerol: Sterilize by autoclave. Chill on ice before use.
14. Electroporation cuvettes: 1 mm gap between electrodes.
15. Electroporation cuvettes: 2 mm gap between electrodes.
16. Electroporation apparatus.
17. 45% glycerol: Dilute glycerol with deionized water and sterilize by autoclave.

2.4 Agroinfection of Plants with PMMoV Having Randomly Mutagenized CP Genes

1. Seeds: *N. sylvestris* seeds.
2. Plastic petri dish: 90 mm in diameter.
3. Filter papers: Round sheets of qualitative filter paper, 70 mm in diameter.
4. Planting pots: Small plastic pots containing soil (*see* **Note 1**).
5. Liquid fertilizer: Hyponex or equivalent.
6. L broth R20G20K50: Liquid L broth containing with 50 mg/L kanamycin, 20 mg/L rifampicin, and 20 mg/L gentamycin.
7. Deep-well plates: Autoclavable plastic 96-well plates, 40 mm depth with tight-sealed silicone rubber lids.
8. Shaking incubator for deep-well plates: M-BR-022UP (Taitech) or equivalent.
9. 45% glycerol: Sterilized by autoclave.
10. 96-well plates with lids.
11. Eight-channel pipette with disposable tips.

12. Liquid reservoir for eight-channel pipette: Sterilize by autoclave.
13. Replicators for 96-well plate: Sterilize by autoclave.
14. L broth R20G20K50 rectangular plates: Made in sterile rectangular plastic dishes (140 mm × 100 mm × 14.5 mm).
15. Infiltration buffer: 10 mM K-MES pH 5.5 (2-(*N*-morpholino) ethanesulfonic acid, adjusted to pH 5.5 with KOH), 10 mM $MgCl_2$. Sterilize by autoclave.
16. 20 mg/mL acetosyringone: Dissolved in dimethyl sulfoxide.
17. Inoculation loops: Sterile, disposable plastic inoculation loops for 1 μL (about 1 mm pore diameter).
18. Autoclaved toothpicks.

2.5 Analysis of the Resistance Breaker Candidates

1. Primers.

PMF1:	GTTAGAATTGGGCAGAACTCGG
PMR1:	GTTATCGTACTCGCCACGGA

2. Leaf punch: 6 mm in diameter.
3. RNA extraction kit.
4. Reverse transcriptase.
5. Random hexamer primer.
6. High-fidelity PCR enzyme.
7. 1% agarose gel in TAE buffer.
8. Agarose gel electrophoresis apparatus.
9. PCR purification kit.
10. Sequencing facility.
11. Mortar and pestle.
12. 10 mM Hepes buffer: 10 mM 4-(2-hydroxyethyl)-1-piperazineethanesulfonic acid, adjust pH 7.4 with 1 N NaOH.
13. Carborundum: carborundum powder of 600 mesh.

3 Methods

3.1 Modification of PMMoV Agroinfection Construct

The PMMoV agroinfection construct, pBTPIW, will be introduced with novel unique restriction sites, KpnI and SmaI, just downstream of the CP gene. The resulting plasmid is named pBT-PIW-KS (*see* **Note 2**).

3.1.1 Introduction of Novel Unique Restriction Site into the Construct

1. Amplify 684-bp and 273-bp DNA fragments by PCR using pBTPIW as a template, high-fidelity DNA polymerase, and primer pairs of PMSnaF and PMCP-A65R, and PMCP-A65F and RzDn-Apa, respectively. Recommended reactions comprise 1 ng template DNA and 200 nM each of the primers in

the manufacturer's recommended buffer condition and will be carried out under the following conditions: 94 °C, 2 min; 25 cycles of 95 °C for 10 s, 50 °C for 10 s, and 68 °C for 1 min; and 68 °C for 3 min. Separate both DNA fragments on a 1% agarose gel and gel-purify using a kit.

2. Combine the DNA fragments from **step 1** by PCR. A recommended reaction comprises 1 ng each of template DNA fragments and 200 nM each of the primers in the manufacturer's buffer condition and will be carried out under the following conditions: 94 °C, 2 min; 95 °C for 10 s, 50 °C for 1 min, and 68 °C for 1 min; 24 cycles of 95 °C for 10 s, 50 °C for 10 s, and 68 °C for 1 min; and 68 °C for 3 min. Gel-purify the 914-bp DNA fragment.
3. Clone the fragment using a PCR cloning kit according to the manufacturer's protocol, prepare the plasmid using a miniprep kit according to the manufacturer's protocol, and confirm the absence of any PCR error by sequencing (*see* **Note 3**).
4. Digest 2 μg of plasmid DNA of selected clone and 1 μg of pBTPIW with SnaBI and ApaI. Gel-purify the 885-bp fragment from the former reaction. Treat the latter reaction with an alkaline phosphatase according to the manufacturer's protocol (*see* **Note 4**).
5. Clone the DNA fragment having KpnI and SmaI sites into pBTPIW. Mix about 20 ng of the gel-purified DNA fragment and about 50 ng of digested and alkaline phosphatase-treated pBTPIW, and ligate them using a ligation kit according to the manufacturer's protocol. Precipitate the ligation products by adding an appropriate amount of coprecipitant, 0.1 volume of sodium acetate pH 5.2, and 2.5 volume of ethanol, and centrifuging at about 20,000 × *g* for 30 min at 4 °C. After rinsed with 70% ethanol, air-dry briefly the DNA pellet. Dissolve the DNA in 5 μL TE.
6. Transform the electrocompetent *E. coli* with the concentrated and desalted ligation product. Transfer the mixture of electrocompetent cells and the ligation product to a 1-mm gap electroporation cuvette, and apply a pulse with the following setting: 1.5 kV, 25 μFd, and 200 Ω. Plate the bacteria onto LB-K50 plates and culture at 28 °C for about 40 h.
7. Pick up some colonies and culture them overnight at 28 °C in 3 mL LB medium containing 50 mg/L kanamycin.
8. Collect the bacteria from the entire culture by two rounds of centrifugation in a 1.5 mL microfuge tubes. Resuspend the bacteria in 150 μL suspension buffer by vigorous vortex. Add 250 μL lysis solution, mix gently by inversion and let stand at room temperature for 5 min. Add 200 μL neutralization solution,

mix well by inversion and let stand at room temperature for 5 min (*see* **Note 5**).

9. Centrifuge at about 20,000 × *g* for 5 min at room temperature and transfer the supernatant to a new 1.5 mL tube containing 0.5 mL chloroform. Mix vigorously by inversion, centrifuge at about 10,000 × *g* for 2 min at room temperature and transfer 0.5 mL upper aqueous phase to a new 1.5 mL tube.
10. Add 1 mL ethanol and mix vigorously by inversion. Centrifuge at about 20,000 × *g* for 3 min, discard supernatant, rinse the pellet with 70% ethanol and let it air-dried.
11. Dissolve the air-dried pellet in 20 μL TER and use 5 μL for restriction analysis with KpnI.

3.1.2 Modification of the Construct for the Insertion of Mutant CP Genes

1. Amplify a 194-bp DNA fragment encompassing the upstream region of CP gene and the mutated initiation codon of CP gene (Fragment A), using pBTPIW as a template, a primer pair of PMSnaF and PMCP-InitMu-R, and high-fidelity DNA polymerase. A recommended reaction comprises 1 ng template DNA and 200 nM each of the primers in the manufacturer's buffer condition and will be carried out under the following conditions: 94 °C, 2 min; 25 cycles of 95 °C for 10 s, 50 °C for 10 s, and 68 °C for 1 min; and 68 °C for 3 min. Gel-purify the 194-bp DNA fragment.
2. Prepare a DNA fragment (Fragment B) by annealing PMCP-InitMu-F and PMCP5-XbA65R and end-filling using a high-fidelity DNA polymerase. A recommended reaction comprises 200 nM each of the oligonucleotides in the manufacturer's buffer condition and will be carried out under the following conditions: 94 °C, 2 min; 95 °C for 10 s, 50 °C for 1 min, and 68 °C for 1 min. Gel-purify the 82-bp DNA fragment.
3. Combine the Fragment A and B by PCR as in **step 1** but using the primers PMSnaF and PMCP5-XbA65R. A recommended reaction comprises 1 ng each of template DNA fragments and 200 nM each of the primers in the manufacturer's buffer condition and will be carried out under the following conditions: 94 °C, 2 min; 95 °C for 10 s, 50 °C for 1 min, and 68 °C for 1 min; 24 cycles of 95 °C for 10 s, 50 °C for 10 s, and 68 °C for 1 min; and 68 °C for 3 min. Gel-purify the 250-bp DNA fragment.
4. Clone the fragment using a PCR cloning kit according to the manufacturer's protocol, prepare the plasmid using a miniprep kit according to the manufacturer's protocol and confirm the absence of any PCR error by sequencing.
5. Digest 2 μg of plasmid DNA of selected clone, and 1 μg of pBTPIW-KS with SnaBI and Acc65I. Gel-purify the 221-bp fragment from the former reaction. Treat the latter reaction

with an alkaline phosphatase according to the manufacturer's protocol.

6. Introduce the 221-bp into pBTPIW-KS digested with SnaBI and Acc65I in the same way as **steps 5–11** in Subheading 3.1.1.
7. Select an appropriate clone by restriction analyses with SnaBI and KpnI, and XbaI. Propagate the clone in 100 mL culture and purify the plasmid, pBTPIWΔCP, using a midiprep kit.

3.2 Random Mutagenesis of PMMoV CP Gene by an Error-Prone PCR

1. Amplify the wild-type PMMoV CP gene by PCR using pBT-PIW as a template, a primer pair of RM-PMMoV-F and RM-PMMoV-R, and high-fidelity DNA polymerase. Clone the 504-bp PCR product into a plasmid using a PCR cloning kit. Confirm the absence of any PCR error by sequencing and use it as a template for error-prone PCR.
2. Perform error-prone PCR using the template from **step 1**, a primer pair of ScXbATG-F and TAAKpSc-R, and a recombinant Taq DNA polymerase (*see* **Note 6**). Set up the reaction to give the final concentrate of 6 mM $MgCl_2$, 0.6 mM $MnCl_2$, 1 mM each of dCTP and dTTP, 0.2 mM each of dATP and dGTP, in Taq buffer (*see* **Note 7**). The reaction will be carried out under the following conditions: 94 °C, 2 min; 35 cycles of 94 °C for 30 s, 56 °C for 30 s, and 72 °C for 1 min; 72 °C, 3 min.
3. Confirm the amplification of CP gene fragment by an agarose gel electrophoresis. Purify the PCR products using a PCR purification kit. Take a small portion of purified PCR product to clone using a TA cloning kit. Evaluate the mutation rates by sequencing.

3.3 Construction of PMMoV Agroinfection Constructs Harboring Randomly Mutagenized CP Gene

1. Digest the purified error-prone PCR products from **step 3** of Subheading 3.2 and 1 μg of pBTPIWΔCP with XbaI and KpnI. Gel-purify the 480-bp mutagenized CP gene fragments. Treat the digested pBTPIWΔCP with alkaline phosphatase.
2. Clone the mutagenized CP gene fragments into pBTPIWΔCP at XbaI and KpnI sites. Mix the half (about 100 ng) of gel-purified mutagenized CP gene fragments with 500 ng of digested and alkaline phosphatase-treated pBTPIW, and ligate them as in **step 5** of Subheading 3.1.1 in a tenfold scaled-up reaction. Precipitate the ligation products by adding 0.1 volume of sodium acetate pH 5.2 and 2.5 volume of ethanol, and centrifuging at about 20,000 × *g* for 30 min at 4 °C. After rinsing with 70% ethanol, air-dry briefly the DNA pellet. Dissolve the DNA in 5 μL TE.
3. Transform the electrocompetent *E. coli* with the concentrated and desalted ligation product. Plate the bacteria onto LB-K50 plates and culture at 28 °C until the colonies grow up to 0.5 mm in diameter.

4. Add 2 mL of LB medium to each plate and suspend the *E. coli* colonies. Collect the *E. coli* suspension from the plates and extract plasmid using a miniprep kit (*see* **Note 8**).
5. Prepare electrocompetent cells of Agrobacterium. Inoculate a single Agrobacterium colony formed on L broth R20G20 plate into 100 mL of L broth R20G20 in a 500 mL flask, culture overnight at 30 °C with vigorous shaking. Wash the Agrobacterium cells three times by spinning at 4000 × *g* for 5 min at 4 °C and resuspending the pellet in 50 mL of ice-cold 10% glycerol. Finally, resuspend the cells in ice-cold 1 mL of 10% glycerol, aliquot it into 40 μL, and store at −80 °C.
6. Add up to 200 ng purified plasmid to an electrocompetent Agrobacterium cells, transfer to a 2-mm gap electroporation cuvette, and apply a pulse with the following setting: 2.5 kV, 25 μF, and 400 Ω. Plate the Agrobacterium onto L broth R20G20K50 plates and culture for 2 days at 28 °C.

3.4 Agroinfection of Plants with PMMoV Harboring Randomly Mutagenized CP Genes

3.4.1 Storage of the Library of PMMoV Harboring Randomly Mutagenized CP Genes

1. Pick up colonies of transformed Agrobacterium from the plates from Subheading 3.3, **step 6**. Add 0.2 mL each of L broth R20G20K50 to the wells of deep-well plates using an eight-channel micropipette and liquid reservoir.
2. Touch a colony with an autoclaved toothpick and inoculate into the L broth R20G20K50 in the deep-well plates. Culture at 28 °C overnight using a shaking incubator.
3. Transfer 0.1 mL each of the culture to 96-well plates, each well of which was added with 0.1 mL 45% glycerol, using an eight-channel micropipette. Mix well by pipetting, place the lid, and store at −80 °C.

3.4.2 Plant Growth and Inoculation

1. Sow seeds of *N. sylvestris* onto two sheets of filter paper wet with tap water and place in a plastic dish. Culture the plate in a plant growth room or growth chamber at 25 °C with 16 h light (50–100 μmol m^{-2} s^{-1}) and 8 h dark cycle.
2. Transfer the tiny seedlings to a potting soil within a week. Grow the plants under the same condition until the first two true leaves fully expand, which may take a few weeks.
3. Inoculate L broth R20G20K50 rectangular plates with the clones of the library for PMMoV with mutagenized CP gene using a replicator and culture for 2 days at 28 °C.
4. Prepare 0.1 mL per clone of the infiltration buffer containing 30 mg/L acetosyringone and aliquot it into 1.5 mL microfuge tubes. Suspend a loopful mass of Agrobacterium grown on the L broth R20G20K50 rectangular plates using an inoculation loop in the infiltration buffer. Let stand at room temperature in the dark for at least 2 h (*see* **Note 9**).

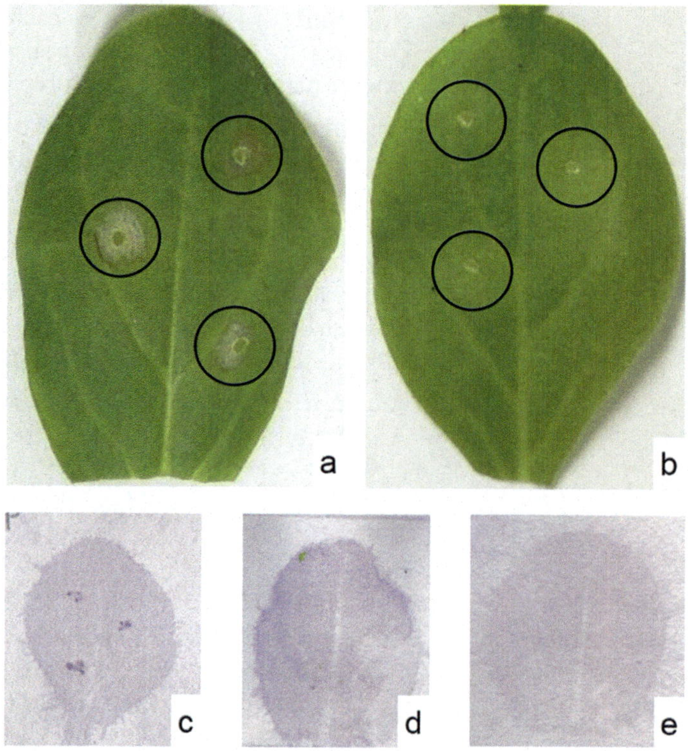

Fig. 3 Agroinfected *N. sylvestris* leaves. Photographs of *N. sylvestris* leaves inoculated with Agrobacterium harboring pBTPIW (**a**) and a mutant (**b**). Hammer blot images of *N. sylvestris* leaves inoculated with Agrobacterium harboring pBTPIW (**c**) and mutants (**d** and **e**). Note that some resistance-escaping mutants accumulate CP (**d**) while others do not (**e**)

5. Inoculate *N. sylvestris* with the Agrobacterium by pricking the leaves with an autoclaved toothpick dipped in the bacterial suspension. To ensure the reliability of the assay, the pricking inoculation should be made in two or more different points on a leaf and in two or more independent leaves. Grow the inoculated plants under the same growth condition mentioned above.
6. Examine the plant response at 7–10 days postinoculation.

3.5 Analysis of the Resistance Breaker Candidates

After inoculation mentioned above, some plants will show necrotic symptoms around the inoculated points (Fig. 3a), while others will show mild chlorosis or no symptoms (Fig. 3b). The clones which do not induce necrotic symptoms will be considered candidates of resistance breakers and analyzed further. Because nonsense mutations and frame-shift mutations abolish the coding capacity of the CP gene and thereby the ability to induce resistance response, it is important to check the coding capacity of the CP gene or the accumulation of CP. One may immunodetect the CP by a hammer blot technique (Fig. 3c–e; *see* Chapter 10 for the protocol), but we have experienced lower accumulation of CP in some candidates, which gave ambiguous results in the immunodetection. Therefore, here we describe the method to examine the sequence of the CP gene from the distal parts of the inoculated points.

1. Detach the inoculated leaves, place on a paper towel, and cut out the inoculated points and surrounding areas of 6 mm in diameter. Collect 0.1 g of remaining tissue for RNA extraction using an RNA extraction kit.

   (Optional) The remaining tissue can be used for preparing virus inoculum for the virus propagation or further analysis. The asymptomatically infected tissue of 0.1 g can be ground with a mortar and a pestle with 1 mL of 10 mM Hepes buffer and small amount of carborundum. The homogenate should be cleared by centrifugation at 10,000 × *g* for 10 min at 4 °C. The supernatant can be used as inoculum or stored at −80 °C.
2. Synthesize cDNA from 1 μg of the total RNA using 2.5 μM (final concentration) random hexamer primer and reverse transcriptase according to the manufacturer's protocol. Incubate the reaction for 10 min at 30 °C, 1 h at 42 °C, and 5 min 95 °C.
3. Amplify the 597-bp cDNA fragment encompassing the entire coding regions of PMMoV CP gene using 1 μL of cDNA from **step 2**, the primer set of PMF1 and PMR1, and a high-fidelity DNA polymerase. The reaction condition should be 94 °C, 1 min; 25 cycles of 94 °C for 30 s, 60 °C for 30 s, and 72 °C for 1 min; and 72 °C for 3 min (*see* **Note 10**).
4. Examine the amplification by an agarose gel electrophoresis. Purify the PCR product using a PCR purification kit according to the manufacturer's protocol, and examine the sequence to identify the mutations that are possibly responsible for the escape from the resistance.

4 Notes

1. Different brands of commercial soil mix can be used. We use Super Mix A (Sakata seed; Yokohama, Japan).
2. The plasmid pBTPIW-SK can be obtained from the corresponding author.
3. We usually perform colony PCR to screen the clones having PCR-amplified DNA fragment, purify the colony PCR products, and directly sequence them to select a clone.
4. Alkaline phosphatase should be inactivated before the plasmid is used for ligation.

5. Miniprep of pBTPIW and derivative plasmids is not efficient with commercial miniprep kits; therefore, we prepare the plasmids by a classic method for screening purpose.
6. High-fidelity DNA polymerase should not be used for the error-prone PCR.
7. The concentration of $MgCl_2$ and $MnCl_2$ should experimentally be optimized through the evaluation of mutation rates by cloning and sequencing. The presented condition gave the average mutation rate of 3.8 in 20 independent clones, 6 of which had no mutation.
8. Although the recovery of pBTPIW and derivative plasmids is not efficient with commercial miniprep kits, we use a miniprep kit in this step because the electroporation-mediated Agrobacterium transformation requires highly pure plasmid DNA.
9. The bacterial suspension can be kept overnight.
10. The primers PMF1 and PMR1 work well to amplify PMMoV CP gene and have been used for the virus diagnosis or the analysis of resistance-breaking strains [26]. For diagnostic purpose, high-fidelity DNA polymerase is not necessary.

Acknowledgments

This work was supported in part by JSPS KAKENHI Grant Numbers 21380032, 24658044, 26292026, and 15K14664.

References

1. Kobayashi K, Sekine K-T, Nishiguchi M (2014) Breakdown of plant virus resistance: Can we predict and extend the durability of virus resistance? J Gen Plant Pathol 80:327–336
2. Whitham S, McCormick S, Baker B (1996) The N gene of tobacco confers resistance to tobacco mosaic virus in transgenic tomato. Proc Natl Acad Sci U S A 93:8776–8781
3. Janzac B, Fabre F, Palloix A, Moury B (2009) Constraints on evolution of virus avirulence factors predict the durability of corresponding plant resistances. Mol Plant Pathol 10:599–610
4. Fabre F, Bruchou C, Palloix A, Moury B (2009) Key determinants of resistance durability to plant viruses: insights from a model linking within- and between-host dynamics. Virus Res 141:140–149
5. Kang BC, Yeam I, Jahn MM (2005) Genetics of plant virus resistance. Annu Rev Phytopathol 43:581–621
6. Ishibashi K, Masuda K, Naito S et al (2007) An inhibitor of viral RNA replication is encoded by a plant resistance gene. Proc Natl Acad Sci U S A 104:13833–13838
7. Ishibashi K, Naito S, Meshi T, Ishikawa M (2009) An inhibitory interaction between viral and cellular proteins underlies the resistance of tomato to nonadapted tobamoviruses. Proc Natl Acad Sci U S A 106:8778–8783
8. Yamaji Y, Maejima K, Komatsu K et al (2012) Lectin-mediated resistance impairs plant virus infection at the cellular level. Plant Cell 24:778–793
9. Sugawara K, Shiraishi T, Yoshida T et al (2013) A replicase of potato virus X acts as the resistance-breaking determinant for JAX1-mediated resistance. Mol Plant-Microbe Interact 26:1106–1112
10. Boukema IW (1982) Resistance to a new strain of TMV in Capsicum chacoense Hunz. Capsicum Newsl 1:49–51
11. Boukema IW (1980) Allelism of genes controlling resistance to TMV in Capsicum L. Euphytica 29:433–439

12. Boukema IW (1984) Resistance to TMV in Capsicum chacoense Hunz. Is governed by allele of the L-locus. Capsicum Newsl 3:47–48
13. Tomita R, Sekine K-T, Mizumoto H et al (2011) Genetic basis for the hierarchical interaction between Tobamovirus spp. and L resistance gene alleles from different pepper species. Mol Plant-Microbe Interact 24:108–117
14. Berzal-Herranz A, de la Cruz A, Tenllado F et al (1995) The CapsicumL3 gene-mediated resistance against the tobamoviruses is elicited by the coat protein. Virology 209:498–505
15. de la Cruz A, Lopez L, Tenllado F et al (1997) The coat protein is required for the elicitation of the Capsicum L2 gene-mediated resistance against the tobamoviruses. Mol Plant-Microbe Interact 10:107–113
16. Gilardi P, Garcia-Luque I, Serra MT (2004) The coat protein of tobamovirus acts as elicitor of both L2 and L4 gene-mediated resistance in Capsicum. J Gen Virol 85:2077–2085
17. Matsumoto K, Sawada H, Matsumoto K et al (2008) The coat protein gene of tobamovirus P0 pathotype is a determinant for activation of temperature-insensitive L1a-gene-mediated resistance in Capsicum plants. Arch Virol 153:645–650
18. Culver JN, Dawson WO (1989) Tobacco mosaic virus coat protein: an elicitor of the hypersensitive reaction but not required for the development of mosaic symptoms in *Nicotiana sylvestris*. Virology 173:755–758
19. Saito T, Meshi T, Takamatsu N, Okada Y (1987) Coat protein gene sequence of tobacco mosaic virus encodes a host response determinant. Proc Natl Acad Sci U S A 84:6074–6077
20. Culver JN (2002) Tobacco mosaic virus assembly and disassembly: determinants in pathogenicity and resistance. Annu Rev Phytopathol 40:287–308
21. Sekine K-T, Tomita R, Takeuchi S et al (2012) Functional differentiation in the leucine-rich repeat domains of closely related plant virus-resistance proteins that recognize common avr proteins. Mol Plant-Microbe Interact 25:1219–1229
22. Burdon JJ, Barrett LG, Rebetzke G, Thrall PH (2014) Guiding deployment of resistance in cereals using evolutionary principles. Evol Appl 7:609–624
23. Ahlquist P, French R, Janda M, Sue Loesch-Friest AL (1984) Multicomponent RNA plant virus infection derived from cloned viral cDNA. Proc Natl Acad Sci U S A 81:7066–7070
24. Grimsley N, Hohn B, Hohn T, Walden R (1986) 'Agroinfection,' an alternative route for viral infection of plants by using the Ti plasmid. Proc Natl Acad Sci U S A 83:3282–3286
25. Grdzelishvili VZ, Chapman SN, Dawson WO, Lewandowski DJ (2000) Mapping of the tobacco mosaic virus movement protein and coat protein subgenomic RNA promoters in vivo. Virology 275:177–192
26. Hamada H, Tomita R, Iwadate Y et al (2007) Cooperative effect of two amino acid mutations in the coat protein of pepper mild mottle virus overcomes L^3 -mediated resistance in Capsicum plants. Virus Genes 34:205–214

Chapter 6

A Cell-Free Replication System for Positive-Strand RNA Viruses for Identification and Characterization of Plant Resistance Gene Products

Kazuhiro Ishibashi and Masayuki Ishikawa

Abstract

Plant cells have lytic vacuoles, which contain ribonucleases and proteinases. The vacuoles are fragile and easily collapsed upon homogenization of plant tissues or cells. Thus, with a few exceptions, plant cell extracts are contaminated by vacuole-derived lytic enzymes and unsuitable for biochemical analyses. Here, we describe a method for removing the vacuoles from intact tobacco BY-2 protoplasts and for cell-free translation and replication of genomic RNA of positive-strand RNA viruses using the extract of evacuolated protoplasts. We also describe a method for the identification and functional characterization of a plant resistance gene product using this system.

Key words BYL, Tomato mosaic virus, Tm-1, Translation, Replication

1 Introduction

Upon host infection, plant viruses replicate their genomes within a cell, move to neighboring cells, and then systemically spread, causing diseases in plants. The lack of in vitro systems recapitulating these processes has limited investigations of the molecular mechanisms of plant virus infection. A breakthrough in this regard is the development of a cell-free translation and replication system for plant positive-strand RNA viruses using plant cell extracts [1]. In this system, input viral genomic RNA is translated to produce replication proteins, and viral RNA is replicated through the action of these replication proteins. Using this system, the early steps of viral infection (e.g., cotranslational template selection by viral replication proteins) have been elucidated [2, 3].

Plants have evolved mechanisms to fight viruses. Many genes that confer resistance to specific viruses, namely, resistance genes, encode nucleotide-binding site and leucine-rich repeat (NB-LRR)-containing proteins [4, 5]. These proteins elicit a defense reaction

Kappei Kobayashi and Masamichi Nishiguchi (eds.), *Antiviral Resistance in Plants: Methods and Protocols*, Methods in Molecular Biology, vol. 2028, https://doi.org/10.1007/978-1-4939-9635-3_6,

upon recognition of viral infection. Other resistance genes encode diverse protein products, suggesting that they evolved independently. The mechanisms of non-NB-LRR protein-mediated resistance are diverse, ranging from inhibiting viral replication or spread to tolerating a low level of virus multiplication without obvious symptoms.

The tomato *Tm-1* gene inhibits the multiplication of tomato mosaic virus (ToMV) within protoplasts [6]. Resistance-breaking ToMV mutants have amino acid substitutions in the helicase domain of replication proteins [7], suggesting that the Tm-1 protein recognizes replication proteins and inhibits viral RNA replication. We found that an extract of *Tm-1* tomato cells contains an activity to inhibit ToMV RNA replication in a cell-free system. By purifying this inhibitory activity, we identified the *Tm-1* gene product as a protein inhibitor of ToMV RNA replication [8]. The cell-free system was also useful for elucidating the mechanism of action of Tm-1 [9, 10].

2 Materials

2.1 Maintenance of Suspension Cultured Cells

1. Culture medium for BY-2 cells: 1× Murashige and Skoog Plant Salt Mixture (Wako Pure Chemical Industries, Ltd., Osaka, Japan), 3% (w/v) sucrose, 1 mg/L thiamine–HCl, 0.2 mg/L 2,4-dichlorophenoxyacetic acid, 100 mg/L myo-inositol, and 200 mg/L KH_2PO_4; adjust pH to 5.8 with KOH. Aliquot 95 mL of the medium in 300-mL Erlenmeyer flasks and sterilize by autoclaving.
2. Culture medium for GCR237 tomato cells: 1× Murashige and Skoog Plant Salt Mixture (Wako), 3% (w/v) sucrose, 0.1 mg/L thiamine–HCl, 0.5 mg/L pyridoxine–HCl, 0.5 mg/L nicotinic acid, 2 mg/L glycine, 2 mg/L 1-naphthaleneacetic acid, 0.2 mg/L 2,4-dichlorophenoxyacetic acid, 0.2 mg/L *trans*-zeatin, and 100 mg/L myoinositol; adjust pH to 5.8 with KOH. Aliquot 75 mL of the medium in 300-mL Erlenmeyer flasks and sterilize by autoclaving.
3. Rotary shaker.

2.2 Preparation of Evacuolated BY-2 Protoplast Lysate (BYL)

1. Protoplast wash buffer: 12.5 mM CH_3COONa, 5 mM $CaCl_2$, and 0.37 M mannitol; adjust pH to 5.8 with acetic acid.
2. Protoplast isolation enzyme solution: 1% (w/v) Cellulase Onozuka RS (Yakult Pharmaceutical Industry, Co., Ltd., Tokyo, Japan) and 0.1% (w/v) Pectolyase Y23 (Kyowa Chemical Industry Co., Ltd., Kagawa, Japan) in protoplast wash buffer; clarify by centrifugation at 1000 × *g* for 5 min.
3. Swinging-bucket centrifuge rotor (Beckman JS24.15 or equivalent).

4. Percoll solution: 70%, 40%, 30%, or 0% (v/v) Percoll, 0.7 M mannitol, 20 mM $MgCl_2$, and 5 mM PIPES–KOH, pH 7.0.
5. Gradient-making device.
6. Dounce homogenizer.
7. TR buffer: 30 mM HEPES–KOH, pH 7.4, 80 mM CH_3COOK, 1.8 mM $(CH_3COO)_2Mg$, 2 mM DTT, and one tablet of cOmplete, Mini, EDTA-free Protease Inhibitor Cocktail (Roche Diagnostics, Basel, Switzerland) per 10 mL.

2.3 Translation and Replication of ToMV RNA Using BYL

1. Viral RNA: Either purified from virions or synthesized by in vitro transcription from full-length complementary DNA clone.
2. 10× translation substrate mixture: 7.5 mM ATP, 1 mM GTP, 250 mM creatine phosphate, 250 mM each of amino acids, and 0.8 mM spermine. Store at −30 °C.
3. Creatine kinase solution: Creatine kinase (Roche) dissolved to 10 mg/mL in 20 mM HEPES–KOH, pH 7.4, 50% (v/v) glycerol. Store at −30 °C.
4. RNasin Plus RNase Inhibitor (Promega Corporation, Madison, WI, USA).
5. 5× RdRP substrate mixture: 5 mM ATP, 5 mM GTP, 5 mM UTP, 125 μM CTP, 50 mM DTT, 0.5 mg/mL actinomycin D, 150 mM creatine phosphate, 17 mM $(CH_3COO)_2Mg$, and 79 MBq/mL [α-^{32}P]CTP.
6. 10× Loening buffer: 0.4 M Tris, 0.2 M CH_3COONa, and 20 mM EDTA-2Na; adjust pH to 7.2 with acetic acid.
7. 2.4% polyacrylamide–8 M urea gel: 2.5 mL 10 × Loening buffer, 2 mL 30% (w/v) acrylamide–1.5% (w/v) bisacrylamide solution, 12.01 g urea, 170 μL 10% (w/v) ammonium persulfate, and 20 μL TEMED for a 138 mm (W) × 100 mm (H) × 1 mm (D) slab gel.
8. Gel-loading dye: 525 mg of urea dissolved in 700 μL 6× dye (15% [w/v] Ficoll type 400, 6× Loening buffer, 0.25% bromophenol blue, and 0.25% xylene cyanol).
9. Gel dryer.
10. Imaging device for autoradiography.

2.4 Immunoprecipitation

1. Immunoprecipitation buffer: 30 mM HEPES–KOH, pH 7.4, 80 mM CH_3COOK, 1.8 mM $(CH_3COO)_2Mg$, 2 mM DTT, 0.2 mM ATP, and one tablet of cOmplete EDTA-free Protease Inhibitor Cocktail (Roche) per 50 mL.
2. Anti-FLAG M2 antibody-conjugated agarose (Sigma-Aldrich, St. Louis, MO, USA).
3. 3× FLAG peptide (Sigma-Aldrich).

3 Methods

3.1 Tobacco BY-2 Suspension-Cultured Cells

1. Culture BY-2 cells at 26 °C on a rotary shaker in the dark.
2. For routine maintenance, transfer 0.8–2 mL of BY-2 cell suspension into 95 mL new media using sterile pipettes every 7 days (*see* **Note 1**).
3. For BYL preparation, transfer 5–10 mL of 7-day-cultured BY-2 cell suspension into two flasks of 95 mL new media and further culture for 3 or 4 days.

3.2 Preparation of Evacuolated BY-2 Protoplast Lysate (BYL)

1. Transfer 50 mL of BY-2 cell suspension into 50-mL conical tubes (4 tubes). Collect the cells by centrifugation at 800 × *g* for 1 min at room temperature. Discard the supernatants.
2. Add protoplast isolation enzyme to fill the tubes, gently mix, and rotate slowly at 25 °C for 2 h.
3. Check for protoplast formation under microscope.
4. Collect the protoplasts by centrifugation at 800 × *g* for 5 min. Discard the supernatants, and wash the protoplasts with protoplast wash buffer four times.
5. Prepare 8 mL of 0–30% Percoll linear gradient (*see* **Note 2**) in centrifuge tubes for Beckman JS24.15 rotor (6 tubes) (Fig. 1 (1)).

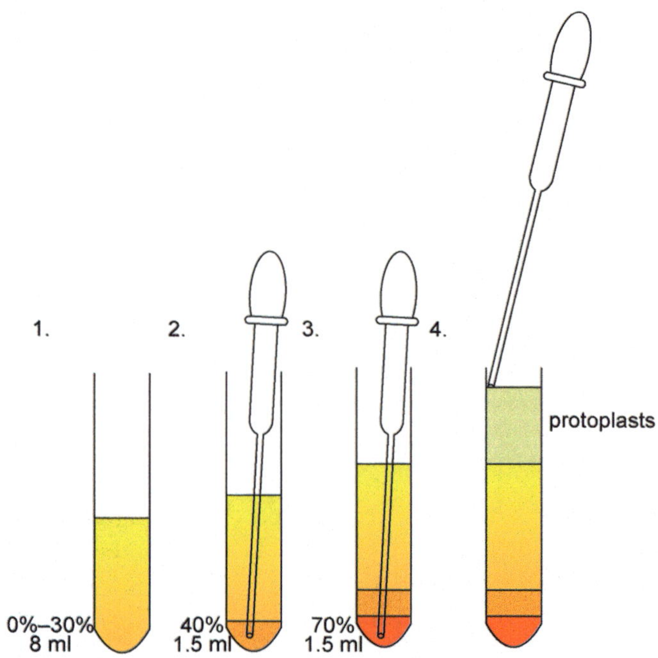

Fig. 1 Preparation of linear stepwise Percoll gradient for evacuolated protoplast isolation

6. Underlay 1.5 mL of 40% Percoll solution and then 1.5 mL of 70% Percoll solution, using Pasteur pipettes to make a stepwise gradient (Fig. 1 (2 and 3)).
7. Suspend the washed protoplasts in ca. 8 mL of 0% Percoll solution and apply the suspensions onto the gradient (Fig. 1 (4)).
8. Centrifuge at 10,000 × *g* at 25 °C for 1 h.
9. Collect evacuolated protoplasts from the interface between the 40% and 70% Percoll layers (*see* **Note 3**) into a 15-mL conical tube.
10. Add wash buffer to the evacuolated protoplasts to dilute Percoll. Collect the evacuolated protoplasts by centrifugation at 800 × *g* for 5 min. Discard the supernatants and wash the evacuolated protoplasts with protoplast wash buffer 3 times.
11. Suspend the evacuolated protoplasts in 4 volumes of ice-cold TR buffer and homogenize with a Dounce homogenizer.
12. Centrifuge the homogenate at 500 × *g* at 4 °C for 10 min and recover the supernatant (BYL). Store at −80 °C.

3.3 In Vitro Translation and Replication of ToMV RNA in Membrane-Containing BYL

1. Assemble the translation mixture on ice (25 μL BYL, 5 μL 10× translation substrate mixture, 1 μL creatine kinase solution, 1 μL RNasin, and 100 ng ToMV RNA; add TR buffer to 50 μL). Incubate the mixture at 25 °C for 1 h.
2. Aliquot 40 μL of translation mixture, add 10 μL of 5× RdRP substrate mixture, and incubate at 25 °C for 1 h.
3. Purify radiolabeled RNA by phenol extraction and ethanol precipitation with ammonium acetate. Dissolve the purified RNA in 30 μL H_2O.
4. Mix 3 μL of RNA with 5 μL of gel-loading dye. Separate RNA by 2.4% polyacrylamide–8 M urea gel electrophoresis in 1× Loening buffer. Dry the gel on filter paper with a gel dryer. Detect RNA by autoradiography.

3.4 Uncoupled In Vitro Translation and Replication of ToMV RNA

1. Centrifuge BYL at 30,000 × *g* for 15 min at 4 °C. Recover the supernatant (membrane-depleted BYL: mdBYL). Suspend the membrane-containing pellet in 1/4 volume (relative to the volume of BYL before centrifugation) of TR buffer by gentle pipetting and keep on ice (P30BYL).
2. Assemble the translation mixture on ice as described in Subheading 3.3, **step 1** using mdBYL instead of BYL. Incubate at 25 °C for 1 h.
3. Aliquot 30 μL of translation mixture, add 10 μL of P30BYL, and incubate at 15 °C for 2 h.

4. Add 10 μL of 5× RdRP substrate mixture and incubate at 25 °C for 1 h.
5. Purify and analyze the radiolabeled RNA as described in Subheading 3.3, **steps 3** and **4**.

3.5 Inhibition of ToMV RNA Replication by the Resistance Protein Tm-1

1. Prepare evacuolated protoplast extract from GCR237 (*Tm-1/Tm-1*) tomato cells as with BYL.
2. For identification of the Tm-1 protein, fractionate the *Tm-1* tomato cell extract by column chromatography. Replace the solvent of each fraction with TR buffer by ultrafiltration for desalting and concentration.
3. Assemble the translation mixture for ToMV RNA as described in Subheading 3.3, **step 1**. Replace TR buffer with crude tomato cell extract or fractions. To distinguish the inhibition of ToMV RNA replication by Tm-1 from nonspecific inhibition of translation and/or replication, perform in vitro translation and replication reaction in two tubes: one for wild-type (wt) ToMV (*Tm-1*-sensitive), and the other for a resistance-breaking mutant for comparison. Specific inhibition of wt ToMV RNA replication is assumed to represent the activity of Tm-1.
4. Once cloned, Tm-1 protein can be prepared by any of following methods: (1) expression of the recombinant protein in *Escherichia coli* (*see* **Note 4**); (2) in vitro translation using mdBYL from in vitro transcribed mRNA for Tm-1 (*see* **Note 5**); (3) preparation of mdBYL from transgenic BY-2 cell lines that express Tm-1 (*see* **Note 6**).
5. Tm-1 inhibits ToMV RNA replication when added after translation and before incubation with P30BYL in the uncoupled translation and replication system. In this case, nonspecific inhibitory effects on translation, often shown by tomato crude cell extracts, can be avoided. For this experiment, mix ToMV RNA-translated mdBYL and Tm-1 protein at an appropriate ratio and incubate at 25 °C for 20 min, and continue the reactions as described in Subheadings 3.4, **steps 3–5**.

3.6 Detection of a Protein–Protein Interaction Between Tm-1 and ToMV Replication Proteins

1. Prepare ToMV replication proteins and FLAG-tagged Tm-1 protein separately by in vitro translation in mdBYL at 25 °C for 1 h. Up to 10 μg of template RNAs can be used in a 50 μL translation mixture to maximize protein production. Replication proteins of resistance-breaking mutants serve as a negative control that does not bind Tm-1.
2. Mix the ToMV replication protein-containing and Tm-1-FLAG-containing translation mixtures and incubate at 25 °C for 20 min (*see* **Note 7**). Dilute the mixture with immunoprecipitation buffer to 1 mL, and add appropriate amounts of

prewashed anti-FLAG antibody-conjugated agarose. Rotate the mixture at 4 °C for 2 h.

3. Spin down the agarose by centrifugation at 3000 × *g* for 1 min. Discard supernatant. Wash the pellet with immunoprecipitation buffer four times.
4. To elute the bound protein, add immunoprecipitation buffer containing 0.1 mg/mL 3× FLAG peptide and incubate on ice for 30 min. Centrifuge at 3000 × *g* for 1 min and analyze the supernatant by immunoblotting.

4 Notes

1. The growth rate of different lineages of BY-2 cells may differ, and may change over time. To keep cell conditions suitable for BYL preparation, it is important to adjust the appropriate transfer volume.
2. Buoyant density of BY-2 protoplasts also varies among cell lineages. If the concentration of Percoll in the linear gradient is inappropriate for the cells used, protoplasts form a thick layer after centrifugation and the yield of evacuolated protoplasts is very low. We recommend finding the best concentration of Percoll in the linear gradient, for example, by testing 10–30% or 10–40%.
3. Vacuole-rich vesicles and nonevacuolated protoplasts remain in the upper layers. To prevent contamination of these cells in evacuolated protoplasts, carefully remove them by Pasteur pipette. It is often difficult to remove cells near the top of the gradient. In such cases, gently overlay protoplast wash buffer and remove the cells with the buffer.
4. The N-terminal fragments of Tm-1 (amino acids 1–431 or 1–201), but not the full-length Tm-1 protein, can be expressed in a soluble form in *E. coli*. The purified N-terminal fragments of Tm-1 inhibit ToMV RNA replication in vitro [11].
5. Translation in BYL depends on the 5′ cap structure on the mRNA. Therefore, prepare mRNA by in vitro transcription in the presence of m^7G cap analog, or by adding the cap structure to transcribed uncapped RNA using capping enzymes.
6. Transformation of BY-2 cells, a simple and well-established method, has been described elsewhere [12].
7. The binding of Tm-1 and ToMV replication proteins is ATP-dependent. ATP (or nonhydrolyzable analogs) should be supplied when using purified proteins.

References

1. Komoda K, Naito S, Ishikawa M (2004) Replication of plant RNA virus genomes in a cell-free extract of evacuolated plant protoplasts. Proc Natl Acad Sci U S A 101(7):1863–1867. https://doi.org/10.1073/pnas.0307131101
2. Kawamura-Nagaya K, Ishibashi K, Huang Y-P, Miyashita S, Ishikawa M (2014) Replication protein of tobacco mosaic virus cotranslationally binds the 5′ untranslated region of genomic RNA to enable viral replication. Proc Natl Acad Sci U S A 111(16):E1620–E1628. https://doi.org/10.1073/pnas.1321660111
3. Ishibashi K, Ishikawa M (2016) Replication of tobamovirus RNA. Annu Rev Phytopathol 54(1):55–78. https://doi.org/10.1146/annurev-phyto-080615-100217
4. De Ronde D, Butterbach P, Kormelink R (2014) Dominant resistance against plant viruses. Front Plant Sci 5. https://doi.org/10.3389/fpls.2014.00307
5. Kobayashi K, Sekine K-T, Nishiguchi M (2014) Breakdown of plant virus resistance: can we predict and extend the durability of virus resistance? J Gen Plant Pathol 80(4):327–336. https://doi.org/10.1007/s10327-014-0527-1
6. Motoyoshi F, Oshima N (1977) Expression of genetically controlled resistance to tobacco mosaic mirus infection in isolated tomato leaf mesophyll protoplasts. J Gen Virol 34(3):499–506. https://doi.org/10.1099/0022-1317-34-3-499
7. Meshi T, Motoyoshi F, Adachi A, Watanabe Y, Takamatsu N, Okada Y (1988) Two concomitant base substitutions in the putative replicase genes of tobacco mosaic virus confer the ability to overcome the effects of a tomato resistance gene, *Tm-1*. EMBO J 7(6):1575–1581
8. Ishibashi K, Masuda K, Naito S, Meshi T, Ishikawa M (2007) An inhibitor of viral RNA replication is encoded by a plant resistance gene. Proc Natl Acad Sci U S A 104(34):13833–13838. https://doi.org/10.1073/pnas.0703203104
9. Ishibashi K, Ishikawa M (2013) The resistance protein Tm-1 inhibits formation of a tomato mosaic virus replication protein-host membrane protein complex. J Virol 87(14):7933–7939. https://doi.org/10.1128/jvi.00743-13
10. Ishibashi K, Ishikawa M (2014) Mechanisms of tomato mosaic virus RNA replication and its inhibition by the host resistance factor Tm-1. Curr Opin Virol 9(Supplement C):8–13. https://doi.org/10.1016/j.coviro.2014.08.005
11. Ishibashi K, Kezuka Y, Kobayashi C, Kato M, Inoue T, Nonaka T, Ishikawa M, Matsumura H, Katoh E (2014) Structural basis for the recognition–evasion arms race between tomato mosaic virus and the resistance gene Tm-1. Proc Natl Acad Sci U S A 111(33):E3486–E3495. https://doi.org/10.1073/pnas.1407888111
12. An G (1985) High efficiency transformation of cultured tobacco cells. Plant Physiol 79(2):568–570. https://doi.org/10.1104/pp.79.2.568

Chapter 7

Plant Protein-Mediated Inhibition of Virus Cell-to-Cell Movement: Far-Western Screening and Biological Analysis of a Plant Protein Interacting with a Viral Movement Protein

Nobumitsu Sasaki, Yasuhiko Matsushita, and Hiroshi Nyunoya

Abstract

Cell-to-cell movement via plasmodesmata is a crucial step for plant RNA viruses to determine their host ranges. Many viruses including *Tomato mosaic virus* (ToMV) encode one or more movement proteins (MPs) that are indispensable for cell-to-cell movement. During movement processes, MPs are thought to interact directly with many plant proteins that may be involved in supporting or inhibiting cell-to-cell movement of viruses. In order to understand the molecular mechanisms that regulate viral spread positively or negatively, it is important to discover such MP-interacting plant proteins and analyze their functions in viral cell-to-cell movement in efficient ways. In this chapter, we provide protocols of a radioisotope-based far-western screening strategy to construct a λ phage cDNA library from a nonhost *Brassica campestris* (syn. *rapa*) for ToMV and identify plant proteins that bind directly to the ^{32}P-labeled probe of ToMV MP, and subsequently a biolistic bombardment method to examine whether a plant protein selected have a function as an inhibitory factor that can interfere with virus cell-to-cell movement.

Key words Far-western screening, λ Phage cDNA library, Protein–protein interaction, Biolistic bombardment, Movement protein, Cell-to-cell movement, Inhibitory effect, Fluorescent protein

1 Introduction

The infection cycle of plant viruses depends on the cellular machinery of plants [1, 2]. Some plant factors are required for systemic virus replication and cell-to-cell movement while other factors are involved in defense induction against virus attacks or inhibition of the function of virus components [3, 4]. To understand virus–host interactions at the molecular level, it is essential to identify and characterize host factors that are involved in virus infection.

Kappei Kobayashi and Masamichi Nishiguchi (eds.), *Antiviral Resistance in Plants: Methods and Protocols*, Methods in Molecular Biology, vol. 2028, https://doi.org/10.1007/978-1-4939-9635-3_7,

Tomato mosaic virus (ToMV) is a well-studied, single-stranded RNA virus [5] and encodes four proteins: the 126K and 183K proteins, the movement protein (MP), and the coat protein (CP). The 126K and 183K proteins are necessary for virus replication. MP is required for cell-to-cell movement through plasmodesmata (cytoplasmic channels). Since CP is dispensable for cell-to-cell movement, the CP gene is replaceable with a reporter gene, such as green fluorescent protein (GFP), allowing for easy detection of ToMV-infected cells [6].

Far-western blotting is an effective method for detecting direct protein–protein interaction in vitro. Whereas conventional western blotting uses an antibody to detect specifically a protein of interest, far-western blotting employs a probe protein (nonantibody) not only to prove an interaction with a protein of interest, but also to screen or identify unknown proteins that can bind directly to the probe protein. For efficient screening, the use of λ-phage cDNA libraries is reported in many studies. The screening of λ-phage cDNA libraries allows us to express and isolate a cDNA clone for a target protein from a phage plaque. Probe proteins can be labeled directly with radioisotopes (^{35}S or ^{125}I) or fused with the consensus 5 amino acids sequence (RRASV) for protein kinase A (PKA) phosphorylation and labeled with [γ-^{32}P]ATP. Reporter molecules (e.g., biotin and DIG) and epitope tags (e.g., GST and His) are also available to mark probe proteins for immunological detection. So far, we have reported successful λ-phage cDNA screenings using the λGEX5 vector [7] and RRASV-fused probe proteins including ToMV MP [8–11].

After a probe protein is confirmed to bind to a target protein, the cDNA clone for the target protein can be recombined to different types of expression plasmids that are selected depending on the purposes of analysis. A transient gene expression system using a biolistic particle gun is one of methods to examine biological roles of a target protein in plant cells [12]. The target protein can be expressed as fusions with a fluorescent protein in order to investigate their subcellular localization by fluorescence microscopy. In addition, by coexpression of an infectious virus clone that is modified to encode another fluorescent protein, we can examine effectively whether over-expression of the target fusion of interest inhibits the viral movement from the initially bombarded cell to neighboring cells.

In this chapter, as an example of a ToMV-MP interacting target protein identified by our radioisotope-based far-western screening (Fig. 1), we show a putative transcription coactivator KELP that was derived from *Brassica campestris*, a nonhost for ToMV, and proven to bind to GST-fused MPs of ToMV and other plant viruses [8]. We also show how we examined the inhibitory effect of KELP on the cell-to-cell movement of ToMV by using the biolistic bombardment method [13, 14].

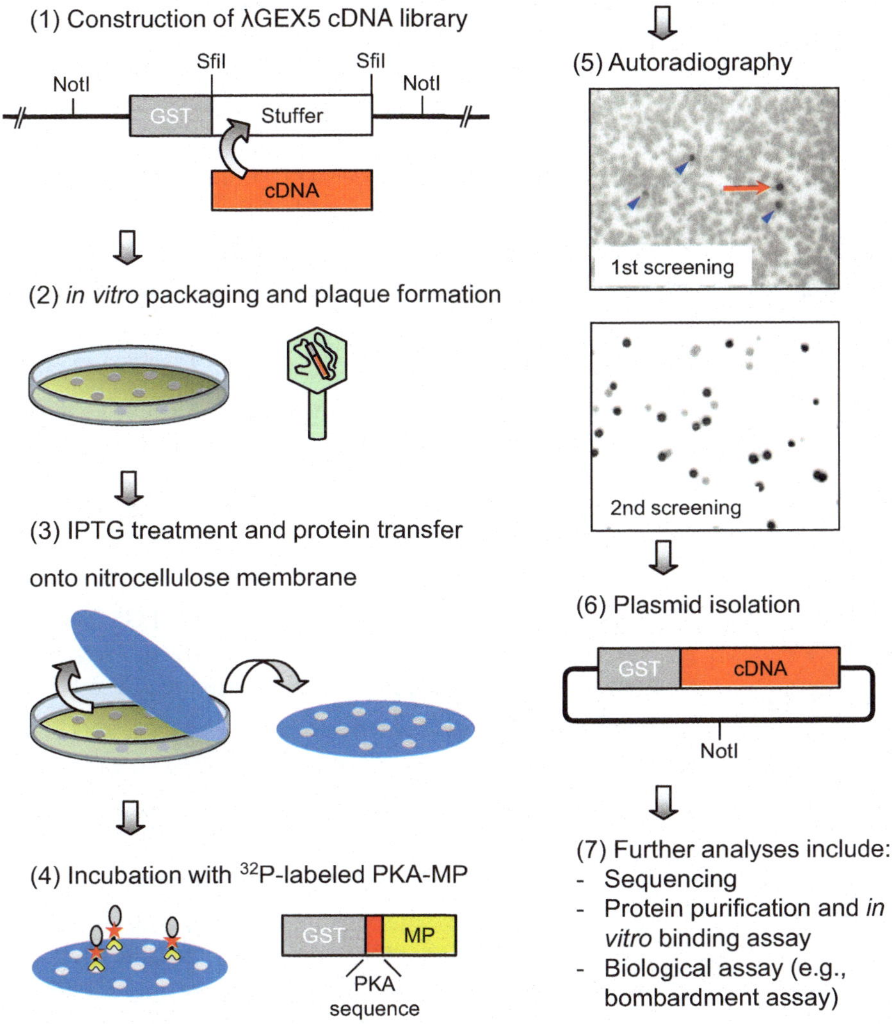

Fig. 1 Schematic representation of far-western screening and subsequent analyses. (1) λGEX5 cDNA library is constructed as follows: after isolation of mRNA from a plant of interest (*see* Subheading 3.1.1), cDNA is synthesized and ligated with phosphorylated adaptors (*see* Subheading 3.1.2). At the same time, λGEX5 is purified (*see* Subheading 3.1.3) and digested with SfiI for vector preparation (*see* Subheading 3.1.4). cDNA is ligated to the SfiI-cut λGEX5 vector (*see* Subheading 3.1.5). (2) in vitro packaging (*see* Subheading 3.1.6) and titration (*see* Subheading 3.1.7) of phages are performed. (3) Protein production is induced by IPTG treatment and transferred to nitrocellulose membrane (*see* Subheading 3.3.1). (4) ^{32}P-labeled PKA-MP probe is prepared by induction (Subheading 3.2.1), purification (*see* Subheading 3.2.2) and phosphorylation (*see* Subheading 3.2.3) of the recombinant protein and incubated with the blotted nitrocellulose membrane (*see* Subheading 3.3.1). (5) Autoradiography is performed after the first and second screenings (*see* Subheading 3.3.1). A red arrow and blue arrowheads shown in the first screening are positive signals from phage plaques. The plaque indicated by the red arrow is selected for the second screening. (6) Plasmids are isolated from positive phage clones by using NotI recognition sites in the λGEX5 vector (*see* Subheading 3.3.2). (7) For further analyses of isolated plasmids, we can do sequencing of the insert clone (*see* Subheading 3.3.2), purify a target protein and confirm its ability to bind MP in vitro (*see* Subheading 3.3.3), and perform a bombardment assay (*see* Subheadings 3.4.1–3.4.4)

2 Materials

2.1 Construction of a λ Phage cDNA Library

2.1.1 Isolation of Brassica campestris mRNA and cDNA Synthesis

1. *Brassica campestris* (*S9/S9*) was used to make a λ phage cDNA library from which we isolate the cDNA clone for KELP. However, this protocol is applicable to any plants of interest.
2. Liquid nitrogen.
3. Mortars and pestles are covered with tinfoil, autoclaved and dried in an oven.
4. QuickPrep Micro mRNA Purification Kit (GE Healthcare).
5. UV spectrophotometer: NanoDrop 2000 UV (Thermo Fisher Scientific).

2.1.2 Ligation of Phosphorylated Oligonucleotide Adapters to cDNA

1. TimeSaver cDNA Synthesis Kit (GE Healthcare), which contains PEG buffer, ATP, T4 DNA Ligase, T4 Polynucleotide Kinase, and spun columns.
2. Oligonucleotide adapters (5′-AGGTGCTGG-3′ and 5′-CCAGCACCTGCA-3′), which can be self-annealed to form the cohesive end of a DNA sequence digested by SfiI.
3. 15 mL and 50 mL centrifuge tubes.
4. Heat block: Dry Thermo Unit, DTU-1B (TAITEC).
5. Phenol–chloroform–isoamyl alcohol (25:24:1).

2.1.3 Purification of the λGEX5 Phage DNA

1. *Escherichia coli* (*E. coli*) strain BB4 (Stratagene).
2. LB (Luria–Bertani) medium: 2.5% Luria–Bertani Broth, Miller and 10 mM $MgSO_4$. After autoclave, 20% maltose solution is prepared by filter sterilization separately and added to $MgSO_4$ containing LB medium at the final concentration of 0.2%.
3. λGEX5 phage: 1.0×10^9 pfu/mL (*see* **Note 1**).
4. Top agarose: 2.5% Luria–Bertani Broth, Miller; 10 mM $MgSO_4$; and 0.5% agarose. Autoclave. The top agarose is prepared by heating the solid top agarose in a microwave oven for a few minutes and then being kept at a 50 °C water bath until just before use.
5. LDB (Lambda Dilution Buffer): 35 mM Tris–HCl (pH 7.4), 100 mM NaCl, 10 mM $MgSO_4$, and 0.01% gelatin.
6. Petri dish (9 cm) (IWAKI).
7. L-1 solution: RNase A (20 mg/mL), DNase (6 mg/mL), 10 mM EDTA, 100 mM Tris–HCl, 300 mM NaCl, BSA (0.2 g/mL), adjusted to pH 7.5.
8. L-2 solution: 30% polyethylene glycol (PEG) 6000, 3 M NaCl.

9. L-3 solution: 100 mM Tris–HCl, 100 mM NaCl, 25 mM EDTA, adjusted to pH 7.5.
10. L-4 solution: 4% sodium dodecyl sulfate (SDS).
11. L-5 solution: 2.55 M potassium acetate (pH 4.8).
12. JETSTAR midiprep kit (VERITAS), which contains E4 solution (600 mM NaCl, 100 mM sodium acetate (pH 5.0), 0.15% Triton X-100), E5 solution (800 mM NaCl, 100 mM sodium acetate (pH 5.0)), E6 solution (1.25 M NaCl, 100 mM Tris–HCl (pH 8.5)), and JETSTAR columns.
13. 80% ethanol: Mix ethanol with sterile RO water.

2.1.4 Preparation of the SfiI-Cut λGEX5 Vector

1. SfiI and 10× M buffer (TOYOBO).
2. SizeSep™ 400 spun column (GE Healthcare).
3. Phenol–chloroform–isoamyl alcohol (25:24:1).
4. Chloroform.
5. Rotator: Rotary Mixer NRC-20D (NISSIN).
6. 50× TAE buffer (Nacalai Tesque): Diluted to 1× with autoclaved RO water.
7. 0.3% agarose gel: Dissolve 0.3 g agarose (Certified Molecular Biology Agarose, Bio-Rad Laboratories) in 100 mL of 1× TAE buffer by using a microwave.
8. 3 M sodium acetate (pH 7.0).
9. 10× Loading Buffer (Takara Bio).

2.1.5 Ligation Between the SfiI-Cut λGEX5 Vector and cDNA

1. Ligation buffer, ATP, and T4 DNA ligase: Components in the TimeSaver cDNA Synthesis Kit (GE Healthcare) are used.

2.1.6 In Vitro Packaging of the λGEX5 cDNA Library

1. Gigapack III Gold Packaging Extract (Agilent Technologies).
2. SM buffer: 100 mM NaCl, 10 mM $MgSO_4$, 50 mM Tris–HCl (pH 7.5), 0.01% gelatin.

2.1.7 Titration and Amplification of the λGEX5 cDNA Library

1. Chloroform (Nacalai Tesque).
2. 14 mL polystyrene round-bottom tubes (BD Biosciences).
3. Wash buffer: 10 mM Tris–HCl (pH 7.5), 15 mM NaCl, 0.1% Tween 20, 1 mM dithiothreitol (DTT).
4. Petri dish (9 cm) (IWAKI).
5. Blocking solution: Block Ace (Dainippon Seiyaku).
6. PBS-T buffer: 137 mM NaCl, 2.68 mM KCl, 10.14 mM Na_2HPO_4, 1.76 mM KH_2PO_4, 0.2% Triton X-100.
7. Imaging plate: BAS-MP (Fujifilm).
8. Bio-imaging analyzer: BAS-1500 (Fujifilm).

2.2 Preparation of ^{32}P-Labeled Recombinant Probe Protein

2.2.1 Induction of Expression of GST-Fused Recombinant Protein in E. coli

1. *E. coli* strain XL10-Gold transformant carrying pGEX-PKA-30K. pGEX-PKA-30K contains the cDNA clone encoding ToMV MP (30K) that is fused with GST and the PKA sequence at the N-terminus, was constructed as described previously [8].
2. LB-Amp medium: 2.5% Luria–Bertani Broth, Miller. After autoclave, 100 mg/mL ampicillin solution is prepared by filter sterilization separately and added to LB medium at the final concentration of 25 μg/mL.
3. 0.2 M IPTG.

2.2.2 Purification of GST-Fused Recombinant Protein

1. NETN: 50 mM Tris–HCl (pH 8.0), 150 mM NaCl, 1 mM EDTA, 0.5% IGEPAL (Sigma).
2. NETN-PD: NETN containing 1 mM phenylmethanesulfonyl fluoride (PMSF) and 1 mM dithiothreitol (DTT).
3. Sonicator: ULTRASONIC DISRUPTOR UD-201 (TAITEC).
4. Sepharose 4B (GE Healthcare).
5. Glutathione Sepharose 4B (GE Healthcare).
6. Rotator: Rotary Mixer NRC-20D (NISSIN).

2.2.3 ^{32}P-Labeling of the PKA Sequence in GST-Fused Recombinant Protein

1. [γ-^{32}P] ATP.
2. cAMP-dependent protein kinase (PKA), catalytic subunit (5,000 U/mL), and 10× PKA buffer (NEB).
3. 50 mM Tris–HCl (pH 8.0) buffer containing 10 mM glutathione.
4. Scintillation counter: Liquid Scintillation Counter System 1409 (Wallac).

2.3 Far-Western Screening

2.3.1 Primary and Secondary Screenings

1. *E. coli* strain BB4 (Stratagene).
2. LB medium containing 10 mM $MgSO_4$ and 0.2% maltose: *see* Subheading 2.2.1.
3. LDB: *see* Subheading 2.2.1.
4. LB agar plate containing 10 mM $MgSO_4$: *see* Subheading 2.2.1.
5. Plastic square dish, 15 cm × 12 cm (EIKEN CHEMICAL).
6. Nitrocellulose membrane.
7. IPTG.
8. Blocking solution: Block Ace (Dainippon Seiyaku).
9. PBS-T buffer: 137 mM NaCl, 2.68 mM KCl, 10.14 mM Na_2HPO_4, 1.76 mM KH_2PO_4, 0.2% Triton X-100.
10. Filter paper (ADVANTEC).
11. Imaging plate: BAS-MP (Fujifilm).

12. Bioimaging analyzer: BAS-1500 (Fujifilm).
13. Petri dish (9 cm).

2.3.2 Isolation of Plasmids from Positive Phage Clones

1. NotI and 10× NotI buffer.
2. Agarose, electrophoresis grade.
3. T4 DNA ligase and ligation buffer.
4. *E. coli* strain XL10-Gold (Stratagene).
5. HiYield Plasmid Mini Kit (RBC Bioscience).
6. PstI and 10× M buffer.
7. pGEX2 primer (5′-CTGGTTCCGCGTGGATCCCC-3′).
8. pGEX5 primer (5′-ATTTCCCCGAAAAGTGCCAC-3′).
9. ABI DNA sequencing analyzer (Thermo Fisher Scientific/ Applied Biosystems).

2.3.3 In Vitro Binding of a ^{32}P-Labeled Probe Protein and a Target Protein

1. 2× sample buffer: 25 mM Tris–HCl, 2% sodium dodecyl sulfate (SDS), 10% glycerol, 0.1% bromophenol blue, and 5% 2-mercaptoethanol.
2. 10% polyacrylamide gel (*see* **Note 2**).
3. Gel electrophoresis apparatus: pageRun (ATTO).
4. PVDF membrane: Immobilon-P (Millipore).
5. Filter paper (ADVANTEC).
6. Methanol.
7. Blotting buffer: 25 mM Tris–HCl (pH 8.0), 192 mM glycine, 0.05% sodium dodecyl sulfate (SDS), and 5% methanol.
8. Semidry blotter: TransBlot SD (Bio-Rad laboratories).
9. 50 mM Tris–HCl (pH 8.0) buffer containing 20% 2-propanol.
10. Buffer A (50 mM Tris–HCl (pH 8.0) and 5 mM 2-mercaptoethanol).
11. Guanidine hydrochloride.
12. Tween 40.
13. PBS-T buffer: *see* Subheading 2.3.1.

2.4 Biolistic Bombardment

2.4.1 Preparation of Gold Particles

1. Gold particles (1 μm) (Bio-Rad Laboratories).
2. Ethanol.
3. 75% ethanol: Mix ethanol with sterile RO water.
4. 50% glycerol: Mix glycerol with sterile RO water and filtrate with a 0.22 μm filter.

2.4.2 Coating Gold Particles with DNA

1. pTH2-KELP-DsRed, pART7-DsRed, and piL.erG3L, which encode DsRed-fused KELP, DsRed, and an infectious clone of ToMV-erGFP, respectively, were constructed as described previously [13] (*see* **Note 3**).

2. 2.5 M $CaCl_2$: Dissolve in sterile RO water and filtrate with a 0.22 μm filter.
3. 0.1 M spermidine (Sigma).

2.4.3 Delivery of DNA-Coated Gold Particles to Plant Leaves

1. *Nicotiana benthamiana*: Use tenth to twelfth true leaves (counted from the cotyledons) of about 8-week-old plants.
2. PDS-1000/He (Bio-Rad Laboratories).
3. Rupture disks (900 or 1100 psi) (Bio-Rad Laboratories).
4. Petri dish (9 cm) (IWAKI).

2.4.4 Observation by Fluorescence Microscopy

1. Glass slide (IWAKI).
2. Cover glass (IWAKI).
3. Epifluorescence microscopy: BX50 (Olympus).

3 Methods

3.1 Construction of a λ Phage cDNA Library

3.1.1 Isolation of Brassica campestris mRNA and cDNA Synthesis

1. The parts of the stigma (e.g., 160 mg) of *Brassica campestris* were collected and ground under liquid nitrogen to a fine powder using a mortar and pestle.
2. Isolate mRNA from the ground sample by using the QuickPrep Micro mRNA Purification Kit or another commercially available kit, according to the manufacturer's instruction.
3. Measure A_{260} and A_{280} of the mRNA solution by using the NanoDrop 2000 UV spectrophotometer (Thermo Fisher Scientific), and determine the concentration of the mRNA. The purity of the isolated mRNA should be assessed by calculating a ratio of A_{260}/A_{280} (*see* **Note 4**).

3.1.2 cDNA Synthesis and Ligation of Phosphorylated Oligonucleotide Adapters to cDNA

1. cDNA is synthesized from 1 to 5 μg of the purified mRNA by using the TimeSaver cDNA Synthesis Kit, according to the manufacturer's instruction. The final effluent of cDNA is stored at −30 °C until the next step.
2. Mix 100 μM of oligonucleotide adapters (5′-AGGTGCTGG-3′ and 5′-CCAGCACCTGCA-3′) in a 1.5 mL microtube, anneal at 65 °C for 30 min, and cool to room temperature.
3. 3.0 μL of the annealed adapters were mixed gently with the total effluent (derived from 3.0 μg of RNA used for cDNA synthesis) together with 30 μL of PEG buffer, 1 μL of 15 mM ATP, and 1 μL of T4 DNA Ligase.
4. Incubate at 16 °C for 1 h, heat-inactivate at 65 °C for 10 min, and chill on ice.
5. Add 1.5 μL of 75 mM ATP and 1 μL of T4 polynucleotide kinase, and mix gently.

6. Incubate at 37 °C for 30 min, heat-inactivate at 65 °C for 10 min and cool at room temperature.
7. Add 140 μL phenol–chloroform–isoamyl alcohol, vortex, and centrifuge for 1 min at 20,000 × *g*.
8. Apply the upper aqueous layer to a spun column with the cap. Remove the cap from the spun column and place it in a 15 mL centrifuge tube. Centrifuge for 2 min at 400 × *g*. Place the column upright in a rack.
9. Slowly apply the upper aqueous layer of the cDNA sample to the center of the flat surface at the top of the bed in the column. Be careful not to allow the sample to flow around the sides of the bed.
10. Place a 1.5 mL microtube in the bottom of a 15 mL centrifuge tube. Place the loaded spun column inside the centrifuge tube, with the tip of the column inside the microtube.
11. Centrifuge for 2 min at 400 × *g*, collect the effluent, and stored at −30 °C until the next step.

3.1.3 Purification of the λGEX5 Phage DNA

1. Incubate the *E. coli* strain BB4 with shaking at 37 °C overnight in 15 mL of LB medium containing 10 mM $MgSO_4$ and 0.2% maltose. Mix 15 mL of the overnight culture, 7.5 mL of 1× LDB, and 20 μL of λGEX5 phage (e.g., 1.0 × 10^9 pfu/mL), and then incubate at 37 °C for 15 min.
2. Mix 300 μL of the phage-inoculated *E. coli* culture and 3 mL of top agarose, which is kept at 55 °C, pour onto an LB agar plate (9 cm petri dish), and allow for solidification. Prepare a total of 50 LB agar plates. Incubate 37 °C for approximately 7 h before plaques merge. Add 4 mL of 1× LDB onto each plate, shake at 4 °C overnight, and collect lysates from plates. By this way, you may obtain approximately 190 mL of the lysate in total (e.g., 1.7 × 10^{13} pfu).
3. Transfer 50 mL of the lysate to a 50 mL centrifuge tube, add 50 μL of chloroform and mix gently. Centrifuge at 4 °C for 10 min at 1,800 × *g*.
4. Transfer the supernatant to a 50 mL tube and incubate at 37 °C for 30 min.
5. Add 100 μL of L-1 solution, mix gently, and incubate at 37 °C for 30 min.
6. Add 10 mL of chilled L-2 solution, mix, and divided to 5 mL aliquots. Incubate on ice for 30 min. Centrifuge at 13,000 × *g* for 10 min at 4 °C.
7. Discard the supernatant, add 3 mL of L-3 solution, and resuspend the pellet gently.
8. Add 3 mL of L-4 solution, mix, incubate at 70 °C for 20 min, and immediately cool on ice.

9. Add 3 mL of L-5 solution, mix, and centrifuge at 13,000 × *g* for 10 min at 4 °C.
10. Slowly apply the supernatants from the two tubes to a JETSTAR column, which is equilibrated with 10 mL of E4 solution. Allow them to run by gravity flow and discard the flow-through. Wash the column with 10 mL of E5 solution and discard the flow-through. Repeat this washing step once more. Apply 5 mL of E6 solution to the column and collect the effluent to a 15 mL centrifuge tube.
11. Add 3.5 mL of isopropanol, mix, and centrifuge at 13,000 × *g* for 30 min at 4 °C. Discard the supernatant, add 5 mL of 80% ethanol, and wash the centrifuge tube. Remove the 80% ethanol thoroughly, dry the pellet, and dissolve it in 200 μL of sterile RO water.
12. Measure A_{260} and A_{280} of the λGEX5 DNA solution by using a UV spectrophotometer and calculate the DNA concentration. Evaluate the purity of the isolated DNA by calculating a ratio of A_{260}/A_{280} (*see* **Note 5**).

3.1.4 Preparation of the SfiI-Cut λGEX5 Vector

1. Mix 30 μg of the λGEX5 phage DNA, 30 μL of 10× M buffer, 4 μL of SfiI and sterile RO water in a total of 300 μL in a 1.5 mL microtube.
2. Incubate at 50 °C for 2 h. Heat 65 °C for 10 min. During these incubations, prepare a SizeSep™ 400 spun column (*see* **Note 6**).
3. Add 300 μL of phenol–chloroform and mix well with a rotator for 5 min at room temperature. Centrifuge at 15,000 × *g* for 10 min and collect approximately 300 μL of the upper aqueous layer.
4. Add 300 μL of chloroform and mix well with a rotator for 5 min at room temperature. Centrifuge at 15,000 × *g* for 10 min and collect the upper aqueous layer as much as possible.
5. Remove the caps from the spun column and place it in a 15 mL centrifuge tube. Centrifuge for 2 min at 400 × *g*. Place the column upright in a rack.
6. To remove a 420-bp stuffer region of the λGEX5 vector, slowly apply the upper aqueous layer of the cDNA sample to the center of the flat surface at the top of the bed in the column. Be careful not to allow the sample to flow around the sides of the bed.
7. Place a 1.5 mL microtube in the bottom of a 15 mL centrifuge tube. Place the loaded spun column inside the centrifuge tube, with the tip of the column inside the microtube. Centrifuge for 2 min at 400 × *g* and collect the effluent.

8. Measure A_{260} and A_{280} and calculate the concentration of the λGEX5 DNA solution and the value of A_{260}/A_{280} by the NanoDrop 2000 UV spectrophotometer (*see* **Note 5**).
9. Optionally, the digested DNA fragments are checked by gel electrophoresis. Mix 1 μL of the effluent, 9 μL of TAE buffer, and 1 μL of 10× Loading Buffer, incubate at 65 °C for 2 min, and run a 0.3% agarose gel at 100 V in TAE buffer.

3.1.5 Ligation Between the SfiI-Cut λGEX5 Vector and cDNA

1. Mix 14 μL (i.e., 1.5 μg) of the SfiI-cut λGEX5 vector, 30 μL of the adapter-ligated cDNA solution, 4.4 μL of 3 M sodium acetate, and 121 μL of cold (−30 °C) ethanol in a 1.5 mL microtube. Chill at −70 °C for 15 min to coprecipitate the vector and cDNA.
2. Centrifuge at 15,000 × *g* for 10 min. Remove the supernatant carefully, and dry the pellet briefly.
3. Dissolve the pellet in 8 μL of ligation buffer by pipetting. Add 1 μL of 1 mM ATP and 1 μL of T4 DNA ligase, vortex gently and spin briefly. Incubate at 16 °C for 4 h.

3.1.6 In Vitro Packaging of the λGEX5 cDNA Library

1. Add 1–4 μL of the ligation reaction to the tube containing the packaging extract of the Gigapack® III Gold Packaging Extract, which should be stored at −80 °C and thawed by fingers just before use. Mix well by pipetting as air bubbles should not be introduced. Centrifuge at 20,000 × *g* for a few seconds.
2. Incubate at room temperature for 2 h.
3. Add 500 μL of SM buffer and 20 μL of chloroform and mix gently.
4. Centrifuge at 20,000 × *g* for a few seconds. Store at 4 °C until ready to proceed.

3.1.7 Titration and Amplification of the λGEX5 cDNA Library

1. On the day before of titrating the phage library, start incubation of the *E. coli* BB4 strain with shaking at 37 °C overnight in 3 mL of LB medium containing 10 mM $MgSO_4$ and 0.2% maltose.
2. Make a tenfold serial dilution of the packaging reaction (10^{-1} to 10^{-6}) in SM buffer. Mix 10 μL of each dilution with 100 μL of SM buffer and 200 μL of the overnight BB4 culture (OD_{600} = approximately 1.4) in a 14 mL polystyrene round-bottom tube and incubate at 37 °C for 15 min.
3. Add 3 mL of top agarose, mix, and pour onto the LB agar plate containing 10 mM $MgSO_4$ (9 cm petri dish). Incubate at 37 °C overnight.
4. Count the plaques on the plates and determine the titer in plaque-forming units per mL (pfu/mL) of the phage library.

In our previous study, the titer of our phage library (approximately 500 μL) was recorded to be 5.0×10^5 pfu/mL.

5. In order to amplify the phage library, mix 80 μL of the phage library (corresponding to 4.0×10^4 pfu), 170 μL of SM buffer, and 500 μL of the overnight BB4 culture (OD_{600} = approximately 1.4), and incubate at 37 °C for 15 min.
6. Add 7.5 mL of top agarose, mix, and pour onto the LB agar plate containing 10 mM $MgSO_4$ (15 cm × 12 cm, square dish). Incubate 37 °C for 6–8 h.
7. Pour 7 mL of SM buffer onto the top agarose and incubate with shaking gently at 4 °C overnight.
8. Collect the phage-containing buffer from the plate to a 15 mL centrifuge tube, centrifuge at 1,800 × *g* for 10 min at 4 °C, and transfer approximately 7 mL of the supernatant to a 15 mL centrifuge tube. Add 100 μL of chloroform, mix, and store the tube at 4 °C until ready to proceed.
9. Determine a titer of an amplified phage library by repeating **steps 2–4**. Prepare a tenfold serial dilution from 10^{-5} to 10^{-8}, because the titer after library amplification should be much higher than that of the original phage library.

3.2 Preparation of ^{32}P-Labeled Recombinant Probe Protein

3.2.1 Induction of Expression of GST-Fused Recombinant Protein in E. coli

1. Grow the transformed *E. coli* cell carrying pGEX-PKA-30K in 50 mL of LB-Amp liquid medium in a 200 mL flask. Incubate with shaking at 37 °C overnight.
2. Mix 450 mL of LB-Amp liquid medium and the overnight culture in a 1000 mL flask, and continue to incubate at 25 °C until the optical density of 600 nm reaches approximately 0.5–0.8.
3. Add 500 μL of 0.2 M IPTG, and incubate at 25 °C for 6 h.
4. Divide the *E. coli* culture to two 500 mL centrifuge bottles and centrifuge at 4,220 × *g* for 10 min at 4 °C. Remove the supernatant from each tube. Resuspend the pellet in one tube with 40 mL of the washing buffer, transfer the cell suspension to the other tube, and resuspend the pellet. Transfer the cell suspension to a 50 mL centrifuge tube. The pellet can be stored at −30 °C or −80 °C until needed in the next step.

3.2.2 Purification of GST-Fused Recombinant Protein

1. Disrupt the pellet of the pGEX-PKA-30K-transformed *E. coli* cell in 35 mL of NETN-PD by a sonicator (output = 10; duty = 0). Perform 30-s sonications with 1-min intervals four times. The sample should be kept on ice during sonication.
2. Centrifuge at 13,000 × *g* for 20 min at 4 °C, and collect the supernatant to a 50 mL centrifuge tube.
3. Add 2 mL of the Sepharose 4B suspension, and mix with a rotator for 30 min at 4 °C. Centrifuge at 1,800 × *g* for 1 min

at 4 °C. This step, which is done to remove any proteins interacting nonspecifically with Sepharose 4B beads, can be omitted.

4. Transfer the supernatant to a 50 mL centrifuge tube containing 225 μL of an 80% (v/v) Glutathione Sepharose 4B suspension, and mix with a rotator for 1 h at 4 °C. Centrifuge at 1,800 × *g* for 1 min at 4 °C. Discard the supernatant.
5. Add 15 mL of NETN, vortex, centrifuge at 1,800 × *g* for 1 min at 4 °C, and discard the buffer. Repeat this step twice more.
6. Add the same volume of NETN as that of the Glutathione Sepharose 4B beads in order to make a 50% (v/v) suspension. Transfer the suspension to a 1.5 mL microtube. You may store it at 4 °C until needed in the next step.

3.2.3 ^{32}P-Labeling of the PKA Sequence in GST-Fused Recombinant Protein

1. Transfer 200 μL of the 50% Glutathione Sepharose 4B suspension containing 0.5–1.0 μg of GST-PKA-MP (*see* **Note 7**) to a 1.5 mL microtube. Centrifuge at 2,600 × *g* for 15 s at 4 °C. Remove the buffer thoroughly by using a microsyringe, resulting in 100 μL of Glutathione Sepharose 4B beads.
2. Add 20 μL of [γ-^{32}P] ATP (3.7 MBq), 2 μL (10 U) of the catalytic subunit of the cAMP-dependent protein kinase (PKA), 20 μL of 10× PKA buffer, and 158 μL of sterile RO water. The volume of the labeling reaction is twice as much as that of Glutathione Sepharose 4B beads. Incubate at 30 °C for 30 min by inverting the tube every 5 min. Centrifuge at 2,600 × *g* for 15 s at 4 °C, and remove the supernatant.
3. Add 100 μL of 50 mM Tris–HCl (pH 8.0) buffer containing 10 mM glutathione, vortex, centrifuge at 2,600 × *g* for 15 s at 4 °C, and collect the supernatant as an effluent. Repeat this elution step four more times.
4. Measure cpm of each effluent with a scintillation counter. Use 1.0–2.0 × 10^5 cpm/μL (specific radioactivity of the freshly labeled probe is approximately 1.0 × 10^7 cpm/μg) of the ^{32}P-labeled protein probe.

3.3 Far-Western Screening

3.3.1 Primary and Secondary Screenings

1. Incubate the *E. coli* strain BB4 with shaking at 37 °C overnight in 8.8 mL of LB medium containing 10 mM $MgSO_4$ and 0.2% maltose.
2. On the next day, add 4.4 mL of 1× LDB, and 200 μL of the phage library (1.0 × 10^7 pfu/mL), and mix. Divide the suspension into 660 μL of 20 aliquots by using 15 mL centrifuge tubes, and then incubate at 37 °C for 15 min.
3. Add 6.6 mL of top agarose, mix, and pour onto the LB agar plate containing 10 mM $MgSO_4$ (15 cm × 12 cm, square dish).

Incubate 42 °C for 3–5 h until diameters of plaques reach approximately 0.5 mm.

4. Cut nitrocellulose membranes to fit to the square dish, immerse in 10 mM IPTG solution, dry completely, and incubate at 37 °C before use. Place a nitrocellulose membrane onto the plaque-forming top agarose layer and incubate at 37 °C for 3.5 h to induce protein expression.
5. Peel the nitrocellulose membrane off the top agarose carefully, transfer to a plastic tray containing 10 mL of wash buffer, and wash with shaking at room temperature for 15 min. Discard the buffer, add 30 mL of a blocking solution (Block Ace), and incubate for blocking the membrane at 4 °C for 1 h.
6. Add 5 mL of a probe solution containing ^{32}P-labeled GST-PKA-MP (corresponding to 1–2 × 10^5 cpm/mL) and incubate at 4 °C overnight.
7. Discard the probe solution and add 10 mL of PBS-T buffer per membrane at room temperature for 5 min. Discard the buffer and repeat this washing step three more times. Air-dry the membrane on a filter paper. The plaque-forming agar plates should be stored at 4 °C. Perform **steps 4–7** for the remaining membranes.
8. Place membranes on an imaging plate (BAS-MP) and keep at room temperature overnight.
9. Analyze the imaging plate by a bio-imaging analyzer (BAS-1500) to obtain an autoradiogram of the membranes. If you find positive signals in the primary screening, copy the autoradiogram to a transparent film sheet (e.g., a sheet used for overhead projector) and compare the positive signals on the film sheet with plaques on the plates. These plaques are used for the secondary screening.
10. Pick up a plaque corresponding to a positive signal from a top agarose layer by using a plastic tip for a 20- or 200-μL micropipette. Resuspend in 1 mL of LDB buffer and incubate at 4 °C overnight.
11. Centrifuge at 20,000 × *g* for 20 min at 4 °C, and collect the supernatant (phage lysate) to a 1.5 mL microtube.
12. Mix 200 μL of the BB4 strain overnight culture (prepared in the same procedure as **step 1**), 100 μL of LDB buffer, and 1–5 μL of the phage lysate, and incubate at 37 °C for 15 min.
13. Add 3 mL of top agarose, mix, and pour onto the LB agar plate containing 10 mM $MgSO_4$ (9 cm petri dish). Incubate 42 °C for 3–5 h until diameters of plaques reach approximately 0.5 mm.
14. Perform the same procedures in **steps 4–9** to confirm that selected phage clones in the primary screening contain positive ones.

3.3.2 Isolation of Plasmids from Positive Phage Clones

1. Collect plaques corresponding to positive signals on the membrane in the second screening, incubate the BB4 strain with the phage lysate overnight, and prepare the cell lysate from the overnight bacterial culture, as described in **steps 10–12** of Subheading 3.3.1. The phage DNA was isolated from the cell lysate as described in Subheading 3.1.3.
2. Mix approximately 500 ng of the phage DNA, 1 μL of 10× NotI buffer, 0.5 μL of NotI, and sterile RO water to a total volume of 10 μL. Incubate at 37 °C for 1–2 h, and then heat at 65 °C for 20 min. You may check the digestion of DNA by electrophoresis of 1–2 μL of the reaction on a 1–1.5% agarose gel.
3. For self-ligation of plasmids, mix 5 μL of the digestion reaction, 1 μL of Ligation Buffer, 0.5 μL of T4 DNA ligase, and 3.5 μL of sterile RO water. Incubate at 16 °C for 30 min to overnight.
4. Transfer 100 μL of *E. coli* XL10-Gold competent cells to a 1.5 mL microtube, and add 5 μL of the ligation reaction. Mix gently, and incubate on ice for 20 min.
5. Incubate at 42 °C for 45 s. Immediately after incubation, move the tube back on ice for 1 min, add 1 mL of SOC medium, and incubate at 37 °C for 1 h. Spread 0.1 mL of the cell culture on to an LB-Amp agar plate.
6. Centrifuge the rest of the cell culture at 1,800 × *g* for a few seconds. Discard 900 μL of the supernatant, resuspend the pellet in 100 μL of the medium, and spread the cell culture on another LB-Amp agar plate. Incubate the two plates at 37 °C overnight.
7. Pick one or a few colonies on the LB-Amp agar plate by using a sterile (autoclaved) toothpick and grow in 3–5 mL of LB-Amp liquid medium. Isolate plasmid DNA from the bacterial culture by using the HiYield Plasmid Mini Kit (RBC Bioscience) according to the manufacturer's instruction.
8. Use PstI to digest the isolated plasmid to determine the size of a cDNA insert. There are two PstI recognition sites in the backbone of the plasmid. Mix 0.5–1 μL of the DNA solution, 1 μL of PstI, 2 μL of 10× M buffer in the volume of 20 μL, and incubate 37 °C for 1 h. Run 10 μL of the reaction on the 1–1.5% agarose gel by electrophoresis at constant 100 V for 30 min.
9. Determine the coding sequence of a cDNA clone in the plasmid by using pGEX2 primer (5′-CTGGTTCCGCGTGGATC CCC-3′) or pGEX5 primer (5′-ATTTCCCCGAAAAGTGC CAC-3′), and an ABI DNA sequencing analyzer (Thermo Fisher Scientific/Applied Biosystems). If the above primers cannot cover the whole sequence of the cDNA insert, design appropriate primers for additional sequencing based on the known

sequence information. The cDNA clone may contain the 3′-part of the open reading frame. If this is the case, 5′-RACE should be done to obtain the full-length cDNA clone.

3.3.3 In Vitro Binding of a ^{32}P-Labeled Probe Protein and a Target Protein

It is important to confirm the binding ability of target proteins encoded in positive clones from the phage cDNA library. For this purpose, GST-fused target proteins are separated by SDS-PAGE (sodium dodecyl sulfate–polyacrylamide gel electrophoresis), immobilized on a membrane, and reacted with a ^{32}P-labeled probe protein (GST-PKA-MP). The protocol in this subsection provides how to perform an in vitro binding assay in a denaturing/renaturing system. *See* Subheadings 3.2.1–3.2.3 for details of incubation of transformed *E. coli* cells, induction and purification of recombinant target and probe proteins, and phosphorylation of the probe protein with the PKA sequence.

1. Mix an appropriate volume of a 50% (v/v) suspension of a GST-fused target protein, 2 μL of 2× sample buffer, and sterile RO water in a total volume of 20 μL. Heat at 65 °C for 15 min and chill on ice.
2. Prepare a 10% polyacrylamide gel (7.8 cm × 9 cm × 1 mm) according to the standard protocol. Set up a gel electrophoresis apparatus (e.g., pageRun; ATTO) according to the manufacturer's instruction.
3. Apply 10–15 μL of the protein sample to the polyacrylamide gel and run at constant 20 mA per gel for 60–90 min. Use a protein marker with molecular weight standards.
4. Prepare a PVDF membrane and filter papers during electrophoresis: Cut to the size of the gel. For the membrane, immerse the membrane in 100% methanol for a few seconds, transfer to blotting buffer, and equilibrate by shaking for 5 min. For filter papers, immerse one by one in blotting buffer directly and equilibrate by shaking for 5 min.
5. After electrophoresis, transfer the gel to the blotting buffer and wash gently. Discard the blotting buffer and repeat this washing step once more.
6. Immerse the gel in the blotting buffer for 20 min to replace the electrophoresis buffer.
7. Transfer the gel to a semidry blotter. From the bottom of the blotter unit, place six filter paper, the PVDF membrane, the gel, six filter paper in order.
8. Perform electrophoretic transfer at constant 200 mA for 2 h.
9. Transfer the membrane to 50 mM Tris–HCl (pH 8.0) buffer containing 20% 2-propanol and wash by shaking at 4 °C for 30 min. Discard the buffer and repeat this washing step once more.

10. Transfer the membrane to buffer A (50 mM Tris–HCl (pH 8.0) and 5 mM 2-mercaptoethanol) and incubate by shaking at 4 °C for 30 min. Discard the buffer and repeat this incubation once more.
11. Transfer the membrane to buffer A containing 6 M guanidine hydrochloride and incubate by shaking at 4 °C for 1 h.
12. For renaturing a recombinant protein on the membrane, prepare dilution series (3, 1.5, 0.75, 0.38, and 0.19 M) of the guanidine hydrochloride solution using buffer A containing 0.04% Tween40.
13. Transfer the membrane to 3 M guanidine hydrochloride solution and incubate by shaking at 4 °C for 5 min and discard the solution. Perform the incubation at 4 °C for 5 min by using the solutions containing 1.5, 0.75, 0.38, and 0.19 M guanidine hydrochloride in order.
14. Transfer the membrane to buffer A containing 0.04% Tween40 incubate at 4 °C overnight.
15. Transfer the membrane to a blocking solution and incubate at 4 °C for 1 h.
16. Add 5 mL of a probe solution containing ^{32}P-labeled GST-PKA-MP (corresponding to 1–2 × 10^5 cpm/mL) and incubate at 4 °C for 4–5 h.
17. Discard the probe solution and wash the membranes in 10 mL of PBS-T buffer at room temperature for 5 min. Discard the buffer and repeat this washing step three more times. Air-dry the membrane on a filter paper.
18. Perform autoradiography of the membrane by using an imaging plate (BAS-MP) and a bioimaging analyzer (BAS-1500).

3.4 Biolistic Bombardment

In this sub-section, we describe how to perform biolistic bombardment and examine whether a target protein has an inhibitory effect on viral cell-to-cell movement (Fig. 2). For visualization of KELP expression and virus-infected cells, an expression plasmid encoding DsRed-fused KELP (pTH2-KELP-DsRed) and one containing an infectious ToMV clone with the endoplasmic reticulum-targeting GFP gene instead of the coat protein gene (piL.erG3L) are used, respectively. As a control, an expression plasmid encoding DsRed is used (pART7-DsRed). The procedures of plasmid construction are described in our previous report [13].

3.4.1 Preparation of Gold Particles

1. Weigh and collect 12 mg of gold particles (1 μm) to a 1.5 mL microtube.
2. Add 200 μL of 100% ethanol and vortex for 5 min.
3. Leave for 5 min at room temperature.

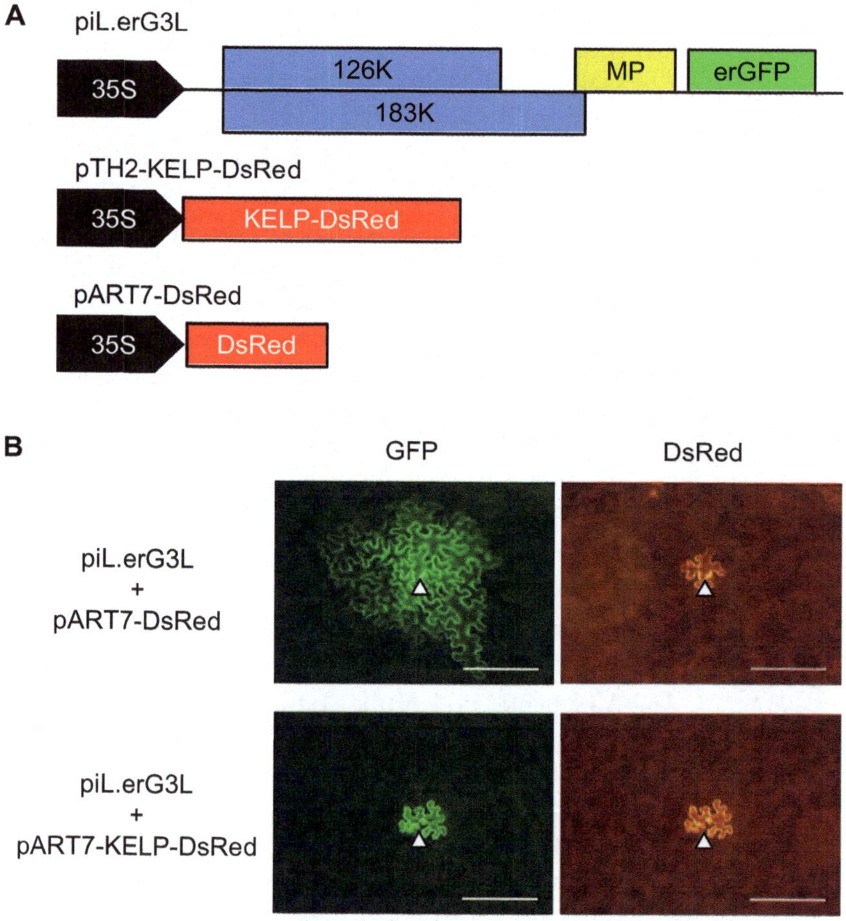

Fig. 2 Diagrams of expression plasmids for bombardment experiments and representative images of bombarded cells. (**a**) All expression cassettes contain the *Cauliflower mosaic virus* 35S promoter (35S). piL.erG3L carries an infectious clone of ToMV-erGFP that encodes the ER-targeting GFP instead of CP. pTH2-KELP-DsRed and pART7-DsRed code for KELP fused with DsRed and unfused DsRed, respectively. (**b**) GFP-expressing infection sites are shown to consist of multiple cells (upper) and a single cell (lower) after bombardment of piL.erG3L with and pART7-DsRed and pART7-BcKELP-DsRed, respectively. A multiple-cell infection site where GFP fluorescence was distributed in cells surrounding an initially bombarded cell indicated successful cell-to-cell movement of the virus. In contrast, a single-cell infection site where GFP fluorescence was confined only to an initially bombarded cell indicated inhibition of virus movement. Images were taken 48 h after bombardment in the same field with filter units for GFP (left) and DsRed (right). Initially bombarded cells are indicated by arrowheads. Scale bars represent 200 μm

4. Spin down at 2,300 × *g* and discard the supernatant.
5. Add 250–300 μL of 75% ethanol, mix by tapping, and then vortex for 3 min.
6. Leave for 5 min at room temperature.
7. Spin down at 2,300 × *g* and discard the supernatant.
8. Add 250–300 μL of sterile RO water, mix by tapping, and then vortex for 1 min.

9. Spin down at 2,300 × g and discard the supernatant.
10. Repeat **steps 8** and **9** twice more. Note that the supernatant should be removed completely.
11. Add 200 μL of 50% sterile glycerol and vortex well. This protocol provides the final density of 60 mg/mL of the gold suspension for five bombardments.
12. Keep at 4 °C until ready to use. The suspension may be available empirically for 4 months.

3.4.2 Coating Gold Particles with DNA

1. Better to adjust the concentration of each plasmid DNA (i.e., pTH2-KELP-DsRed, piL.erG3L, and pART7-DsRed) to 1 μg/μL in advance. Put 1–5 μL of one or more DNA plasmids in total to a 1.5 mL microtube. Use 1 μg per plasmid DNA for coating gold particles.
2. Add 50 μL of the gold particle suspension (60 mg/mL) and mix well by pipetting.
3. Add 50 μL of 2.5 M $CaCl_2$ and quickly and instantly vortex with the tube lid open. Keep vortexing by holding the tube on the vortex mixer, add 10 μL of 0.1 M spermidine, close the tube lid, and vortex vigorously. Leave for 1 min at room temperature.
4. Spin down at 2,300 × g and discard the supernatant.
5. Add 140 μL of 70% ethanol slowly not to disturb the pellet of the gold particles and discard the supernatant by sucking gently.
6. Repeat **step 5** by using 100% ethanol twice more.
7. Add 80 μL of 100% ethanol and mix the suspension by tapping the side of the microtube strongly.

3.4.3 Delivery of DNA-Coated Gold Particles to Plant Leaves

1. Place 12 μL of the DNA-coated gold particle suspension on the center of the macrocarrier and leave with the lid open until the ethanol is evaporated completely.
2. Place the macrocarrier in a macrocarrier holder by using forceps.
3. Open the valves of the He gas and turn on the vacuum pump and the chamber.
4. Using forceps, place a rupture disk (900 or 1,100 psi) into the retainer cap, screw the cap back in place and tighten using a torque wrench.
5. Using forceps, place a stopping screen into the center of the macrocarrier launch assembly unit and then the macrocarrier holder containing the macrocarrier with DNA-coated gold particles upside down. Screw the launch assembly with a coverlid and insert to the top slot of the chamber.

6. Put a detached leaf of *N. benthamiana* with the abaxial side up inside a lid of a petri dish. Set the petri dish at the center of the target shelf.
7. Place the target shelf at the 9-cm level and close the chamber door.
8. Evacuate the chamber up to about 26 in. of mercury by keeping on the VAC position and then hold the chamber vacuum by switching the HOLD position. Press and hold the FIRE switch to bombard the leaf sample until the rupture disk bursts.
9. Release the switch and change to the VENT position. After the vacuum is released completely, open the chamber door and remove the petri dish lid from the chamber.
10. Place a water-wet paper towel inside the petri dish, transfer the bombarded leaf onto the paper towel, and close the dish lid. Cover the dish with a metal tray and keep overnight at room temperature.

3.4.4 Observation by Fluorescence Microscopy

1. Cut the bombarded leaf into a piece of appropriate size and place on a glass slide.
2. Drop a water droplet on the leaf piece and put a coverslip. Tap the cover slip carefully with a fingertip to remove air bubbles and stretch the uneven surface of the leaf piece.
3. Fix the cover slip to the glass slide with adhesive tape.
4. Observe fluorescent cells under an epifluorescence microscopy (Olympus, BX50) equipped with appropriate filter units for GFP and DsRed. Cells showing the ER-localizing GFP signal and the DsRed signal are regarded as those infected by ToMV-erGFP and initially bombarded cells, respectively. The effect of KELP-DsRed to inhibit the cell-to-cell movement of ToMV-erGFP is evaluated by calculation of the ratio of single cell infection sites against total infection sites and comparison with the control experiment using the expressing plasmid for DsRed.

4 Notes

1. The original λGEX5 phage, which was kindly provided by Dr. Rikiro Fukunaga (Osaka University of Pharmaceutical Sciences), was amplified with *E. coli* BB4 strain, according to the **steps 1** and **2** in Subheading 3.1.3.
2. 10% polyacrylamide gels were made by a standard method: The resolving gel (375 mM Tris–HCl (pH 8.8), 10% acrylamide, 0.1% sodium dodecyl sulfate (SDS), 0.1% ammonium persulfate (APS), and 0.04% TEMED) and the stacking gel

(125 mM Tris–HCl (pH 6.4), 4.5% acrylamide, 0.1% SDS, and 0.1% TEMED) are made. We use 30 (w/v) % Acrylamide–Bis Mixed Solution (37.5:1) (Nacalai Tesque).

3. piL.erG3L was kindly provided by Dr. Tetsuo Meshi (National Institute of Agrobiological Sciences).
4. If the ratio is around 2.0, isolated mRNA is regarded as "pure."
5. If the ratio is 1.8–2.0, isolated DNA is regarded as "pure."
6. Invert a spun column several times and remove the top and then the bottom caps in order. Allow the column to drain. Caution should be taken not to let the gel dry. Add 2 mL of the ligation buffer, replace the cap, invert the column several times, remove the cap, and allow the column to drain. Repeat this step twice more, by using 2 mL of the buffer for each wash. When the last of the buffer enters the gel, place the bottom cap and then the top cap. Use at **step 5** of Subheading 3.1.4.
7. The Glutathione Sepharose 4B bead suspension and commercially available protein markers with known concentrations were separated by SDS-PAGE and stained by Coomassie brilliant blue. The amount of GST-PKA-MP in the suspension was estimated roughly by comparison of the band density of GST-PKA-MP with that of a protein marker locating at a similar migration position.

References

1. Heinlein M (2015) Plant virus replication and movement. Virology 479–480:657–671. https://doi.org/10.1016/j.virol.2015.01.025
2. Ishibashi K, Nishikiori M, Ishikawa M (2010) Interactions between tobamovirus replication proteins and cellular factors: their impacts on virus multiplication. Mol Plant-Microbe Interact 23:1413–1419. https://doi.org/10.1094/MPMI-04-10-0102
3. Ishibashi K, Masuda K, Naito S, Meshi T, Ishikawa M (2007) An inhibitor of viral RNA replication is encoded by a plant resistance gene. Proc Natl Acad Sci U S A 104: 13833–13838
4. Liu C, Nelson RS (2013) The cell biology of *Tobacco mosaic virus* replication and movement. Front Plant Sci 4:1–10
5. Ishikawa M, Okada Y (2004) Replication of tobamovirus RNA. Proc Jpn Acad Ser B Phys Biol Sci 80:215–224
6. Tamai A, Meshi T (2001) Tobamoviral movement protein transiently expressed in a single epidermal cell functions beyond multiple plasmodesmata and spreads multicellularly in an infection-coupled manner. Mol Plant-Microbe Interact 14:126–134
7. Fukunaga R, Hunter T (1997) MNK1, a new MAP kinase-activated protein kinase, isolated by a novel expression screening method for identifying protein kinase substrates. EMBO J 16:1921–1933
8. Matsushita Y, Deguchi M, Youda M, Nishiguchi M, Nyunoya H (2001) The tomato mosaic tobamovirus movement protein interacts with a putative transcriptional coactivator KELP. Mol Cells 12:57–66
9. Matsushita Y, Miyakawa O, Deguchi M, Nishiguchi M, Nyunoya H (2002) Cloning of a tobacco cDNA coding for a putative transcriptional coactivator MBF1 that interacts with the tomato mosaic virus movement protein. J Exp Bot 53:1531–1532
10. Konagaya K, Matsushita Y, Kasahara M, Nyunoya H (2004) Members of 14-3-3 protein isoforms interacting the resistance gene

product N and the elicitor of *Tobacco mosaic virus*. J Gen Plant Pathol 70:221–231
11. Yoshioka K, Matsushita Y, Kasahara M, Konagaya K, Nyunoya H (2004) Interaction of tomato mosaic virus movement protein with tobacco RIO kinase. Mol Cells 17:223–229
12. Crawford KM, Zambryski PC (2000) Subcellular localization determines the availability of non-targeted proteins to plasmodesmatal transport. Curr Biol 10:1032–1040
13. Sasak N, Ogata T, Deguchi M, Nagai S, Tamai A, Meshi T, Kawakami S, Watanabe Y, Matsushita Y, Nyunoya H (2009) Overexpression of putative transcriptional coactivator KELP interferes with *Tomato mosaic virus* cell-to-cell movement. Mol Plant Pathol 10:161–173. https://doi.org/10.1111/j.1364-3703.2008.00517.x
14. Sasaki N, Odawara T, Nagai S, Yoshimura K, Matsushita Y, Nyunoya H (2012) Interference with initial and short-distance cell-to-cell movement of *Tomato mosaic virus* in transgenic tobacco plants with high expression of BcKELP, a virus movement protein interactor from *Brassica campestris*. Physiol Mol Plant Pathol 78:38–44. https://doi.org/10.1016/j.pmpp.2012.01.003

Chapter 8

Transfection of Protoplasts Prepared from *Arabidopsis thaliana* Leaves for Plant Virus Research

Naoi Hosoe, Takuya Keima, Yuji Fujimoto, Yuka Hagiwara-Komoda, Masayoshi Hashimoto, Kensaku Maejima, Shigetou Namba, and Yasuyuki Yamaji

Abstract

Plant viruses use numerous host factors for efficient replication of the viral genome. Protoplasts, plant cells from which cell walls are removed, are the useful system to analyze the virus translation and replication in vivo. Here, we report a protocol for preparation of protoplasts from *Arabidopsis thaliana* leaves and transfection of plasmids to the protoplasts. Protoplasts isolated from the loss-of-function mutant of viral host factor(s) would be helpful to analyze the function of host factors in virus infection cycles.

Key words *Arabidopsis thaliana*, Protoplast, Virus replication analysis, Plant virus, PEG-meditated transfection

1 Introduction

Plant viruses are obligate intercellular pathogens. After invasion into host plant cells, viruses multiply within the single cells. Then, viruses spread from infected to noninfected cells of the host plant symplastically through plasmodesmata and the phloem system. Since only a limited number of genes are encoded in viral genomes, viruses require numerous host factors to successfully infect their hosts [1–5]. In the case of plant RNA viruses, viruses use host plant proteins, lipids, membranes, or metabolites during the replication of their genomic RNAs [6–10]. Analyzing the loss-of-function mutants of such host factor(s) would provide valuable information how viruses employ sophisticated strategies to infect host plants.

Naoi Hosoe and Takuya Keima contributed equally to this work.

Kappei Kobayashi and Masamichi Nishiguchi (eds.), *Antiviral Resistance in Plants: Methods and Protocols*, Methods in Molecular Biology, vol. 2028, https://doi.org/10.1007/978-1-4939-9635-3_8,

Protoplasts are plant cells whose cell walls are removed by enzyme treatment, and we can easily transfect those "naked cells" with DNAs or RNAs including viral genomic RNAs. By viral RNA transfection experiment, the earliest steps of viral life cycles such as translation and genomic RNA replication can be monitored because the virus propagation is restricted within single protoplasts and intercellular movement does not occur. Thus, the protoplast transfection assay has long been used for studying viral replication. In this chapter, we provide a protoplast transfection protocol, which includes preparation of protoplasts from *Arabidopsis thaliana* and transfection of plasmids to the protoplasts by a polyethylene glycol (PEG) mediated method for virus research. The protocol is the modified version of the previously reported one [11]. Comparative analysis between the protoplasts prepared from leaves of wild-type and of mutant lines would allow us to analyze the function of host factors in viral translation and replication at a single-cell level.

2 Materials

2.1 Preparation of Protoplasts from Expanded Leaves of A. thaliana

1. Plant: *A. thaliana* ecotype Columbia-0 (Col-0) plants maintained in the growth chamber with 16 h light–8 h dark conditions at 22 °C throughout the assays. For the protoplast assay, preparation of healthy 5- or 6-week-old plants is critical (Fig. 1a). Avoid using plants whose leaves are purple in color or dried.
2. Enzyme solution: 5 mM MES (pH 5.6), 1% (w/v) Cellulase Onozuka R-10 (Yakult), 0.25% (w/v) Macerozyme Onozuka R-10 (Yakult), 0.4 M mannitol, 8 mM $CaCl_2$. 50 mL of enzyme solution is enough to digest 5.0 g of expanded leaves from single line *A. thaliana*. Enzyme solution can be stored for at least 3 months at −30 °C. Unused enzyme solution should not be refrozen.
3. W5 solution with 0.1 M mannitol: 1.5 mM MES (pH 5.6), 154 mM NaCl, 125 mM $CaCl_2{\cdot}2H_2O$, 5 mM KCl, 5 mM glucose, 0.1 M mannitol. After scaling up to 1 L with distilled water, autoclave, cool to room temperature, and store at 4 °C.
4. 21% (w/v) sucrose: 20 mL of 21% (w/v) sucrose is enough to isolate protoplasts from single line of *A. thaliana*.
5. Disposable scalpels.
6. Stainless steel sieves.
7. Disposable plastic petri dish.
8. 50 mL conical centrifuge tubes.
9. Disposable serological pipettes.

Fig. 1 Preparation of protoplasts from *A. thaliana* leaves and transfection to the protoplasts. (**a** and **b**) 5- or 6-week-old *A. thaliana* leaves shown in (**a**) was cut into thin slices with about 1 mm in width using scalpels (**b**). (**c** and **d**) Crude protoplast suspension was overlaid on top of the 21% (w/v) sucrose (**c**) and, after the centrifugation, intact protoplasts are located at the interface between the water layer and the sucrose layer indicated by a bracket (**d**). (**e**) Microscopic image of intact protoplasts under a light microscope. (**f**) Fluorescent image of protoplasts transfected with a plasmid encoding GFP under 35S promoter under a fluorescence microscope

10. Wide-opening tip.
11. A shaker.
12. A high-speed centrifuge with a swing rotor.
13. An optical microscope.

2.2 Transfection of Plasmid DNA into Cells

1. Plasmid: A plasmid vector encoding GFP under the control of 35S promoter described previously [12].
2. MaMg solution: 5 mM MES (pH 5.6), 400 mM mannitol, 15 mM $MgCl_2 \cdot 6H_2O$. After scaling up to 100 mL with distilled water, autoclave, cool to room temperature, and store at 4 °C.
3. PEG solution: 40% (w/v) PEG 4000, 400 mM mannitol, 100 mM $Ca(NO_3)_2$. All reagents must be dissolved in distilled water by incubating the tube containing the ingredients in hot water. Cool to room temperature before use.

4. Cell counters.
5. An optical microscope.
6. Disposable serological pipettes.
7. Wide-opening tip.
8. Round-bottom polypropylene tubes.
9. 24-well cell culture plates.
10. A fluorescent microscope.

3 Methods

3.1 Preparation of Protoplasts from Expanded Leaves of A. thaliana

1. If enzyme solution is stored at −30 °C, incubate the solution in water bath with sometimes mixing upside down until all reagents are dissolved. Pour the 50 mL of enzyme solution into a 100 mL flask.
2. On a paper towel, cut 5.0 g of expanded rosette leaves into thin slices with about 1 mm in width using disposable scalpels (Fig. 1b) (*see* **Note 1**). Transfer the leaf slices into the flask containing enzyme solution (*see* **Note 2**).
3. Cover the flask with aluminum foil and incubate at 23 °C for 3 h with gentle shaking at 60 rpm on a shaker.
4. After the incubation, filter the protoplast suspension with a stainless steel sieve and transfer to a 50 mL conical centrifuge tube (*see* **Notes 3** and **4**).
5. Collect the protoplasts by centrifugation using a swing rotor at 50 × *g* for 5 min with slow deceleration at 23 °C (*see* **Note 4**).
6. Remove the supernatant using a disposable serological pipette without disturbing the pellet of protoplasts and resuspend the protoplasts in 50 mL of W5 solution with 0.1 M mannitol (*see* **Note 4**). Check the protoplasts quality using an optical microscope (*see* **Note 5**).
7. Collect the protoplasts by centrifugation using the swing rotor at 50 × *g* for 5 min with slow deceleration at 23 °C.
8. Remove the supernatant using a disposable serological pipette and the protoplasts are resuspended in 10 mL of W5 solution with 0.1 M mannitol.
9. Overlay the protoplast suspension on top of the 20 mL of 21% (w/v) sucrose using tips with wide openings and centrifuge at 80 × *g* for 10 min with slow deceleration at 23 °C (Fig. 1c).
10. Transfer the intact protoplasts at the interface between the water layer and the sucrose layer to a new 50 mL conical centrifuge tube containing 30 mL of W5 solution with 0.1 M mannitol using tips with wide openings and mix them gently

(Fig. 1d). Check the quality of protoplasts using an optical microscope (*see* **Note 5**).

11. Collect the protoplasts by centrifugation at 60 × *g* for 5 min using a swing rotor with slow deceleration at 23 °C.
12. Remove the supernatant using a disposable serological pipette and the protoplasts are resuspended in 30 mL of W5 solution with 0.1 M mannitol.
13. Collect the protoplasts by centrifugation at 50 × *g* for 5 min with slow deceleration at 23 °C.
14. Remove the supernatant using a disposable serological pipette and the protoplasts are resuspended in 10 mL of W5 solution with 0.1 M mannitol.
15. Incubate the protoplasts on ice for 30 min.

3.2 Transfection of Plasmid DNA into Cells

1. Count the number of protoplasts using a cell counter (*see* **Note 6**).
2. Collect the protoplasts by centrifugation using the swing rotor at 50 × *g* for 5 min at 4 °C with slow deceleration.
3. Remove the supernatant and resuspend the protoplasts in MaMg solution at the density of 5.0×10^6 protoplasts/mL.
4. 200 or 300 μL of protoplast suspension (*see* **Note 7**) is mixed with the plasmids (*see* **Note 8**) at the bottom of the round-bottom polypropylene tubes by pipetting once.
5. Add the 300 μL of PEG solution to the mixture and mix gently and immediately.
6. Incubate the protoplasts for 30 min at room temperature.
7. Add 10 mL of W5 solution with 0.1 M mannitol to each tube and resuspend the protoplasts.
8. Collect the protoplasts by centrifugation using the swing rotor at 50 × *g* for 5 min with slow deceleration.
9. Remove the supernatant and resuspend the pellet in the 1 mL of W5 solution with 0.1 M mannitol and incubate for 1 to 3 days in the 24-well cell culture plates.

4 Notes

1. Using sharp scalpels is critical for isolation of protoplasts. When the scalpels become dull by cutting leaves, use a new one. We usually used two scalpels to cut 5.0 g of *A. thaliana* leaves. Digestion of more than 5.0 g of leaves in a flask may decrease the protoplast yield.
2. Sliced leaves are easy to dry, which may decrease the protoplast yield. To avoid this, dip the sliced leaves into the enzyme

solution quickly and shake the flask gently to make the leaves sink into the solution certainly.

3. Collection of the protoplast suspension may become easier by laying a disposable plastic petri dish under the sieve.
4. Long-time incubation of protoplasts in enzyme solution may decrease the yield. To avoid the damage to the protoplasts, these steps should be carried out as quickly as possible.
5. Intact protoplasts are circular in shape (Fig. 1e). If the osmotic pressure of the solution is not proper, the shape of protoplasts may become irregular; for example, high pressure makes the protoplasts shrink, while low pressure makes them burst. If you are not familiar with the assay, check the quality of protoplasts after each step to verify that each step is performed adequately.
6. In our experimental condition, 1.5 mL of protoplast suspension (at the density of 5.0×10^6 protoplasts/mL) is obtained from about 5.0 g of expanded leaves on average. When a plasmid harboring GFP under the control of 35S promoter is transiently expressed, transfection efficiency measured by GFP fluorescence is 24.2 ± 5.29% (Fig. 1f).
7. Enough RNA for northern blot analysis could be obtained from 300 μL of protoplast suspension and for qRT-PCR from 200 μL. For western blot analysis, we use 300 μL of protoplast suspension per each transfection.
8. Each plasmid should be purified by additional ethanol precipitation after usual extraction procedures. We analyzed the plasmid quality and concentration by electrophoresis. For plant virus inoculation, we introduced 10–100 μg of plasmids harboring infectious virus cDNA along with 10 μg of internal control plasmids into 200–300 μL protoplasts. In transient expression assay, 10 μg of plasmids is transfected into 150–200 μL protoplasts.

References

1. Hyodo K, Okuno T (2014) Host factors used by positive-strand RNA plant viruses for genome replication. J Gen Plant Pathol 80:123–135
2. Sanfaçon H (2015) Plant translation factors and virus resistance. Viruses 7:3392–3419
3. Hashimoto M, Neriya Y, Keima T et al (2016) EXA1, a GYF domain protein, is responsible for loss-of-susceptibility to plantago asiatica mosaic virus in *Arabidopsis thaliana*. Plant J 88:120–131
4. Hashimoto M, Neriya Y, Yamaji Y et al (2016) Recessive resistance to plant viruses: potential resistance genes beyond translation initiation factors. Front Microbiol 26:1695
5. Keima T, Hagiwara-Komoda Y, Hashimoto M et al (2017) Deficiency of the eIF4E isoform nCBP limits the cell-to-cell movement of a plant virus encoding triple-gene-block proteins in *Arabidopsis thaliana*. Sci Rep 6:39678
6. Sharma M, Sasvari Z, Nagy PD (2011) Inhibition of phospholipid biosynthesis decreases the activity of the tombusvirus replicase and alters the subcellular localization of replication proteins. Virology 415:141–152
7. Nishikiori M, Mori M, Dohi K et al (2011) A host small GTP-binding protein ARL8 plays crucial roles in tobamovirus RNA replication. PLoS Pathog 7:e1002409

8. Nagy PD, Pogany J (2011) The dependence of viral RNA replication on co-opted host factors. Nat Rev Microbiol 10:137–149
9. Diaz A, Zhang J, Ollwerther A et al (2015) Host ESCRT proteins are required for bromovirus RNA replication compartment assembly and function. PLoS Pathog 11:e1004742
10. Hyodo K, Hashimoto K, Kuchitsu K et al (2017) Harnessing host ROS-generating machinery for the robust genome replication of a plant RNA virus. Proc Natl Acad Sci U S A 114:1282–1290
11. Abel S, Theologis A (1994) Transient transformation of *Arabidopsis* leaf protoplasts: a versatile experimental system to study gene expression. Plant J 5:421–427
12. Chiu W, Niwa Y, Zeng W et al (1996) Engineered GFP as a vital reporter in plants. Curr Biol 6:325–330

Chapter 9

CRISPR/Cas9-Mediated Editing of Genes Encoding rgs-CaM-like Proteins in Transgenic Potato Plants

Zhila Osmani, Shinnosuke Jin, Masafumi Mikami, Masaki Endo, Hiroki Atarashi, Kaien Fujino, Tetsuya Yamada, and Kenji S. Nakahara

Abstract

A tobacco calmodulin-like protein, rgs-CaM, has been shown to interact with viruses in a variety of ways; it contributes to geminivirus infections but is also involved in primed immunity to the cucumber mosaic virus. Sequence similarity searches revealed several calmodulin-like proteins similar to rgs-CaM (rCML) in *Arabidopsis* and other Solanaceae plants, including potato (*Solanum tuberosum*). To analyze the functions of each rCML, mutations were introduced into potato rCMLs using the CRISPR/Cas9 system. Here, we describe our protocol of the CRISPR/Cas9-mediated targeted mutagenesis in stably transformed potato plants.

Key words Calmodulin-like protein, CRISPR/Cas9, Potato plants, rgs-CaM, Targeted mutagenesis

1 Introduction

Calmodulin and calmodulin-like proteins (CML) are calcium (Ca^{2+})-binding proteins at their EF hand motifs and are present in many plant genes. Ca^{2+} is an important secondary messenger acting directly after various biotic and abiotic stresses. They function during stress responses by transducing calcium signaling, probably via interaction with diverse endogenous proteins [1].

A tobacco CML, rgs-CaM, interacts with plant viruses inconsistently, suggesting that it has diverse functions. We have shown that rgs-CaM is involved in antiviral defense; it detects viral RNA silencing suppressors and Ca^{2+} as virus infection cues to induce salicylic acid signaling [2], and it counteracts viral RNA silencing suppressors by directing their degradation via autophagy in a primed tobacco plant (referred to as systemic acquired resistance) [2, 3]. However, rgs-CaM has also been shown to suppress RNA silencing and contribute to geminivirus infections [4, 5]. Sequence similarity searches

Kappei Kobayashi and Masamichi Nishiguchi (eds.), *Antiviral Resistance in Plants: Methods and Protocols*, Methods in Molecular Biology, vol. 2028, https://doi.org/10.1007/978-1-4939-9635-3_9,

of rgs-CaM with *Arabidopsis* and Solanaceae plant genes suggest that there are several CMLs that are similar to rgs-CaM (rCML) in each species, including potato (*Solanum tuberosum*), and each of the rCMLs may have a distinct function. To analyze each function of potato rCMLs, functional analysis using knockout mutants of each of the rCMLs is required. While progeny from seeds would show a range of genetic and phenotypic varieties, the major potato cultivars are tetraploid and are propagated vegetatively using tubers. Therefore, making homozygous mutants using conventional random mutagenesis and genetics is difficult. Recently developed genome editing techniques, including the bacterial clustered regularly interspaced short palindromic repeats (CRISPR)/associated protein 9 nuclease (Cas9) system, enabled us to make a knockout mutant potato. This was done by inducing mutations into a particular gene on all four chromosomes in regenerated plants of the stably transformed potato prepared in advance [6, 7]. CRISPR/Cas9 produces a double-stranded break at the targeted site by expressing the Cas9 nuclease and single guide RNA (sgRNA), and thus mutations are frequently induced when the break site is repaired. We induced site-directed mutations in the potato rCML genes using the CRISPR/Cas9 system. Here, we describe our protocol for CRISPR/Cas9-mediated site-directed mutagenesis in stably transformed potato plants.

2 Materials

2.1 Construction of a Binary Vector for CRISPR/Cas9

1. Plasmids necessary for constructing the binary vector for modifying the rCML gene using CRISPR/Cas9: pUC19_AtU6oligo (Fig. 1b) and pZK_gYSA_FFCas9, which were made in advance [8–10].
2. Synthetic oligonucleotides for sgRNA: Select a 20-nucleotide sequence of the target rCMV just upstream of PAM (*see* **Note 1** and Fig. 1a) and synthesize the following oligonucleotides for integration into pUC19_AtU6oligo (Fig. 1b): 5′-ATTG followed by the 20-nucleotide sequence and 5′-AAAC followed by the antisense of the 20-nucleotide sequence.
3. Synthetic oligonucleotides for PCR: AtU6h, 5′-TGGGAATCTGAAAGAAGAGAAGCA-3′; AtU6c, 5′-TTCGACTCTAGCTGTATATAAACTCAGC-3′; rCMLh, 5′-GATGGAAATGTGTCACCACCGGAGC-3′ (Fig. 1a); and rCMLc, 5′-CCCATCAACTCACTCTCTTTATTCC-3′ (Fig. 1a).
4. Restriction enzymes: BbsI, EcoRI, I-SceI.
5. DNA ligase: T4 DNA ligase with 2× rapid ligation buffer (Promega).
6. *Escherichia coli*: K12 strain such as DH5α or JM109 depending on the strains available in your laboratory. Prepare competent cells according to your laboratory protocols.

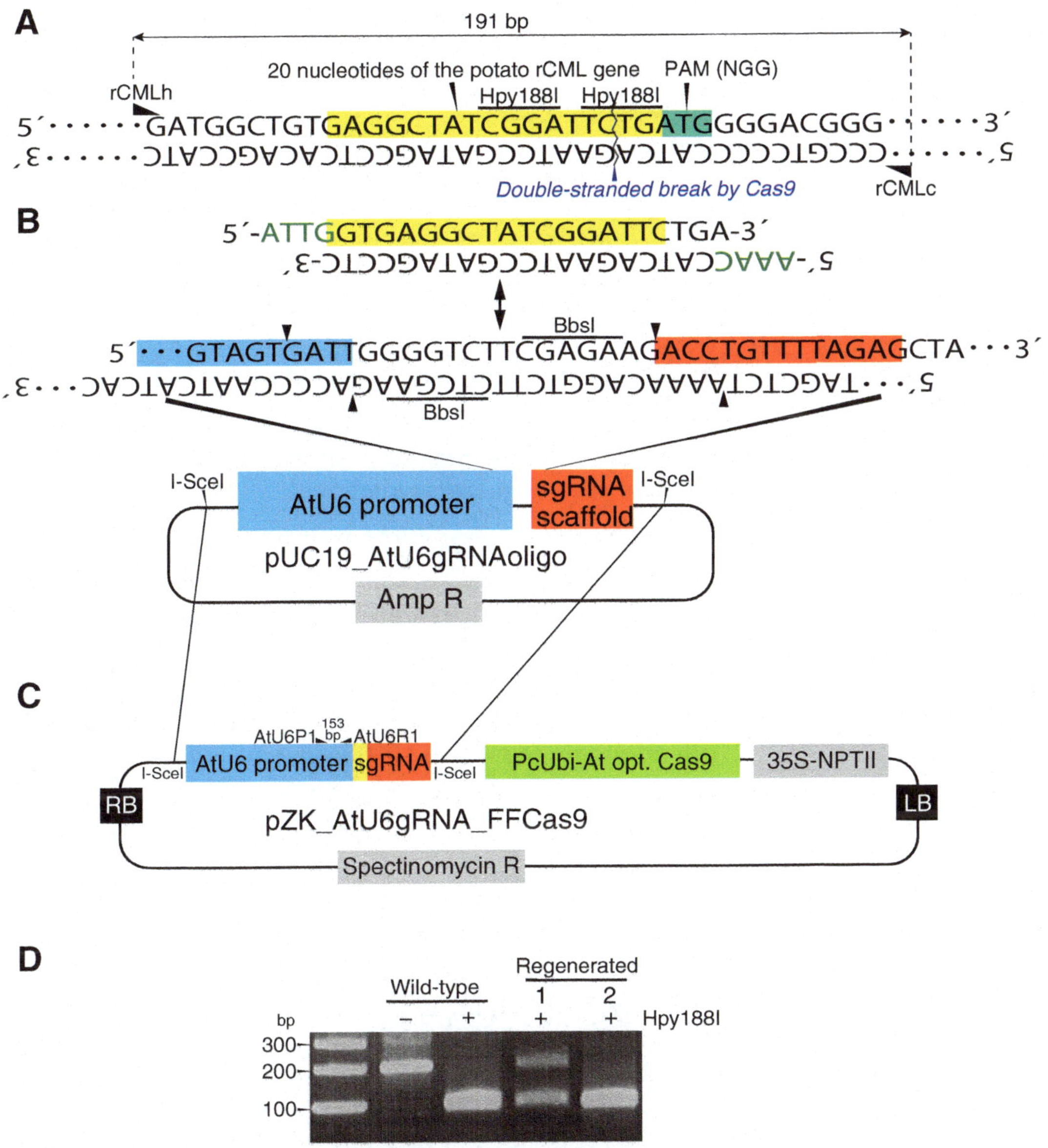

Fig. 1 Construction of the binary vector for the rCML gene editing by CRISPR/Cas9 via *Agrobacterium*-mediated transformation of potato plants and analysis of the CRISPR-mediated editing. (**a**) The 20-nucleotide sequence highlighted in yellow just upstream of the PAM sequence was selected for CRISPR/Cas9 targeting the rCML. (**b**) A pair of synthetic oligonucleotides of 20 nucleotides with adaptive sequences were annealed and cloned into pUC19_AtU6sgRNAoligo using the restriction enzyme, BbsI. (**c**) The partial sequence between the I-SceI sites of pUC19_AtU6sgRNAoligo, which included the *Arabidopsis* U6 promoter, the 20-nucleotide sequence followed by the sgRNA scaffold sequence, was subcloned into the binary vector. (**d**) The modification around the 20-nucleotide sequence by CRISPR/Cas9 in regenerated potato plants can be confirmed by cleaved amplified polymorphic sequence analysis. PCR fragments with a pair of primers, rCMLh and rCMLc (**a**), were digested with Hpy188I (**a**) and fractionated in agarose gel electrophoresis. The PCR fragments of regenerated plant 1 were not completely digested by Hpy188I, suggesting that the rCML gene in the plant was modified

7. Agarose gel electrophoresis.
8. Gel stain: GelRed (Biotium, Fremont, CA, USA).
9. *Agrobacterium tumefaciens*: LBA4404 (Takara Bio, Otsu, Japan).
10. YEP medium: Add 1 g yeast extract, 1 g tryptone peptone, and 0.5 g NaCl to 100 mL water. Adjust pH of the YEP medium to 7.0. After autoclaving, the reagent can be stored at room temperature.
11. $CaCl_2$: Sterilize 20 mM $CaCl_2$ with a 0.22 μm filter.
12. 2× PCR master mix: 2× Go-Taq master mix (Promega).

2.2 Transformation of Potato Plants

1. Soil: Sterile light well-draining soil in pots.
2. 70% ethanol
3. 1% sodium hypochlorite: Mix 100 mL of 10% sodium hypochlorite with 900 mL of sterile water.
4. Glass tube: Glass culture tubes with cotton wool plugs.
5. Filter paper.
6. MS salts stock (10×): Dissolve an MS salts pack in 100 mL water. Store at −21 °C.
7. MS vitamins stock (100×): Add 5 mg nicotinic acid, 5 mg pyridoxine–HCl, 1 mg thiamine–HCl, and 20 mg glycine to 100 mL water. Store at −21 °C.
8. B5 vitamins stock (100×): Add 10 mg nicotinic acid, 10 mg pyridoxine–HCl, and 100 mg thiamine–HCl to 100 mL water. Store at −21 °C.
9. 1000 mg/L gibberellic acid (GA3) stock: Add 20 mg gibberellic acid to 20 mL 50% ethanol. Then filter sterilize and store at 4 °C.
10. 1000 mg/L 1-Naphthalene-acetic acid stock: Add 20 mg 1-naphthalene-acetic acid to 1 mL 1 N NaOH. Dissolve well and then make up to 20 mL with water. Store at 4 °C.
11. 1000 mg/L trans-zeatin-riboside stock: Weigh 20 mg trans-zeatin-riboside and add to 1 mL 1 N NaOH. Dissolve completely and then make up to 20 mL with water. Then filter sterilize and store at −20 °C (*see* **Note 2**).
12. 250 mg/mL cefotaxime stock: Add 1 g cefotaxime to 4 mL water and mix well. Then filter sterilize and store at −20 °C (*see* **Note 3**).
13. 100 mg/mL kanamycin stock: Add 1 g kanamycin to 10 mL water. Dissolve it completely, sterilize by filtration, and store at −20 °C.
14. 100 mg/mL spectinomycin: Add 1 g spectinomycin to 10 mL water. Dissolve it completely, sterilize by filtration, and store at −20 °C.

15. 50 mg/mL rifampicin stock: Weigh 0.5 g of rifampicin. Add to 10 mL of 100% methanol. Dissolve completely and store at −20 °C (*see* **Note 4**).
16. 20 mM acetosyringone stock: Mix 7.8 mg acetosyringone with 2 mL DMSO. Then filter sterilize and store at −20 °C (*see* **Note 5**).
17. *Vir* gene induction medium: Dissolve 2 mL MS salts stock, 200 μL MS vitamins stock, and 600 mg glucose in 20 mL water. Adjust the pH of the induction medium to 5.8. After autoclaving, store at −20 °C.
18. Liquid propagation medium: Prepare by mixing one pack of MS salts mixture, MS vitamins, and sucrose (Table 1). Adjust the pH of the culture medium to 5.8. Then autoclave the medium (*see* **Note 6**).
19. Root medium: Prepare by dissolving one pack of MS salts, 10 mL B5 vitamins, 0.1 g myoinositol, and 30 g sucrose in 1 L water (Table 1). Adjust the pH of this medium to 5.8. After autoclaving and briefly cooling the medium, add 800 μL cefotaxime and 2 mL kanamycin.
20. Coculture medium: Prepare by mixing MS salts, B5 vitamins, adenine sulfate, glucose, mannitol, GA3, 1-naphthalene-acetic acid, trans-zeatin-riboside (*see* **Note 7**), and agar. Adjust the pH of this medium to 5.7 (Table 1; *see* **Notes 8** and **9**).
21. Shoot regeneration medium: Prepared using the ingredients and methods of the coculture medium but 800 μL cefotaxime and 2 mL kanamycin are added for 1 L of medium after autoclaving and briefly cooling the medium (Table 1). Adjust the pH of this medium to 5.7 (*see* **Note 10**).

2.3 Detection of Mutation

1. TPE buffer: 100 mM Tris–HCl pH 9.5, 500 mM KCl, and 10 mM EDTA.
2. High fidelity PCR enzyme: KOD-FX neo with reaction buffer (Toyobo, Japan).
3. Restriction enzyme: Hpy188I (NEB).
4. Agarose gel electrophoresis.

3 Methods

3.1 Construction of a Binary Vector for CRISPR/Cas9

3.1.1 Plasmid Construction

1. Anneal a pair of the synthetic oligonucleotides for sgRNA to form a double-stranded DNA with cohesive ends (Fig. 1b). Mix 1 μL of each of 100 μM oligonucleotides with 48 μL distilled water. Incubate the mixture for 5 min at approximately 95 °C in boiling water and then gradually lower its temperature to room temperature.

Table 1
The components used in coculture, shoot regeneration, root induction, and shoot elongation medium for transformation of the Desiree cultivar

Components	Concentration in medium Propagation	Coculture	Shoot regeneration	Root induction and shoot elongation
MS salts	1×	1×	1×	1×
Nicotinic acid	0.5 mg/L	1 mg/L	1 mg/L	1 mg/L
Pyridoxine–HCl	0.5 mg/L	1 mg/L	1 mg/L	1 mg/L
Thiamine–HCl	0.1 mg/L	10 mg/L	10 mg/L	10 mg/L
Glycine	2 mg/L	–	–	–
Myoinositol	100 mg/L	100 mg/L	100 mg/L	100 mg/L
Adenine sulfate	–	40 mg/L	40 mg/L	–
Glucose	–	20 g/L	20 g/L	–
Mannitol	–	20 g/L	20 g/L	–
Sucrose	30 g/L	–	–	30 g/L
GA3	–	0.05 mg/L	0.05 mg/L	–
1-Naphthalene-acetic acid	–	0.02 mg/L	0.02 mg/L	–
Trans-zeatin-riboside	–	3 mg/L	3 mg/L	–
Agar	–	8 g/L	8 g/L	–
Cefotaxime	–	–	300 mg/L	300 mg/L
Kanamycin	–	–	100 mg/L	100 mg/L

2. Ligate the annealed oligonucleotides with the linearized pUC19_AtU6oligo. Digest 1 μg of pUC19_AtU6oligo with 10 U of BbsI (NEB) for 2 h. Fractionate the linearized pUC19_AtU6oligo by agarose gel electrophoresis, and purify it using a kit. Then adjust its concentration to 30 ng/μL.
3. Clone the pUC19_AtU6oligo by integrating the selected sequence of the target rCML gene with *E. coli*. Gently mix 2.5 μL of the annealed oligonucleotide, 1.5 μL of the linearized pUC19_AtU6oligo, 5 μL of 2× rapid ligation buffer, and 1 μL of T4 DNA ligase. Incubate the mixture for 30 min at room temperature. Transform *E. coli* with the ligated pUC19_AtU6oligo according to your laboratory protocol. Select the transformants with the pUC19_AtU6oligo plasmid under LB medium containing ampicillin and amplify them to extract the plasmid vectors. Confirm that the nucleotide sequences

correspond to the synthesized oligonucleotides for sgRNA in cloned plasmids by determining their nucleotide sequences.

4. Subclone the sgRNA expression cassette in pUC19_AtU6oligo to make the binary vector pZK_AtU6gRNA_FFCas9. Replace the sgRNA expression cassette between the I-SceI restriction enzyme with that of the binary vector pZK_gYSA_FFCas9 (Fig. 1c). Digest pUC19_AtU6oligo containing the sgRNA sequence and the pZK_gYSA_FFCas9 binary vector with I-SceI and fractionate the sgRNA expression cassette of the pUC19_AtU6oligo and the linearized binary vector with agarose gel. Ligate and clone the expression cassette and the binary vector with T4 DNA ligase as described above. Confirm that the sgRNA expression cassette is appropriately integrated into the binary vector by digesting it with EcoRI. Note that this is inapplicable if the selected 20-nucleotide sequence has GAATTC, which is digested by EcoRI. You will see digested fragments of 8890, 7978, and 1187 bp. If pZK_gYSA_FFCas9 is digested with EcoRI, only fragments of 8890 and 8876 bp will be present.

3.1.2 Preparation of Competent Cells of A. tumefaciens LBA4404

1. Streak glycerol stock of *A. tumefaciens* to an LB plate containing rifampicin and streptomycin and incubate the plate for a few days at 28 °C.
2. Scrape a colony of *A. tumefaciens* and inoculate it into 2 mL of YEP with rifampicin and streptomycin. Incubate the medium overnight at 28 °C with shaking.
3. Transfer the overnight culture to 100 mL of YEP with the same antibiotics and incubate the culture at 28 °C with shaking until its OD_{600} = 0.8–1.0.
4. Place the culture on ice for 10 min and harvest cells by centrifugation at 1200 × *g* for 5 min at 4 °C.
5. Add 6 mL of 20 mM $CaCl_2$, suspend and spin down the cells as above, then discard the supernatant, and finally suspend cells in 2 mL of 20 mM $CaCl_2$.
6. Dispense the suspended cells into 100 μL aliquots in 1.5 mL sterile microcentrifuge tubes and store competent cells in a deep freezer at −80 °C until use.

3.1.3 Transforming A. tumefaciens LBA4404 with the pZK_AtU6gRNA_FFCas9 Binary Vector

1. Add 1–3 μg of the binary vector to the 1.5 mL microcentrifuge tube containing 100 μL of competent cells and let the tube stand on ice for 30 min.
2. Dip the tube in liquid nitrogen for 20 s and then incubate it in a water bath at 37 °C for 5 min.
3. Add 900 μL YEP into the tube and shake it at 28 °C for 4 h.
4. Spread the culture on LB plates containing rifampicin, streptomycin, and spectinomycin and incubate them overnight at 28 °C for 2–3 days.

5. Select transformants that include the binary vector by colony PCR. Prepare 10 μL of the PCR mixture containing 5 μL of 2× PCR master mix and 0.75 μM of each of the primers, AtU6h and AtU6c. Scrape a colony of *A. tumefaciens* and inoculate it into the PCR mixture. Incubate the mixture under the following conditions: at 94 °C for 2 min followed by 27 cycles of 94 °C for 30 s, 60 °C for 30 s, and 72 °C for 2 min with the last step at 72 °C for 7 min. Fractionate and visualize the PCR products using 1.5–2% agarose gel electrophoresis stained with GelRed (Biotium, Fremont, CA, USA).

3.2 Transformation of Potato Plants

3.2.1 Planting the Potato Tubers

Choose a sterile light well-draining soil to facilitate development of roots and tubers. Make small holes in the soil (about 6 in. deep) and put each potato tuber in a hole with their eyes facing up. Then cover the tubers with soil. When the plants grow, mound soil around the stems because the developing tubers tend to lose their surrounding soil, resulting in exposure of their surface to light. Keep the plants in a greenhouse with 16 h of light, 8 h of darkness, and at a temperature of 25 ± 1 °C (*see* **Notes 11** and **12**) (Fig. 2a).

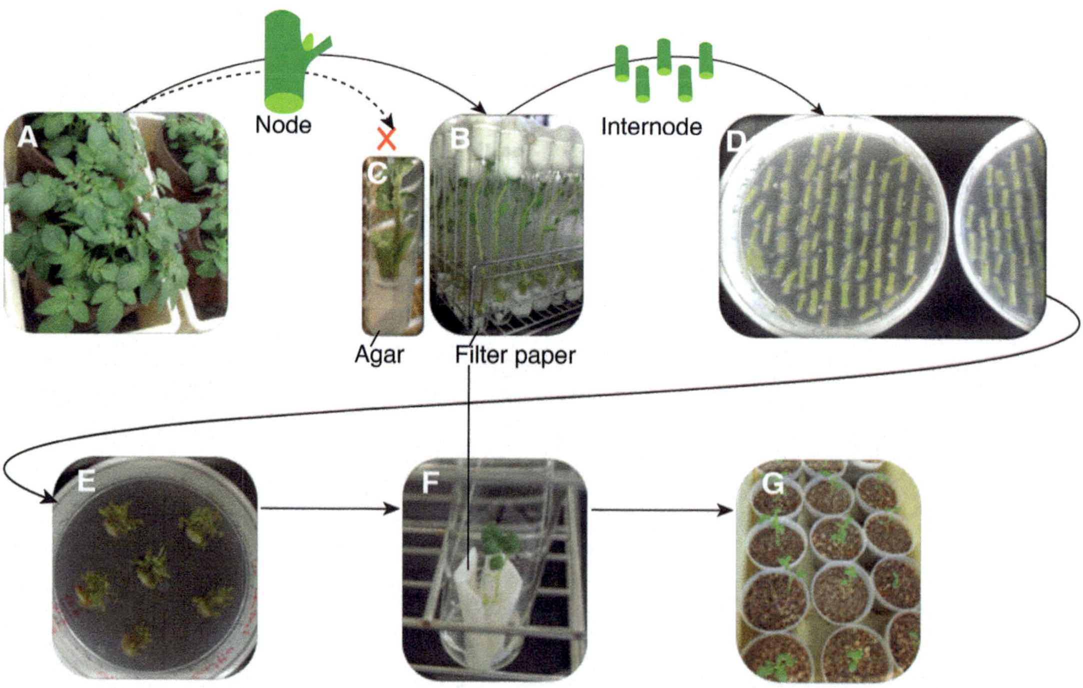

Fig. 2 Schematic flowchart of *Agrobacterium*-mediated transformation of potato plants for genome editing via CRISPR/Cas9. (**a**) Potato plants were grown in a greenhouse and nodes were excised. (**b**) Nodes were sterilized and planted in glass tubes with a piece of filter paper to fix the nodes. (**c**) Nodes should not be grown on agar-medium (*see* **Note 14**). (**d**) Internodes were inoculated with *A. tumefaciens* by soaking in its cultured medium and were then placed on agar-medium with appropriate antibiotics. After shoots with roots were newly developed on internodes (**e**), they were placed into glass tubes with pieces of paper (**f**) and then potted (**g**)

3.2.2 Generation and Propagation of Potato Tissue Culture Plantlets

Carry out all steps under sterile conditions. All equipment and water that are used for making stocks should also be sterile.

1. Put paper bridges inside the glass tubes (Fig. 2b, f; *see* **Note 13**).
2. Pour 5 mL of the liquid propagation medium into each glass tube (*see* **Note 14**).
3. Put glass tubes with cotton wool plugs in the holder boxes and cover with a large bag, then autoclave and dry (*see* **Note 15**).
4. After 3–4 weeks (Fig. 2a), cut stems of the young plants and trim away all leaves (*see* **Note 16**).
5. Transfer the stems to 70% ethanol in a sterile bottle and rinse with sterile distilled water for 30 s.
6. Remove ethanol and wash twice with sterile distilled water for 2–3 min to remove all traces of ethanol.
7. Add 1% sodium hypochlorite and rinse for 3–5 min.
8. Remove sodium hypochlorite and wash three times with sterile distilled water.
9. Transfer the stem pieces onto sterile filter paper using forceps and dry their surface.
10. Cut the auxiliary buds from the sterile stems with a scalpel (at least 2.5–4 cm long) and remove all leaves and damaged cell layers.
11. Transfer the auxiliary bud pieces with long forceps and culture one auxiliary bud per glass tube containing liquid propagation medium. Seal tubes with cotton wool plugs (*see* **Note 17**) (Fig. 2b).
12. Place in a growth chamber at 24 °C under a 16-h–8-h light–dark regime for 3–4 weeks.

3.2.3 *Agrobacterium*-Mediated Stable Transformation of Potato Plants

Carry out all steps under sterile conditions and ensure all equipment is sterile.

1. After 3–4 weeks, remove the sterile plantlets from the glass tubes using long forceps and put them on sterile filter paper.
2. Cut approximately 4–6 mm-long internodal stem segments from sterile plantlets as explants (*see* **Note 18**) (Fig. 2b).
3. Prepare an LB plate with *Agrobacterium* transformed with a plasmid of the pZK_AtU6gRNA_ FFCas9 binary vector. Scrape a colony of *A. tumefaciens* and add to 10 mL YEP with 10 μL spectinomycin and 20 μL rifampicin for plasmid maintenance (with your strain). *Agrobacterium* cells are grown until OD_{600} is approximately 0.6. Depending on the inoculum and the strain used, this may take up to 2 days (*see* **Note 19**).
4. Pellet the cells by centrifugation at 2500 × *g* for 10 min. Remove the supernatant.

5. Resuspend the cells with 200 μL acetosyringone (20 mM) and 20 mL of the Vir gene induction medium. Shake at 200 rpm at 28 °C for 2 h.
6. Put internodal stems or leaf segments in this solution. Shake at 200 rpm at 28 °C for 15 min for leaves and 20 min for internodes. Remove the solution.
7. Put internodal stem or leaf segments on sterile filter paper and dry briefly.
8. Then place them on the coculture medium.
9. Keep at 28 °C in the dark for 3 days (*see* **Note 20**).
10. After 3 days, wash the internodal stems or leaf segments with sterile liquid MS medium twice. Then put them on filter paper to dry briefly.
11. Put the internodal stems or leaf segments on the shoot regeneration medium (*see* **Note 21**).
12. Keep them in a growth chamber with 16-h–8-h light–dark conditions at 22 ± 1 °C for about 2 months (*see* **Note 22**).
13. After observing signs of regeneration (Fig. 2e), transfer the explants to the root medium. Keep the explants in glass tubes with this medium for approximately 10 days.
14. After the plants begin to root, transfer the young and small plants to a sterile light well-draining soil. Cover the pots with plastic and make a few holes in the plastic for air exchange. Keep in a growth chamber with 16-h–8-h light–dark at 22 ± 1 °C for about 1 week. Then remove the cover (*see* **Note 23**).

3.3 Detection and Identification of Mutations That Were Induced by CRISPR/Cas9

1. Extract genomic DNAs from regenerated plants (*see* **Note 24**). Grind approximately 50 mg of plant tissues (leaves) in TPE buffer. Incubate the ground sample at 75 °C for 5 min and centrifuge it at 20,000 × *g* for 1 min. Place 25 μL of the supernatant into a new microcentrifuge tube and add 175 μL of TPE buffer.
2. Amplify the DNA fragment that includes the region targeted by CRISPR/Cas9 (Fig. 1a). Incubate the PCR mixture (20 μL) containing 1× reaction buffer, 400 μM dNTP, 6 pmol rCMLh primer (Fig. 1a), 6 pmol rCMLc primer (Fig. 1a), 0.4 U high fidelity PCR enzyme, and 1 μL (0.1–10 ng) of extracted DNA at 94 °C for 2 min, followed by 30 cycles of 98 °C for 10 s, 60 °C for 30 s, and 68 °C for 2 min with the last step at 68 °C for 7 min. Fractionate and visualize PCR products using 1.5–2% agarose gel electrophoresis.
3. After confirming specific amplification of the expected DNA fragment (in this case, 190 bp, Fig. 1a), digest the DNA

fragment with Hpy188I. Prepare the reaction mixture (10 μL), including 1× CutSmart buffer, 10 U Hpy188I, and 3 μL of the PCR mixture, and incubate it at 37 °C for 2 h. Fractionate and visualize digested-PCR products using 1.5–2% agarose gel electrophoresis (*see* **Note 25**) (Fig. 1d).

4. Confirm the CRISPR/Cas9-mediated editing of the rCML gene by cloning and sequencing of the PCR product that has been amplified with the primer pair, rCMLh and rCMLc (*see* **Note 26**).

4 Notes

1. You can select an appropriate 20-nucleotide sequence for your target genes at the following addresses: http://crispr.hzau.edu.cn/CRISPR2/ or http://www.genome.arizona.edu/crispr/
2. Trans-zeatin-riboside stock should be covered with foil to keep it from being exposed to light.
3. Do not prepare higher concentrations of this stock. Higher concentrations do not filter through a 0.2 μm syringe filter.
4. Rifampicin may precipitate in the freezer. Dissolve again before use.
5. Use resistant filters (PTFE) because this solution does not filter through a 0.2 μm syringe filter.
6. After autoclaving, it is best to store it at room temperature for 2 days to check for contamination. Then cultivate the plants. If it is necessary to store the medium, it can be placed inside a refrigerator and then used to cultivate the seedlings.
7. Trans-zeatin-riboside can be used at concentrations of 2.5, 3, and 4 mg/L depending on cultivars. The best concentration of trans-zeatin-riboside for the Desiree cultivar is 3 mg/L [11].
8. Antibiotics are not used in this medium.
9. Add GA3 and trans-zeatin-riboside after autoclaving and briefly cooling the medium.
10. After autoclaving, when the MS medium temperature reaches approximately 40° C, add selected antibiotics (kanamycin). Then pour 15–20 mL of the medium into different petri dishes.
11. The plant must not mature and should not be allowed to begin flowering. Thus, cut the upper leaves weekly to prevent flowering. The best time for harvesting the stem for propagation is between weeks 2 and 3.

12. The harvested stems cannot be stored. This means that the next steps should follow immediately after harvesting the stems.
13. A paper bridge can be made with any kind of paper that can absorb the MS medium and is firm enough not to dip into the liquid MS medium.
14. Potato is known to display a number of physiological disorders called intumescences or edema [12] that result in blister-like protrusions on the leaves (Fig. 2c). It is important to note that this is not a disease per se, but rather a disorder, which means that the problem may be resolved through investments in laboratory equipment. The problem may occur at several levels and in all cases is usually caused by some form of air pollution, insufficient ventilation of cultures, bad design of growth chambers, or problems with equipment used for air-cooling. In our research, using cotton wool plugs instead of plastic and liquid medium instead of solid medium were very effective in resolving this issue.
15. After cooling the medium, remove the cover bag under sterile conditions and then culture the nodes in the tubes.
16. Unless they are very small, all leaves should be removed.
17. Leave the bottom part of the auxiliary bud on the paper bridge so that the medium is just in contact with the nodes, not immersing them.
18. In this case internodes are better than leaves, because the internodes are damaged less during the different stages of the protocol.
19. If the overnight culture cells are left to grow for too long they will enter the "death phase" and begin to die off. You can add 100 μL of the overnight culture(s) to an Erlenmeyer flask containing 150 mL new YEP medium with the appropriate antibiotics. Grow again to reach an OD_{600} of approximately 0.6.
20. Initially, you can cultivate leaving only small distances between plants (Fig. 2d), then in the next steps increase the distance, allowing space for the explants to feed and grow (Fig. 2e).
21. In the final two stages of subculturing, you can use a smaller amount of cefotaxime, for example 800 μL for 1 L medium, because plant growth improves.
22. Subculturing should be performed every 10 days.
23. During this week, punch more holes in the plastic to allow the plants to adapt to the outside environment.
24. We simply prepare extracts for genomic PCR from potato, tomato, and *Arabidopsis* plants according to this protocol. However, genomic DNAs can be extracted according to your laboratory protocol.

25. CRISPR/Cas9-mediated modification does not always disrupt the restriction enzyme site (in this case, the *Hpy*188I sites; Fig. 1a). The target rCML gene may be mutated in regenerated plants, in which case the disruption of the *Hpy*188I sites may not be detected.
26. PCR products can be cloned using the TA cloning kit with pGEM-Teasy vector (Promega). In most cases the target gene is mutated heterologously by CRISPR/Cas9. We usually determine the sequences of five or more clones.

Acknowledgments

We thank Takehiro Ohki for technical advice. This work was supported in part by the Japan Society for the Promotion of Science (JSPS). KAKENHI grant numbers 16H04879 and 16KK0167 (to K.S.N.).

References

1. Zhu X, Dunand C, Snedden W, Galaud JP (2015) CaM and CML emergence in the green lineage. Trends Plant Sci 20:483–489
2. Jeon EJ, Tadamura K, Murakami T, Inaba JI, Kim BM, Sato M et al (2017) rgs-CaM detects and counteracts viral RNA silencing suppressors in plant immune priming. J Virol 91:e00761-00717
3. Nakahara KS, Masuta C, Yamada S, Shimura H, Kashihara Y, Wada TS et al (2012) Tobacco calmodulin-like protein provides secondary defense by binding to and directing degradation of virus RNA silencing suppressors. Proc Natl Acad Sci U S A 109:10113–10118
4. Chung HY, Lacatus G, Sunter G (2014) Geminivirus AL2 protein induces expression of, and interacts with, a calmodulin-like gene, an endogenous regulator of gene silencing. Virology 460–461:108–118
5. Li F, Huang C, Li Z, Zhou X (2014) Suppression of RNA silencing by a plant DNA virus satellite requires a host calmodulin-like protein to repress RDR6 expression. PLoS Pathog 10:e1003921
6. Wang S, Zhang S, Wang W, Xiong X, Meng F, Cui X (2015) Efficient targeted mutagenesis in potato by the CRISPR/Cas9 system. Plant Cell Rep 34:1473–1476
7. Sawai S, Ohyama K, Yasumoto S, Seki H, Sakuma T, Yamamoto T et al (2014) Sterol side chain reductase 2 is a key enzyme in the biosynthesis of cholesterol, the common precursor of toxic steroidal glycoalkaloids in potato. Plant Cell 26:3763–3774
8. Fauser F, Schiml S, Puchta H (2014) Both CRISPR/Cas-based nucleases and nickases can be used efficiently for genome engineering in *Arabidopsis thaliana*. Plant J 79:348–359
9. Kanazashi Y, Hirose A, Takahashi I, Mikami M, Endo M, Hirose S et al (2018) Simultaneous site-directed mutagenesis of duplicated loci in soybean by using a single guide RNA. Plant Cell Rep 37(3):553–563. https://doi.org/10.1007/s00299-00018-02251-00293
10. Mikami M, Toki S, Endo M (2015) Comparison of CRISPR/Cas9 expression constructs for efficient targeted mutagenesis in rice. Plant Mol Biol 88:561–572
11. Kashani K, Javaran MJ, Mohebodini M, Moieni A, Deh Abadi MS (2012) Regeneration and *Agrobacterium*-mediated transformation of three potato cultivars (*Solanum tuberosum* cv. Desiree, Agria and Marfona) by human proinsulin gene. Aust J Crop Sci 6:1212–1220
12. Craver JK, Miller CT, Williams KA (2014) Characterization and comparison of lesions on ornamental sweetpotato "Blackie", tomato "Maxifort", interspecific geranium "Caliente Coral", and bat-faced cuphea "Tiny Mice". J Am Soc Hortic Sci 139:603–615

Chapter 10

A Simplified Method to Engineer CRISPR/Cas9-Mediated Geminivirus Resistance in Plants

Zahir Ali, Syed Shan-e-Ali Zaidi, Manal Tashkandi, and Magdy M. Mahfouz

Abstract

Throughout the world, geminiviruses cause devastating losses in economically important crops, including tomato, cotton, cassava, potato, chili, and cucumber; however, control mechanisms such as genetic resistance remain expensive and ineffective. CRISPR/Cas9 is an adaptive immunity mechanism used by prokaryotes to defend against invading nucleic acids of phages and plasmids. The CRISPR/Cas9 system has been harnessed for targeted genome editing in a variety of eukaryotic species, and in plants, CRISPR/Cas9 has been used to modify or introduce many traits, including virus resistance. Recently, we demonstrated that the CRISPR/Cas9 system could be used to engineer plant immunity against geminiviruses by directly targeting the viral genome for degradation. In this chapter, we describe a detailed method for engineering CRISPR/Cas9-mediated resistance against geminiviruses. This method may provide broad, durable viral resistance, as it can target conserved regions of the viral genome and can also be customized to emerging viral variants. Moreover, this method can be used in many crop species, as it requires little or no knowledge of the host plant's genome.

Key words CRISPR/Cas9, Virus resistance, Plant virus, Geminivirus, Viral immunity

1 Introduction

Viruses belonging to the Geminiviridae family infect important crops of several plant families throughout the world, including Cucurbitaceae (gourds, squash, watermelon, and melon), Euphorbiaceae (cassava), Solanaceae (tobacco, petunia, pepper, tomato, and potato), Malvaceae (okra and cotton), and Fabaceae (cowpea, mung bean, common bean, lima bean, and soybean) [1–3]. Conventional breeding strategies for improving plant resistance to diseases are expensive, time-consuming, labor-intensive, and often ineffective, especially in the case of rapidly mutating viral diseases [4]. Other strategies seek to boost the plant's cellular immunity against viruses [5]. Several novel approaches have been used to attempt to enhance resistance against viruses, including

Kappei Kobayashi and Masamichi Nishiguchi (eds.), *Antiviral Resistance in Plants: Methods and Protocols*,
Methods in Molecular Biology, vol. 2028, https://doi.org/10.1007/978-1-4939-9635-3_10,

RNA interference-mediated targeting of viral protein production, expression of mutant proteins that impair viral replication, and targeting of host genes that cause symptoms [6].

The Clustered Regularly Interspaced Palindromic Repeats (CRISPR)/CRISPR-associated protein-9 (Cas9) nuclease (CRISPR/Cas9) system, derived from an adaptive immunity mechanism that bacteria and Archaea use to defend against invading nucleic acids such as phages and plasmids, has received special interest as a genome-engineering tool because of its ease, simplicity, efficiency, and reproducibility of use. Although established recently, CRISPR/Cas9 has been used in important model and crop plants such as rice, wheat, maize, tomato, potato, tobacco, cotton, soybean, and *Arabidopsis thaliana* [7]. Recent studies have used CRISPR/Cas9 to engineer virus resistance in plants, either by directly targeting and cleaving the viral genome, or by modifying the host plant genome to introduce genetic viral immunity [8–16].

Recently, we demonstrated that CRISPR/Cas9 can efficiently target the geminivirus genome, leading to interference and degradation of the viral genome and resulting in a significant reduction in viral symptoms [12]. To establish immunity against geminiviruses, we first generated *Cas9*-overexpressing (*Cas9*-OE) *Nicotiana benthamiana* transgenic lines and developed a *Tobacco rattle virus* (TRV)-based system for the delivery of the single guide RNA (sgRNA) into the *Cas9*-OE *N. benthamiana* lines [17, 18]. Engineered TRV genomic RNAs RNA1 and RNA2 were transformed into *Agrobacterium* and agro-infiltration was used to systemically infect the optimized TRV for sgRNA delivery into the *Cas9*-OE *N. benthamiana* plants. After the systemic establishment of TRV in the *Cas9*-OE *N. benthamiana* plants, the plants were infected with different geminiviruses. Plants with a complete CRISPR/Cas9 system exhibited robust resistance against *Tomato yellow leaf curl virus* (TYLCV), *Beet curly top virus* (BCTV), and *Merremia mosaic virus* (MeMV) [11, 12]. Other studies in *N. benthamiana* have demonstrated that the CRISPR/Cas system can interfere with infection by *Bean yellow dwarf virus* (BeYDV) and *Beet severe curly top virus* (BSCTV) or even plant RNA viruses *Turnip mosaic virus* (TuMV) [13–15]. Moreover, we have also demonstrated that this technology can be used to target and cleave *Cotton leaf curl Kokhran virus* (CLCuKoV) and showed that targeting the conserved nonanucleotide sequence can simultaneously target multiple begomoviruses (CLCuKoV, TYLCV, TYLCSV, MeMV, BCTV-Worland, and BCTV-Logan), conferring broad-spectrum geminivirus resistance [11, 19]. These studies showed that *N. benthamiana* plants expressing the CRISPR/Cas9 system displayed reduced viral titers, which abolished or significantly reduced disease symptoms.

Further studies have refined the system and defined key parameters for successful induction of virus resistance. For example,

Cas9 expression has been correlated with the levels of virus suppression, indicating the need to use a single background transgenic line with optimum expression of *Cas9* and the sgRNA for practical applications [14]. Furthermore, catalytically inactive Cas9 (dCas9) can be used to mediate virus interference, thereby eliminating concerns about off-target activities in the plant genome [13, 20]. Our work, therefore, returns CRISPR/Cas9 to its origins as an immunity mechanism and our findings highlight the potential for using *Cas9* overexpression to induce viral immunity without time-intensive tissue culturing practices. Here, we provide a stepwise method for the development of plant lines with CRISPR/Cas9-mediated immunity against Geminiviruses.

2 Materials

2.1 Designing and Cloning the sgRNA

1. TYLCV sequence data (*see* Subheading 3.1 and **Note 1**).
2. CRISPR-P: http://crispr.hzau.edu.cn/CRISPR2/ [21, 22] or CCTop: https://crispr.cos.uni-heidelberg.de/ [21, 22] (*see* Subheading 3.1 and **Note 2**).
3. Phusion polymerase: Phusion high-fidelity DNA polymerase (New England Biolabs).
4. dNTP mixture: Mixture of an equal concentration (e.g., 2.5 mM) of dATP, dCTP, dGTP, and dTTP.
5. TYLCV-IR-T1 and SPDK-gRNA-R (*see* Subheading 3.1 for primers designing and Table 1 for the sequences).
6. Plasmid with sgRNA scaffold sequence (*see* **Note 3**).
7. Thermal cycler.
8. TRV (*Tobacco rattle virus*) vector RNA2 construct: A modified pYL156 having Pea early browning virus (PEBV) subgenomic promoter in the upstream of insertion site. Provided by Savithramma Dinesh-Kumar, University of California at Davis.
9. Restriction enzymes: *Xba*I and *Xma*I or other restriction enzymes as per design of experiment, preferably high-fidelity (New England Biolabs).
10. T4 DNA ligase.
11. *E. coli* competent cells: Any types of competent cells of *Escherichia coli* strains such as DH5α.
12. LB (Luria Bertani) medium: 1% (w/v) tryptone, 0.5% (w/v) yeast extract, 1% (w/v) NaCl. For agar plate medium, add 1.5% (w/v) agar. Sterilize by autoclave.
13. LB agar plates and LB liquid medium containing 50 mg/L kanamycin.
14. Mini-prep kit for plasmids.
15. Sequencing facility.

Table 1
Primers used in the method

Primer name	Sequence (5′–3′)	Use
TYLCV2.3-IR-T-F	AATTGGGAAAGTGCTTCCTCT	Semi-qPCR, to amplify TYLCV IR flanking region, to make probe for southern, detection of TYLCV by PCR
TYLCV2.3-IR-T-R	ATAGTCACGGGCCCTTACAACA	
TYLCV-IR-T1	CGAGTCTAGAGGCCATCCGTATA ATATTACGTTTTAGAGCTAGAAA TAGCAAG	To clone TYLCV IR-sgRNA
SPDK-gRNA-R	ACATGCCCGGGAAAAAAAGCACCG ACTCGG	To clone all sgRNAs
Cas9-gw-F	CACCATGGACTATAAGGACCACG	To clone *Cas9*
Cas9-R	TTACTTTTTCTTTTTTGCCTGGC	To clone *Cas9* and as reverse primer in *Cas9* detection RT-PCR
Cas9-seqF6	GCCCTCCAAATATGTGAACTTCC	Forward primer in *Cas9* detection by RT-PCR and Cas9 sequencing
NB-Actin-RT-R	GATGAAGATACTCACAGAAAGA	NB. Actin RT control
NB-Actin-Q-RT-F	TAGCTCTTCTCCACGGATGAGC	NB. Actin RT control
CAS9-SEQ-1860-F	AAATCGAGTGCTTCGACTCCGTGGAAAT	Cas9 sequencing
CAS9-SEQ-2442-F	ACAGAAGAACAGCCGCGAGAGAATGAA	Cas9 sequencing
CAS9-SEQ-2960-F	ACACTAAGTACGACGAGAATGACAAGC	Cas9 sequencing
Cas9-SEQ-3726-F	GATCATCAAGCTGCCTAAGTACTCCCTG	Cas9 sequencing
CAS9SEQ F1	CCCATCAACGCCAGCGGCGTG	Cas9 sequencing
CAS9-SEQ-1064-F	GACATCCTGAGAGTGAACACCGAGATCA	Cas9 sequencing
CAS9SEQ F2	GCTGCGGAAGCAGCGGACCTTC	Cas9 sequencing
CAS9SEQ F3	CGAGGACATTCTGGAAGATATC	Cas9 sequencing
CAS9SEQ F5	GCGAGCAGGAAATCGGCAAGGC	Cas9 sequencing
Cas9-887-SeqR	CAGGTCGAAGTTGCTCTTGAAGTT	Cas9 sequencing
Cas9-397-seq	CCATCTCGTTGCTGAAGATCTCTT	Cas9 sequencing

2.2 Cloning Cas9

1. Phusion polymerase.
2. dNTP mixture.
3. Cas9-gw-F and Cas9-R (Table 1).
4. pX330 plasmid [20].
5. Thermal cycler.

6. pENTR/D-TOPO (Life Technologies).
7. pK2GW7 binary vector Gateway compatible [21].
8. LR clonase (Life Technologies, Invitrogen).
9. *E. coli* competent cells: Any types of competent cells of *Escherichia coli* ccdB-sensitive strains such as DH5α.
10. LB agar plates and LB liquid medium containing 50 mg/L kanamycin.
11. LB agar plates and LB liquid medium containing 50 mg/L spectinomycin.
12. Mini-prep kit.
13. Sequencing facility.

2.3 Generation of N. benthamiana Plants Overexpressing Cas9 (Cas9-OE)

1. Electro-competent cells of *Agrobacterium tumefaciens* GV3101 strain.
2. LB agar plates and LB liquid medium containing 100 mg/L spectinomycin, 30 mg/L gentamycin, and 25 mg/L rifampicin
3. MS medium: 4.4 g/L Murashige and Skoog (MS) basal salt mixture (Sigma-Aldrich or others), 1 mL/L Murashige and Skoog Vitamin Solution (1000×; Sigma-Aldrich or others), 3% (w/v) sucrose, pH 5.8. For agar plates, add 0.8% (w/v) agar (plant tissue culture grade). Sterilize by autoclave.
4. Sterile plastic petri dishes.
5. Autoclavable plant boxes such as Magenta™ vessel GA-7 (Sigma-Aldrich).
6. *Nicotiana benthamiana*: 2-week-old plants grown aseptically on MS agar medium in plant boxes.
7. Transformation medium: MS liquid medium.
8. Sterile filter paper.
9. MS cocultivation medium: MS agar medium containing 1 mg/L 6-benzylaminopurine and 0.1 mg/L 1-naphthaleneacetic acid.
10. MS regeneration and selection medium: MS cocultivation medium supplemented with 50 mg/L kanamycin and 200 mg/L timentin.
11. Root-inducing medium: half-strength MS media [2.2 g/L Murashige and Skoog (MS) basal salt. Mixture, 0.5 mL/L Murashige and Skoog Vitamin Solution (1000×), 1.5% (w/v) sucrose, pH 5.8] added with 0.8% (w/v) agar (plant tissue culture grade), sterilized by autoclave, added with 50 mg/L kanamycin, and poured into plant boxes.
12. Soil, pots, and plastic domes for plant growth.

2.4 Cas9 Confirmation and Expression Analysis

1. MS medium seedling selection plates: the same as root-inducing medium (*see* Subheading 2.3, **item 11**).
2. Soil and pots.
3. RNA extraction kit such as RNeasy mini kit.
4. SuperScript 3 reverse transcriptase (Life Technologies, Invitrogen).
5. Oligo-dT primer.
6. Phusion polymerase.
7. dNTP mixture.
8. Cas9-seqF6 and Cas9-R primers (Table 1).
9. Thermal cycler.
10. Agarose gel electrophoresis.
11. Protein extraction buffer: 100 mM Tris–HCl, pH 7.5, 150 mM NaCl, 0.5% NP-40, 1 mM EDTA, 3 mM DTT and protease inhibitors (1 mM PMSF, 5 μg/mL leupeptin, 1 μg/mL aprotinin, 1 μg/mL pepstatin, 5 μg/mL antipain, 5 μg/mL chymostatin, 2 mM Na_2VO_3, 2 mM NaF and 50 μM MG132).
12. SDS-PAGE system.
13. PVDF membrane.
14. Electric blotter.
15. TBS-T: 50 mM Tris–HCl, 0.138 M NaCl, 0.0027 M KCl, 0.05% Tween 20, pH 8.0. Dissolve Tris Buffered Saline, with Tween® 20, pH 8.0 powder (Sigma-Aldrich) in 1 L distilled water.
16. 5% skim milk: 5% (w/v) skim milk powder in TBS-T.
17. 1% skim milk: 1% (w/v) skim milk powder in TBS-T.
18. Anti-FLAG antibody (Sigma-Aldrich).
19. Goat anti-mouse IgG-HRP secondary antibody (Santa Cruz).
20. Enhanced Chemiluminescence (ECL)-detection reagent (Thermo Scientific).
21. Chemiluminescence imaging system.

2.5 TRV-Mediated sgRNA Delivery for Cas9 Against TYLCV

1. *N. benthamiana Cas9-OE* plants: The transgenic plants should be grown according to the relevant law and regulations in each nation/state/province. Preferably grown in a greenhouse on the soil.
2. TRV vectors: pYL192 for TRV-RNA1 (Provided by Savithramma Dinesh-Kumar, University of California at Davis), and TRV-RNA2:TYLCV (constructed in Subheading 3.1).
3. Infectious clone of TYLCV: Infectious clones should be constructed by each of the research groups or obtained from other groups in the same nation/state/province because of the phytosanitary regulations.

4. Electrocompetent cells of *Agrobacterium tumefaciens* GV3101 strain.
5. LB agar plates and LB liquid medium containing 100 mg/L spectinomycin, 30 mg/L gentamycin, and 25 mg/L rifampicin.
6. Infiltration medium: 10 mM MES (2-(*N*-morpholino)ethane-sulfonic acid), 10 mM $CaCl_2$. Adjust pH to 5.7 with KOH and sterilize by autoclave. Then add 200 μM acetosyringone (add 1/500 volume of 100 mM stock solution in dimethyl sulfoxide).
7. Spectrophotometer.
8. 1 mL needleless syringes.

2.6 Mutation Detection by T7EI Assay, Restriction Site Loss Assay, and Sanger Sequencing

1. DNA extraction buffer:100 mM Tris–HCl, 1 mM EDTA, 100 mM NaCl, 100 mM LiCl, 100 mM β-mercaptoethanol, and 0.4% RNase.
2. PCI: Mixture of phenol–chloroform–isoamyl alcohol at a ratio of 25:24:1 saturated with 10 mM Tris–HCl, pH 8.0, 1 mM EDTA.
3. Phusion polymerase.
4. dNTP mixture.
5. TYLCV2.3-IR-T-F and TYLCV2.3-IR-T-R primers (Table 1).
6. Thermal cycler.
7. PCR purification kit.
8. NanoDrop spectrophotometer.
9. T7EI: T7 endonuclease I supplied with NEBuffer 2(New England Biolabs).
10. SspI restriction enzyme.
11. Agarose gel electrophoresis.
12. pJET2.1 (Thermo Scientific) or other PCR cloning vector for blunt-ended PCR products.
13. T4 DNA ligase.
14. *E. coli* competent cells.
15. LB agar plates and LB liquid medium containing 50 mg/L ampicillin.
16. Mini-prep kit.
17. Sequencing facility.

2.7 Viral Quantification with Semiquantitative PCR, Rolling Circle Amplification, and Southern Blot Analysis

1. DNA extracted in Subheading 3.6.1.
2. Phusion polymerase.
3. dNTP mixture.
4. TYLCV2.3-IR-T-F and TYLCV2.3-IR-T-R primers (Table 1).
5. NbActin-F and NbActin-R primers (Table 1).
6. Thermal cycler.

7. Agarose gel electrophoresis.
8. Rolling circle amplification kit: TempliPhi 100 Amplification kit (Sigma-Aldrich or GE Healthcare).
9. NcoI restriction enzyme.
10. DIG DNA labeling and detection kit (Roche Diagnostics GmbH).
11. DNA Molecular Weight Marker II, DIG-labeled (Roche Diagnostics GmbH).
12. Nylon membrane.
13. CDP-Star chemiluminescent substrate (Roche Diagnostics GmbH).
14. Chemiluminescence imaging system such as Alpha Innotech digital imaging system.
15. ImageJ: http://rsb.info.nih.gov/ij.

3 Methods

3.1 Designing and Cloning the sgRNA

1. Select the genomic region of the geminivirus to be targeted. In this example, we selected the intergenic region (IR) of TYLCV (Fig. 1a) (*see* **Note 1**). Target site selection can be done manually by looking for an "NGG" as the protospacer-adjacent motif (PAM); use the 20 bp upstream, avoiding five or more consecutive thymines (T) in a row. Software-assisted selection is possible with CRISPR-P or CCTop [20, 21] (*see* **Note 2**). Manual selection is preferable to target a specific structure, motif, or domain (Fig. 1a).

Fig. 1 CRISPR/Cas9-mediated interference of *Tomato yellow leaf curl virus* (TYLCV) in *N. benthamiana*. (**a**) CRISPR/Cas9 machinery (Cas9, sgRNA) designed for the intergenic region (IR) sequence of TYLCV is shown. Black arrows represent the partially overlapping open reading frames (ORFs), and the IR is represented by double arrowheads in the TYLCV genome. (**b**) Schematic representation of the experimental design. The sgRNA specific for the TYLCV IR sequence engineered to TRV-RNA2 was infiltrated into *Cas9*-OE plants. After establishing CRISPR/Cas9 system targeting the TYLCV IR sequence in plants, TYLCV was agro-infiltrated into the lower leaves. Samples were collected at 5, 10, 15, and 30 days postinfiltration (dpi). PEBV, *Pea early browning virus* promoter. (**c**) Mutation analysis using a restriction site loss assay. The IR (560 bp) sequence containing an SspI recognition site at the target region was analyzed for loss of recognition. The arrow indicates the 269-bp SspI-resistant DNA fragment in samples harboring IR-sgRNA, but not in samples harboring the TRV empty vector used as a control. Note that the 560 bp fragment has two additional SspI sites outside the target sequence (for detail, *see* Fig. 3c in Ref. 11). (**d**) Alignment of Sanger sequencing reads of PCR amplicons cloned from the IR targeted region. PAM and Ssp*I* recognition sequences are represented. (**e**) Semiquantitative PCR of TYLCV genomic DNA. TYLCV infiltration into CRISPR/Cas9 established plants. IR-sgRNA containing plants accumulated low levels of TYLCV compared to the control plants. Actin of *N. benthamiana* (NB-Actin) was used for normalization. (**f**) Rolling-circle amplification (RCA) assay of the TYLCV genome. Plants harboring IR-sgRNA accumulated less TYLCV genomic DNA than those with the TRV empty vector controls. (**g**) Southern blot analysis of TYLCV genomic DNA accumulation. Genomic DNA of TYLCV was detected with DIG-labeled 560-bp IR probe. IR-sgRNA-harboring plants showed lower accumulation of the TYLCV genome than plants inoculated with the TRV empty vector and TYLCV. Arrowhead indicates the expected size of the TYLCV genome

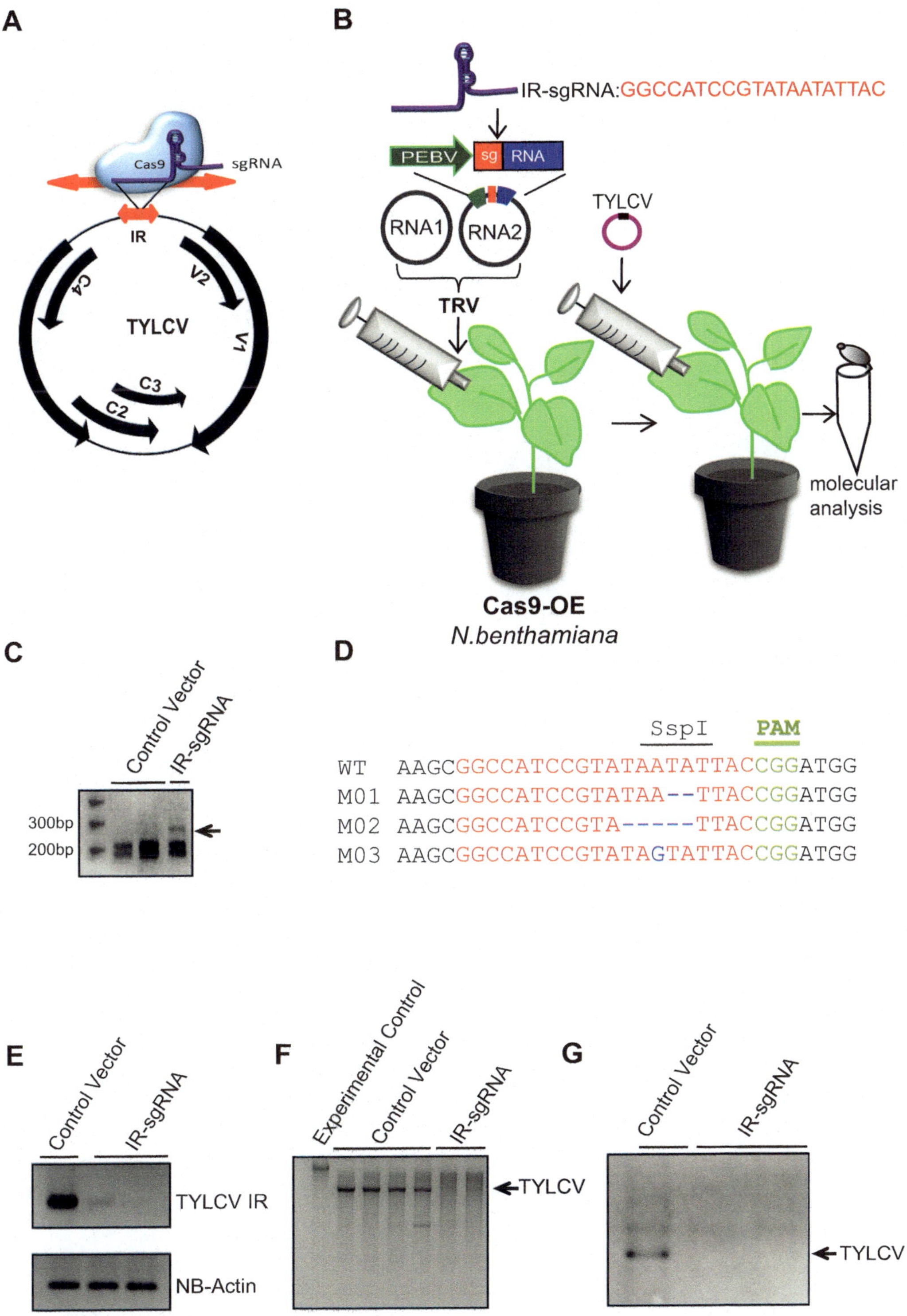

Fig. 1 (*caption on page 174*)

2. Design a forward primer to contain an XbaI recognition site with a four-nucleotide 5′ extension, a 20-nucleotide target sequence, and an 84-nucleotide Cas9-binding sgRNA scaffold, and a reverse primer to contain complementary sequence to the sgRNA 3′end, a 7-T repeat (terminator), and an XmaI recognition site with a four-nucleotide 5′ extension. Perform PCR with these primers to amplify a 125-bp PCR fragment using Phusion polymerase with 5× Phusion HF buffer, and a backbone containing the 84 bp sgRNA Cas9 binding scaffold as a template. The fragment should contain a four-nucleotide extension, six-nucleotide XbaI recognition site, the 20-nucleotide target sequence, the 84-bp Cas9 binding loop for sgRNA including a seven-T repeat, six-nucleotide XmaI recognition site, and a five-nucleotide extension (*see* **Note 3**).

   Clone the amplified fragment using a restriction ligation procedure to have an sgRNA targeting the TYLCV IR region in the downstream of PEBV promoter of the TRV RNA2 vector. Cut both the PCR product and the RNA2 vector with XbaI and XmaI and gel purify.
3. Using T4 ligase, insert the gel-purified 116-bp restriction fragment of the sgRNA for TYLCV target into the TRV RNA2 vector.
4. Transform *E. coli* and plate onto LB agar plate containing 50 mg/L kanamycin.
5. Pick up a few single colonies, grow them in LB medium containing 50 mg/L kanamycin, and extract plasmid DNA using a mini-prep kit.
6. Perform Sanger sequencing to confirm all cloned sequences.

3.2 Cloning of Cas9

1. PCR amplify the complete 3XFlag–NLS-Cas9-NLS cassette with Cas9-GW-F and Cas9-R primers using Phusion polymerase with Phusion 5× HF buffer, and the pX330 plasmid as a template [20].
2. Clone the gel-purified PCR product into pENTR/D-TOPO according to the manufacturer's protocol. Transform *E. coli* and plate onto LB agar plate containing 50 mg/L kanamycin. Extract plasmid DNA from a few clones as mentioned above.
3. Confirm clones with restriction digestion and Sanger sequencing using Cas9-seq series primers (Table 1) for authenticity of clones and the in-frame translational fusions.
4. By LR Clonase reaction under the manufacturer's protocol, move *Cas9* coding sequence into the pK2GW7 binary vector to generate a *35S::Cas9-T35S* construct. Transform *E. coli* and plate onto LB agar plate containing 100 mg/L spectinomycin. Pick up a few single colonies, grow them in LB medium containing 100 mg/L spectinomycin, and extract

plasmid DNA using a mini-prep kit. Confirm clones with restriction digestion. Use this construct for the production of the *N. benthamiana Cas9*-OE line (*see* **Note 4**).

3.3 Generation of *N. benthamiana* Plants Overexpressing Cas9 (Cas9-OE)

1. Introduce the pK2GW7.Cas9 binary vector from the previous step into the *Agrobacterium tumefaciens* GV3101 strain by electroporation. Plate the bacteria onto LB agar plate containing 100 mg/L spectinomycin, 30 mg/L gentamycin, and 25 mg/L rifampicin.
2. Pick up and grow single colonies carrying the T-DNA vector overnight at 28 °C in 5 mL LB medium containing 100 mg/L spectinomycin, 30 mg/L gentamycin, and 25 mg/L rifampicin.
3. Use the overnight culture to inoculate 50 mL of the same medium.
4. Collect the cells at an OD_{600} of 1.0, resuspend in transformation medium to an OD_{600} of 0.4, and pour onto a petri dish.
5. Use a surgical blade to cut leaf tissues from aseptically grown 2-week-old *N. benthamiana* plants into small (approximately 1 cm) leaf disks. Make small incisions on the leaf surface with the scalpel to maximize contact of *Agrobacteria* with the cells.
6. Incubate leaf disks with an *Agrobacterium* culture for 30 min with occasional shaking.
7. Remove extra *Agrobacterium* cells with sterile filter paper and keep the leaf discs for 2 days for cocultivation with *Agrobacterium* in the growth chamber on MS cocultivation media.
8. After 2 days, transfer leaf disks onto MS regeneration and selection medium.
9. Excise shoot tissues appearing after 3–4 weeks and place on the root-inducing medium.
10. After proper roots have formed (2–3 weeks), carefully remove the agar medium from the roots, transfer the plantlets to soil and acclimate them to ambient atmosphere under plastic domes.
11. Grow the plants until seed setting and collect T1 seeds.

3.4 Cas9 Confirmation and Expression Analysis

3.4.1 Plant Growth

1. Grow the T1 generation of *N. benthamiana Cas9-OE* plants on MS medium seedling selection plates.
2. Transfer some kanamycin-resistant plants to soil and grow them further for the analysis.

3.4.2 Confirmation of the Expression of Cas9 in the T1 Plants by Semiquantitative RT-PCR

1. Extract total RNA from the leaves of *N. benthamiana Cas9-OE* plants using the RNAeasy mini kit or equivalent.
2. Synthesize first-strand cDNA using SuperScript 3 reverse transcriptase with oligo-dT primer.

3. Amplify a fragment corresponding to the C-terminal end of Cas9 using a primer pair (Cas9-seqF6 and Cas9-R) with Phusion polymerase with 5× Phusion HF buffer. Input PCR condition as follows: 98 °C for 1 min 30 s, 30 cycles of 98 °C for 15 s, 60 °C for 30 s, and 72 °C for 15 s, followed by 72 °C for 5 min.
4. Analyze PCR products on an agarose gel. The expected size of the PCR product should be 471 bp.

3.4.3 Confirmation of the Presence of the Cas9 Protein by Western Blotting Using Anti-Flag Antibody

1. Extract total proteins from 100 μg leaf tissue of *N. benthamiana Cas9-OE* plants with 100 μL of extraction buffer, clear the extract by centrifugation at 10,000 × *g* for 5 min at 4 °C.
2. Denature the proteins in a sample buffer and separate them on a 6% SDS polyacrylamide gel electrophoresis by a standard protocol.
3. Transfer the proteins onto PVDF membrane using electric blotter by a standard protocol.
4. Perform immunoblot analysis by a standard protocol. Use primary mouse anti-FLAG antibody (1:1000) in 5% skim milk and secondary goat anti-mouse (1:1000) in 1% skim milk.
5. Detect reacting proteins by chemiluminescence using ECL detection reagent and a chemiluminescence imaging system according to the manufacturer's protocol.
6. Select transgenic lines with good Cas9 protein expression and grow them to collect T2 seeds.

3.5 TRV-Mediated sgRNA Delivery for Cas9 Against TYLCV

The complete strategy is provided in (Fig. 1b) and described as follows.

1. Grow *N. benthamiana Cas9*-OE plants on the soil in a greenhouse to the 6–8 leaf stage (2–3 weeks).
2. Introduce vectors containing the TRV RNA1 and TRV RNA2 genomes separately into the *A. tumefaciens* GV3101 strain and spread on LB agar plates containing 50 mg/L kanamycin, 30 mg/L gentamycin, and 25 mg/L rifampicin.
3. Introduce vectors containing the infectious clone of TYLCV (or any geminivirus under experiment) into the *A. tumefaciens* GV3101 strain and spread on the appropriate selection plates.
4. Grow the transformed single colonies for 36–48 h in selective medium to OD_{600} of 1.2, spin down, resuspend and dilute to an OD_{600} of 0.3 with infiltration medium.
5. Incubate cultures at ambient temperature in the dark for 2–4 h.
6. Prior to infiltration, mix bacterial cultures at OD_{600} at a 1:1:0.05 ratio for TRV-RNA1:TRV-RNA2:TYLCV.

7. Infiltrate the bacterial culture mix into the lower (abaxial) side of two lower leaves of 3- to 4-week-old *Cas9*-OE *N. benthamiana* plants using a needleless 1 mL syringe.
8. Alternatively, first infiltrate the mixture of TRV-RNA 1:TRV-RNA2 at a ratio of 1:1 to preestablish the CRISPR/Cas9 system for 5 days and then superimpose TYLCV by agroinfiltration to one lower leaf at an OD_{600} of 0.05.
9. Collect leaf disc samples from infiltrated and upper uninfiltrated leaves at 5, 10, 15, and 30 days postinfiltration and subject to targeted genome modification analysis to determine and quantify the activity of the Cas9 endonuclease mutations resulting from double-strand break repair through the nonhomologous end joining pathway (*see* **Note 5**).

3.6 Mutation Detection by T7EI Assay, Loss of Restriction Site Assay, and Sanger Sequencing

3.6.1 PCR Amplification of DNA Fragments Containing the Target Site

1. Extract genomic DNA from the samples collected as mentioned above by a standard protocol using DNA extraction buffer, PCI extraction, chloroform extraction, and ethanol precipitation. Quantify the DNA using NanoDrop spectrophotometer.
2. PCR amplify the fragment encompassing the target sequence (*see* **Note 6**). In this example, a 560 bp fragment of the IR flanking region was amplified using TYLCV2.3-IR-T-F and TYLCV2.3-IR-T-R primers with Phusion Polymerase with 5× Phusion HF buffer. PCR condition was as follows: 98 °C for 1 min 30 s, 30 cycles of 98 °C for 15 s, 60 °C for 30 s, and 72 °C for 15 s, followed by 72 °C for 5 min.
3. Confirm the amplification of the fragments on an agarose gel, purify the DNA fragments using a PCR purification kit and quantify using NanoDrop. Mutation can be detected by T7EI assay in most cases. A loss of restriction site assay can be used to detect mutations if a restriction enzyme recognition site was included in the target DNA. In both cases, mutations detected should be validated by Sanger sequencing.

3.6.2 Mutation Detection by a T7EI Assay

1. Prepare heteroduplex formation reaction by mixing 200 ng purified PCR products, 1.8 μL NEBuffer 2, and sterile water to 18 μL total volume. Denature and reanneal the PCR products in a thermal cycler to allow heteroduplex formation using a following cycling program: 95 °C for 10 min, 85 °C for 2 min, 75 °C for 3 min, 65 °C for 3 min, 55 °C for 3 min, 45 °C for 3 min, 35 °C for 3 min, 25 °C for 3 min, and 4 °C.
2. Treat the reannealed PCR products with T7EI by adding 0.5 μL of T7EI and 0.2 μL Buffer 2 and 1.3 μL of sterile water, incubate at 37 °C for 1 h. If the thermal cycler is to be used for the incubation reaction, the lid temperature should not be more than 37 °C.

3. Analyze the reaction product on an agarose gel, capture the gel image and quantify the cut and uncut bands using ImageJ. Calculate the molar ratio of cut fragments to the total fragments. The mutation rates is calculated as follows:

$$\text{Mutation rate}\,(\%) = 100 \times \left(1 - (1 - \text{ratio of cut fragment})^{1/2}\right).$$

3.6.3 Mutation Detection by a Loss of Restriction Site Assay

1. For TYLCV, the IR contains an SspI restriction enzyme recognition site, and can be used in restriction enzyme protection analysis. In a 20-μL reaction, 300 ng of the IR flanking PCR products were subjected to SspI restriction and the results are shown in (Fig. 1c). To ensure complete digestion, the reaction should be performed for 8 h at 37 °C. If the thermal cycler is to be used for the incubation reaction, the set temperature should not be more than 37 °C.
2. Analyze the reaction product on an agarose gel, capture the gel image and quantify the cut and uncut bands using ImageJ. Uncut bands represent the modified target sequence, and therefore, the molar ratio of the uncut fragments is the mutation rate.

3.6.4 Validation of Mutations Detected by Sanger Sequencing

1. Ligate the PCR products to pJET2.1 using T4 DNA ligase. To validate the mutations detected by T7EI assay, the original PCR products from Subheading 3.6.1 are used. It is preferable to gel purify the undigested DNA fragments in the loss of restriction site assay and clone them.
2. Transform *E. coli* and spread the bacteria onto the LB agar plate containing 50 mg/L Ampicillin.
3. Pick up 8 or more colonies per sample, grow the bacteria in LB medium containing 50 mg/L Ampicillin and extract plasmid DNA using a mini-prep kit.
4. Sequence the flanking regions of the target site to identify the mutation occurred. The representative results are shown in Fig. 1d.

3.7 Viral Quantification with Semiquantitative PCR, Rolling Circle Amplification, and Southern Blot Analysis

3.7.1 Semiquantitative PCR for TYLCV

1. Amplify the DNA fragment encompassing the IR of TYLCV, using the TYLCV2.3-IR-T-F and -R primers (Table 1) with Phusion polymerase with 5× Phusion HF buffer and the 50 ng genomic DNA extracted in Subheading 3.6.1 from all samples including the controls as templates.
2. As the experimental control, amplify a partial fragment of *N. benthamiana* Actin gene using the NbActin-F and -R primers (Table 1) with Phusion polymerase with 5× Phusion HF buffer and the genomic DNA extracted from all samples including the controls as templates.

3. Run a 20–23 cycle PCR with conditions as follows: 98 °C for 1 min 30 s, 20–23 cycles of 98 °C for 15 s, 60 °C for 30 s, and 72 °C for 15 s, followed by 72 °C for 5 min.
4. Run the PCR product on a 1% agarose gel. Representative results from the uninoculated upper leaves are shown in Fig. 1e.

3.7.2 Rolling Circle Amplification for TYLCV

Amplify the TYLCV genomic DNA following the rolling circle amplification kit protocol. The amplified DNA can be detected on an agarose gel electrophoresis after cutting into unit size molecules by restriction digestion.

1. Incubate the 50 ng genomic DNA for 3 min at 95 °C in sample buffer and place on ice for 5 min.
2. Add enzyme mix and reaction buffer to the samples and incubate at 30 °C for 12–18 h for amplification.
3. Inactivate the enzyme at 65 °C for 15 min.
4. Add *Nco*I to the samples and incubate for 1 h at 37 °C to linearize the amplified DNA.
5. Resolve DNA samples on a 1% agarose gel. Representative results from the uninoculated upper leaves are shown in Fig. 1f.

3.7.3 DIG-Based Southern Blot Analysis for the TYLCV Genome

1. Using DNA DIG labeling and detection kit, PCR amplify a fragment of the IR sequence of the TYLCV genome (an approximately 560 bp fragment) using the TYLCV2.3-IR-T-F and -R primers (Table 1).
2. Resolve 2 μg total genomic DNA from Subheading 3.6.1 on a 1% agarose gel alongside the DNA DIG-labeled size marker-II.
3. Blot the DNA fragments onto a nylon membrane by a standard capillary transfer procedure overnight followed by UV-crosslinking. Block the membrane and allow hybridization with the respective IR-DIG labeled probe.
4. Visualize DNA bands using alkaline phosphatase-conjugated anti-DIG (1:10,000) and CDP-star chemiluminescent substrate in a chemiluminescence imaging system. Representative results from the uninoculated upper leaves are shown in Fig. 1g.
5. Quantify virus titer using ImageJ.

4 Notes

1. In this method, we used TYLCV as an example geminivirus. Any geminivirus or DNA virus can be targeted using the described method. A comprehensive list of all geminiviruses can be found at www.talk.ictvonline.org with respective accession numbers that can be used to download viral genome sequences from www.ncbi.nlm.nih.gov.

2. We highly recommend checking the off-targeting activity of the designed sgRNA using online tools like CRISPR-P or CCTop [20, 21].
3. Alternatively, the sgRNA clones can be custom-synthesized in a plasmid vector using an oligonucleotide synthesis service company such as BlueHeronBio gene synthesis service (BlueHeronBio, Bothell, WA, USA). Each sgRNA, flanked by *Xba*I and *Xma*I restriction sites, should consist of a 116-bp fragment containing the 20-nucleotide target sequence, 77 bp functioning as a Cas9 binding loop followed by a repeat of seven thymine residues as the transcriptional terminator and six-nucleotide recognition sequence of each restriction enzyme (*Xba*I and *Xma*I). The 116 bp sgRNA fragment of each target can be subcloned into the TRV RNA2 vector under the PEBV viral promoter by restriction ligation cloning using XbaI and XmaI restriction enzymes.
4. In this example, we described *N. benthamiana*, but any plant species where the efficiency of CRISPR/Cas9 and TRV has been experimentally demonstrated can be used.
5. Take photographs of plants at each stage for comparison of molecular data with the phenotypic data.
6. Perform PCR reactions using undigested genomic DNA. For the cloning and Sanger sequencing purpose, restriction enzyme predigested genomic DNA can be used to enrich for the modification when the target sequence includes the restriction enzyme recognition site. In this case, design PCR primers such that the amplified regions do not have the recognition sites except for that in the target sequence.

Acknowledgments

We would like to thank members of the laboratory for genome engineering for continuous discussions. This study is supported by King Abdullah University of Science and Technology (KAUST) Office of Sponsored Research (OSR) under Award No. OSR-2015-CRG4-2647.

References

1. Zaidi SS, Martin DP, Amin I et al (2017) *Tomato leaf curl New Delhi virus*: a widespread bipartite begomovirus in the territory of monopartite begomoviruses. Mol Plant Pathol 18(7):901–911. https://doi.org/10.1111/mpp.12481
2. Zaidi SS, Shafiq M, Amin I et al (2016) Frequent occurrence of *Tomato leaf curl New Delhi virus* in cotton leaf curl disease affected cotton in Pakistan. PLoS One 11:e0155520
3. Zubair M, Zaidi SS, Shakir S et al (2017) Multiple begomoviruses found associated with cotton leaf curl disease in Pakistan in early 1990 are back in cultivated cotton. Sci Rep 7:680

4. Whitham SA, Hajimorad MR (2016) Plant genetic resistance to viruses. In: Wang A, Zhou X (eds) Current research topics in plant virology. Springer International Publishing, Cham, pp 87–111
5. Mandadi KK, Scholthof KB (2013) Plant immune responses against viruses: How does a virus cause disease? Plant Cell 25:1489–1505
6. Sahu PP, Prasad M (2015) Application of molecular antiviral compounds: novel approach for durable resistance against geminiviruses. Mol Biol Rep 42:1157–1162
7. Zhang K, Raboanatahiry N, Zhu B et al (2017) Progress in genome editing technology and its application in plants. Front Plant Sci 8:177
8. Zaidi SS, Tashkandi M, Mansoor S et al (2016) Engineering plant immunity: using CRISPR/Cas9 to generate virus resistance. Front Plant Sci 7:1673
9. Zaidi SS, Mansoor S, Ali Z et al (2016) Engineering plants for geminivirus resistance with CRISPR/Cas9 system. Trends Plant Sci 21:279–281
10. Zaidi SS, Tashkandi M, Mahfouz M (2017) Engineering molecular immunity against plant viruses. Prog Mol Biol Transl Sci 149:167–186
11. Ali Z, Ali S, Tashkandi M et al (2016) CRISPR/Cas9-mediated immunity to geminiviruses: differential interference and evasion. Sci Rep 6:26912
12. Ali Z, Abulfaraj A, Idris A et al (2015) CRISPR/Cas9-mediated viral interference in plants. Genome Biol 16:238
13. Baltes NJ, Hummel AW, Konecna E et al (2015) Conferring resistance to geminiviruses with the CRISPR–Cas prokaryotic immune system. Nat Plants 1:15145
14. Ji X, Zhang H, Zhang Y et al (2015) Establishing a CRISPR-Cas-like immune system conferring DNA virus resistance in plants. Nat Plants 1:15144
15. Aman R, Ali Z, Butt H et al (2018) RNA virus interference via CRISPR/Cas13a system in plants. Genome Biol 19:1
16. Ali Z, Mahas A, Mahfouz M (2018) CRISPR/Cas13 as a tool for RNA interference. Trends Plant Sci 23:374–378
17. Ali Z, Abul-Faraj A, Li L et al (2015) Efficient virus-mediated genome editing in plants using the CRISPR/Cas9 system. Mol Plant 8:1288–1291
18. Ali Z, Abul-Faraj A, Piatek M et al (2015) Activity and specificity of TRV-mediated gene editing in plants. Plant Signal Behav 10:e1044191
19. Mahas A, Mahfouz M (2018) Engineering virus resistance via CRISPR-Cas systems. Curr Opin Virol 32:1–8
20. Aouida M, Eid A, Ali Z et al (2015) Efficient fdCas9 synthetic endonuclease with improved specificity for precise genome engineering. PLoS One 10:e0133373
21. Lei Y, Lu L, Liu HY et al (2014) CRISP-P: a web tool for synthetic single-guide RNA design of CRISPR-system in plants. Mol Plant 7:1494–1496
22. Stemmer M, Thumberger T, Del Sol Keyer M et al (2015) CCTop: an intuitive, flexible and reliable CRISPR/Cas9 target prediction tool. PLoS One 10:e0124633

Chapter 11

Computational Workflow for Small RNA Profiling in Virus-Infected Plants

Livia Donaire and César Llave

Abstract

In this chapter we describe a series of computational pipelines for the in silico analysis of small RNAs (sRNA) produced in response to viral infections in plants. Our workflow is primarily focused on the analysis of sRNA populations derived from known or previously undescribed viruses infecting host plants. Furthermore, we provide an additional pipeline to examine host-specific endogenous sRNAs activated or specifically expressed during viral infections in plants. We present some key points for a successful and cost-efficient processing of next generation sequencing sRNA libraries, from purification of high quality RNA to guidance for library preparation and sequencing strategies. We report a series of free available tools and programs as well as in-house Perl scripts to perform customized sRNA-seq data mining. Previous bioinformatic background is not required, but experience with basic Unix commands is desirable.

Key words Plant viruses, Small RNAs, Antiviral silencing, sRNA-seq, Bioinformatic analysis, Next generation sequencing

1 Introduction

RNA silencing is an important antiviral mechanism in plants and other organisms [1]. The accumulation of virus-derived small RNAs (vsRNAs) in infected tissues and their contribution to antiviral silencing have been proven for many plant viruses (reviewed by [2]). Since the advent of modern next generation sequencing (NGS) technics, sequencing the small RNA (sRNA) transcriptome in the plant cell is high throughput and effortless in terms of experimental concerns [3]. At the time of writing, at least four different massive sequencing platforms are available for sRNA profiling (454 GS FLX, Life Sciences-Roche; Illumina GA/HiSeq/MiSeq/NextSeq; Applied Biosystems SOLiD; Ion Torrent, Thermo Fisher Scientifics). All these platforms have standardized protocols for library preparation, which for sRNA profiling involves isolation of sRNAs and attachment to specific adapter sequences. Performance comparison between different

Kappei Kobayashi and Masamichi Nishiguchi (eds.), *Antiviral Resistance in Plants: Methods and Protocols*, Methods in Molecular Biology, vol. 2028, https://doi.org/10.1007/978-1-4939-9635-3_11,

NGS platforms along with the advantages and limitations for each technology has been widely discussed [4]. A necessary step, and a major challenge for most biologists, is the transformation of raw data into genome-scale datasets that can be interpreted and applied at the molecular, cellular and genomic levels. The NGS computational workflow requires a number of sophisticated tools for monitoring, viewing, analyzing, and storing data that are hardly manageable for scientists in the field of biological research [5, 6]. Here we outline four current analytical pipelines for computational profiling of both vsRNAs and endogenous virus-associated sRNAs in virus-infected plants. We do not aim to provide an exhaustive compilation of dedicated software and bioinformatic tools, or to dissect methods for preparation of sRNA sequencing libraries. Rather, we aim to share an elementary guideline for the analysis of sRNA-seq data using in-house and free available bioinformatic resources.

The proposed workflow consists of four multistep pipelines. Pipeline 1 encompasses the analysis of raw data, including data quality tests, size filtering and subtraction of contaminant RNA species such as structural RNAs. Pipeline 2 and 3 are devoted to characterizing vsRNA populations. When the target viral genome is available, it is possible to track the origin of the vsRNAs by mapping sRNA-seq reads onto the viral genome. Pipeline 2 involves the alignment of filtered short reads against a viral reference genome, and it is used to profile sRNA derived from known viruses. As a result, specific signatures such as origin, relative abundance, size distribution, strand orientation, or 5′ nucleotide identity are reported for both total and unique vsRNA sequence reads. For viruses without sequenced genomes, identification is achieved by first assembling overlapping sRNA reads into longer contiguous pieces of cDNA or contigs. Pipeline 3 is thus used for viral discovery and for resolving the genome organization of novel infecting viruses in host plans. Finally, Pipeline 4 allows for the identification and quantification of virus-responsive sRNAs derived from host genomic loci.

2 Materials

2.1 RNA Isolation and sRNA Library Construction

1. High-quality plant RNA isolation kit: *mir*Vana™ miRNA Isolation Kit (Thermo Fisher Scientific) or Macherey-Nagel NucleoSpin miRNA kit (Clontech Laboratories).
2. Agilent 2100 BioAnalyzer.
3. Qubit 3.0 fluorometer.
4. A method for sRNA library preparation.

2.2 In Silico Bioinformatic Analysis

2.2.1 Hardware Requirements

1. Operating system: 64-bit Linux (*see* **Note 1**).
2. Available RAM: at least 4 GB (recommend 16 GB or higher).
3. Free disk space: depends on the number of samples and sequencing depth (20–50 GB per sample).

2.2.2 Software Requirements (See Note 2)

1. Java SE Runtime Environment.

   To run java applications (recommended version 1.6 JREs from Oracle or higher), check the java version installed on your computer by typing in a Terminal:

```
$ java -version
```

   If java is not already installed or the available version is older than the recommended one, install Oracle Java 8 by typing in the command line:

```
$ sudo add-apt-repository ppa:webupd8team/java
$ sudo apt-get update
$ sudo apt-get install oracle-java8-installer
```

2. Perl (version 5.0) and Bioperl.

   Perl is already installed in most Linux distributions. Check that the current Perl version in your computer is 5.0 or higher by typing:

```
$ perl -version
```

   To install Bioperl, type:

```
$ sudo apt-get install bioperl
```

3. SRAtoolkit (version 2.8.2).

   To install the SRAtoolkit [7], type:

```
$ sudo apt-get install sra-toolkit
```

4. FastQC (version 0.11.5).

   To download FastQC [8], type:

```
$ wget https://launchpad.net/ubuntu/+archive/
primary/+files/fastqc_0.11.5+dfsg-3_all.deb
```

   To install it, type:

```
$sudo dpkg -i fastqc_0.11.5+dfsg-3_all.deb
$sudo apt-get install -f
```

5. Adapter Removal (version 2.1.7).

   Download AdapterRemoval [9] from GitHub by typing:

```
$ wget -O adapterremoval-2.1.7.tar.gz https://
github.com/MikkelSchubert/adapterremoval/ar-
chive/v2.1.7.tar.gz
```

   To unpack it, type:

```
$ tar -xvzf adapterremoval-2.1.7.tar.gz
```

   To compile and install it, type:

```
$ cd adapterremoval-2.1.7
/adapterremoval-2.1.7/$ make
/adapterremoval-2.1.7/$ sudo make install
```

   (*see* **Note 3**).

6. Perl scripts from Github.

   Download the Perl script "trim.fastq.length.pl" from https://github.com/MadsAlbertsen/miscperlscripts.

   Likewise, download from https://github.com/ldonaire/Perl-scripts, the eight following Perl scripts: "countfq_20to26.pl", "countfa_20to26.pl", "fqtofqunique.pl", "misis_table_analysis.pl", "nucomp_fa.pl", "nucomp_fq.pl", "rna_to_dna.pl", and "srna_count_comparison.pl".

   Save all of them into a new folder called "perl_scripts" in your working directory (*see* **Note 4**).

7. Python (version 2.7.12) and Python packages.

   Python is already installed in most Linux distributions. To check the version installed in your computer type:

```
$ python --version
```

   First, install pip package management system:

```
$ sudo apt-get install python-pip
```

   Then, install numpy and pyfasta packages as follows:

```
$ sudo python -m pip install numpy
$ sudo python -m pip install pyfasta
```

8. Bowtie (version 1.1.2).

   Install Bowtie [10] by typing:

```
$ sudo apt-get install bowtie
```

9. SAMools (version 1.6).

   To install SAMtools [11], type:

```
$sudo apt-get install samtools
```

10. MISIS (version 2.7).

    Download MISIS [12] from https:/www.fasteris.com/apps/misis.php.

    Unpack it into your working directory:

```
/working_directory$ tar -xvzf Misis.tar.gz
```

11. Integrated Genomics Viewer (IGV) (version 2.4.2).

    Download the binary distribution of IGV [13] from:

    http://software.broadinstitute.org/software/igv/download

    Unpack it into your working directory:

```
/working_directory$ unzip IGV_2.4.3.zip
```

12. eXpress (version 1.5.1).

    Download the Linux 64-bit executable application of eXpress [13] from https://pachterlab.github.io/eXpress/index.html.

    Unpack it into your working directory:

```
/working_directory$ tar -xzvf express-1.5.1-
linux_x86_64.tgz
```

    Give executable permissions by typing:

```
/working_directory$ chmod +x
express-1.5.1-linux_x86_64/express
```

2.2.3 Local Databases

1. Creating a local Rfam index (*see* **Notes 5** and **6**).

   Download Rfam data in fasta format from version 11.0 (or previous):

```
/working_directory$ wget ftp://ftp.ebi.ac.uk/
pub/databases/Rfam/11.0/Rfam.fasta.gz
```

   Unpack it into a new folder called "Rfam_index" in your working directory:

```
/working_directory/Rfam_index$ gunzip Rfam.
fasta.gz
```

   Convert RNA sequences into DNA by using the Perl script "rna_to_dna.pl" (Subheading 2.2.2, **item 6**):

```
/working_directory/Rfam_index$ perl perl_
scripts/rna_to_dna.pl Rfam_index/Rfam.fasta
```

   The output file will be a fasta file called Rfam-DNA.fasta.

   If you want to search for miRNAs in your datasets, it is necessary to first remove pre-miRNA and miRNA sequences from the Rfam fasta file. To do so, create a file containing the fasta IDs of all pre-miRNAs and miRNAs included by using the Unix command "grep" and remove the first ">" using the Unix command "sed":

```
/working_directory/Rfam_index$ grep -E
'miR|MIR|mir' Rfam-DNA.fasta | sed 's/^>//' >
miR_headers.txt
```

   The output file "miR_headers.txt" contains 27,394 fasta IDs of miRNAs and pre-miRNA sequences from different organisms.

   Then, use pyfasta (Subheading 2.2.2, **item 7**) to eliminate those sequences from the fasta file as follows:

```
/working_directory$ pyfasta extract --header
--fasta Rfam_index/Rfam-DNA.fasta --exclude
--file miR_headers.txt > Rfam-modif.fasta
```

   The output file "Rfam-modif.fasta" contains 35,5610 fasta sequences.

   To create a Bowtie index, type:

```
/working_directory$ bowtie-build Rfam_index/
Rfam-modif.fasta Rfam_index/Rfam
```

   The output should be six files with ebwt extension.

2. Creating local plant virus indexes.

   For known viruses, virus reference genomes can be downloaded in fasta format from the NCBI sequence repository (nucleotide database). As an example, our reference genome consists of the *Tobacco rattle virus* isolate PpK20, a positive, single-stranded RNA virus with a bipartite genome. The sequences of TRV1 and TRV2 with the accession numbers NC_003805 and NC_003811, respectively, are downloaded into our working directory and referred to as TRV1.fasta and TRV2.fasta.

To create a Bowtie index of TRV1, type:

```
/working_directory$ mkdir virus_index
/working_directory$ mv TRV1.fasta virus_index
/working_directory$ bowtie-build virus_index/
TRV1.fasta virus_index/TRV1
```

The output should be six files with ebwt extension.

3. Creating a local plant host index.

Genomic or transcriptomic fasta files of the plant hosts can be downloaded from different plant genomic databases such as Solgenomics, TAIR, PlantGDB, and NCBI. For instance, our test experiment is run on the Arabidopsis transcriptome (CDS+UTRs, version 10) available at the TAIR website:

https://www.arabidopsis.org/download_files/Sequences/TAIR10_blastsets/TAIR10_cdna_20101214_updated

Create a new folder called "plant_index" and save the Arabidopsis cDNAs into a file called "TAIR10_cdna_20101214_updated.fasta".

It is highly recommended to shorten the fasta headers to avoid later execution problems. To remove the description and leave only the gen ID use the Unix "awk" command:

/working_directory$ awk '{print /^>/ ? $1 : $0}' plant_index/TAIR10_cdna_20101214_updated.fasta > plant_index/TAIR10cdna.fasta

To create a Bowtie index, type the following:

/working_directory$ bowtie-build plant_index/TAIR10cdna.fasta plant_index/TAIR10cdna

The output should be six files with ebwt extension.

3 Methods

3.1 Viral Small RNA Library Construction and Sequencing

In the following section, we briefly provide some clues and tricks to optimize our sequencing performance. It is not our goal to offer a detailed description of current protocols for library preparation and sequencing, which have been extensively reviewed by others [14, 15].

3.1.1 Purification of High-Quality RNA

Several methods are suitable for high-quality RNA isolation including guanidine isothiocyanate and acid-phenol methods such as TRIzol Reagent (Thermo Fisher Scientific) or TRI reagent (Sigma-Aldrich/Merck). If using commercial kits, ensure that short RNAs (<200 nts) are retained.

3.1.2 RNA Quality Test

To improve NGS library construction and further increase sequencing performance, sRNAs should be size-fractionated from total RNA using gel electrophoresis or silica spin columns. RNA integrity and degradation are determined with the RNA integrity number (RIN) calculated using the Agilent 2100 BioAnalyzer (Agilent Technologies, CA). For plant samples a RIN value higher than eight is usually recommended, but lower values can be acceptable provided that discrete RNA bands are visualized in the gel-like image displayed by BioAnalyzer [16].

The amount of starting material is important, so the RNA source has to be measured preferentially using highly sensitive quantitation assays such as the Qubit 3.0 fluorometer (Thermo Fisher Scientific). The required amount of high-quality RNA is 1–3 μg. Treatment with DNase is advisable as DNA contamination may produce inaccurate measurements of RNA concentration.

3.1.3 Library Preparation

Most sequencing companies offer fast and cost-effective preparation of libraries for ultra-deep sequencing with low input requirements for purified total RNA. Different library preparation options include individual libraries and/or pooled barcoded libraries according to customer preferences and further downstream applications [15]. Multiplexed-sequencing is the preferred option to process a large number of samples in a single flow cell lane, without drastically increasing cost or time. Multiplexing consists in the incorporation of individual barcode sequences to each sample during library preparation so that they can be easily distinguished and sorted during data analysis.

If the NGS libraries are prepared in the lab, standard protocols for preparing directional sRNA libraries are described elsewhere [14, 15]. The directional method is advisable since the sequencing always reads from the 5′ to the 3′ end of the original sRNA species. Alternatively, numerous kits for making sequencing oriented libraries are available commercially. For instance, TruSeq Small RNA Library Prep Kit (Illumina) or SMARTer smRNA-Seq Kit (Clontech Laboratories) allow size selection and multiplexing and are highly recommended. Libraries can be done from microgram to picogram of starting material, but problems arise when the amount of input RNA is too low (200 ng of total RNA) as adapter dimers compete with the desired products in the RT-PCR reaction. Thus, quality assessments of libraries are of primary importance when PCR-based cloning techniques are used and should be done thorough the whole process. The quality of an sRNA library can be assessed by a number of criteria. For instance, gel-based size selection of amplicons is critical due to the presence of adapter dimers and/or higher molecular RNA products created during the ligation steps.

3.1.4 Sequencing

Here we report the analysis of sRNA datasets obtained using massively parallel sequencing from Illumina, which utilizes an optimized four-color DNA sequencing by synthesis (SBS) technology (HiSeq 2500 System). In our test, we pooled four libraries in a multiplex format to obtain approximately 42 M of clean reads per sample. The recommended run type is 50 bp and single-ends (SE) that allow enough read length for accurate adapter trimming. In sRNA-seq, an important factor is the library size, which is the number of clean sequence reads for a given sample. For accurate profiling of sRNAs, the deeper the library is sequenced, the more sRNA species will be detected, including those expressed at very low levels, and the more precise their quantification will be [4].

3.2 Pipeline 1: Raw-Data Analysis

This analysis is intended to parse small RNA reads from the raw sequence data, and to classify and characterize sRNA reads. sRNA reads from 18 to 26 nucleotides in length are parsed and trimmed from raw sequences upon identification of 3′ adaptor sequences. Then low-quality, contaminated or background sequence reads are filtered out. Samples in our test were obtained from healthy, buffer (mock)-inoculated Arabidopsis plants (MOCK8) or from Arabidopsis plants infected with TRV (TRV8).

3.2.1 Obtaining Small RNA-Seq Data in fastq Format

Download raw data directly from the website of the sequencing supplier or from public databases (*see* **Note 7**). The example data can be obtained from the Short Read Archive (SRA) under the accession numbers SRR6156924 and SRR6156925.

1. To directly download it, type from the working directory (here called "At_TRV"):

```
/At_TRV$ wget ftp://ftp-trace.ncbi.nih.gov/
sra/sra-instant/reads/ByRun/sra/SRR/SRR615/
SRR6156924/SRR6156924.sra
```

or

```
/At_TRV$ wget ftp://ftp-trace.ncbi.nih.gov/
sra/sra-instant/reads/ByRun/sra/SRR/SRR615/
SRR6156925/SRR6156925.sra
```

To transform SRA to fastq format use SRAtoolkit (Subheading 2.2.2, **item 3**):

```
/At_TRV$ fastq-dump SRR6156924.sra
```

The standard output should be as follows:

```
Read 43279723 spots for SRR6156924.sra
Written 43279723 spots for SRR6156924.sra
```

or

```
/At_TRV$ fastq-dump SRR6156925.sra
```

The standard output should be as follows:

```
Read 44951890 spots for SRR6156925.sra
Written 44951890 spots for SRR6156925.sra
```

Rename the new created fastq files by typing:

```
/At_TRV$ mv SRR6156924.fastq MOCK8.fq
```

or

```
/At_TRV$ mv SRR6156925.fastq TRV8.fq
```

2. Quality-control checkpoints.

Specific controls should be applied at different stages of the analysis to ensure reproducibility and reliability of the results. Quality control for the raw data involves the analysis of sequence quality, the presence of adaptors, sequence content, GC content, N content, overrepresented k-mers, PCR artifacts, or contaminations by degraded RNA. This can be done using FastQC [8], which is a popular tool to perform quality control on Illumina reads (*see* **Note 7**). To run FastQC (Subheading 2.2.2, **item 4**) from the command line, type

/At_TRV$ fastqc MOCK8.fq TRV8.fq

The output delivered by this program includes several files for each fastq file analyzed. Quality results are showed in the html file that can be opened using a common web browser.

For sRNA reads it is important to pass the per base sequence quality, which should be higher than 30 in the full length of the sRNA sequence.

3.2.2 3′ Adapter Trimming

Adaptor sequences should be trimmed from the raw data to obtain clean reads. This is a critical checkpoint in the workflow when a read length of 50 bp is used in the sequencing setting. In our example, the adapter sequence was already removed by the sequencing supplier, however, whenever required, adaptor sequences can be trimmed out using AdapterRemoval [9] (Subheading 2.2.2, **item 5** and **Note 8**).

3.2.3 Filtering by sRNA Size

vsRNAs are mostly of 20–26 nts in length and derive from Dicer-like-mediated cleavage of double-stranded RNA substrates. To filter reads by length, use the Perl script "trim.fastq.length.pl" (*see* Subheading 2.2.2, **item 6**):

```
/At_TRV$ perl perl_scripts/trim.fastq.length.
pl -i MOCK8.fq -o MOCK8_20to26.fq -m 20 -x 26
```

Options:

```
-i <file.fq>      input file in fastq format
-o <file.fq>      output file in fastq format
-m <integer>      minimum length
-x <integer>      maximum length
```

The standard output (*see* **Note 9**) should be as follows:

```
43279723 sequences evaluated
36331025 sequences within 20 bp to 26 bp saved
```

or

```
/At_TRV$ perl perl_scripts/trim.fastq.length.
pl -i TRV8.fq -o TRV8_20to26.fq -m 20 -x 26
```

The standard output should be as follows:

```
44951890 sequences evaluated
41393135 sequences within 20 bp to 26 bp saved
```

The output is a fastq file containing sRNA sequences from 20 to 26 nts in length for each input file ("MOCK8_20to26.fq" and "TRV8_20to26.fq").

3.2.4 Removing Structural sRNAs Sequences

A significant fraction of sRNA reads in the sequenced set maps to structural non-coding RNA such as ribosomal RNA, transfer RNA, small nucleolar RNA or small nuclear RNA. It is highly recommended to filter out these reads before proceeding with the analysis. To do so, map your reads against a local Rfam database using Bowtie (Subheading 2.2.2, **item 8** and Subheading 2.2.3, **item 1**) by typing

```
/At_TRV$ bowtie Rfam_index/Rfam -q
MOCK8_20to26.fq -k 1 -v 1 --best -t --un
MOCK8noRfam.fq >MOCK_Rfam_maps.csv
```

Options:

`-q <file.fq>`	`input file in fastq format`
`-k <integer>`	`reports <integer> good alignment per read`
`--best`	`hits guaranteed best stratum; ties broken by quality`
`-v <integer>`	`report end-to-end hits with <integer> mismatches`
`-t`	`print wall-clock time taken by search phases`
`--un <file.fq>`	`write unaligned reads to a new file`
`> <file.csv>`	`redirect the mapping results to a file`

The standard output should be as follows:

```
Time loading forward index: 00:00:01
Time loading mirror index: 00:00:01
Time for 1-mismatch full-index search:
00:05:40
# reads processed: 36331025
# reads with at least one reported alignment:
2429303 (6.69%)
# reads that failed to align: 33901722
(93.31%)
```

```
Reported 2429303 alignments to 1 output
stream(s)
Time searching: 00:05:42
Overall time: 00:05:42
```

or

```
bowtie Rfam_index/Rfam -q TRV8_20to26.fq -k 1
-v 1 --best -t --un TRV8noRfam.fq >TRV_Rfam_
map.csv
```

The standard output should be as follows:

```
Time loading forward index: 00:00:01
Time loading mirror index: 00:00:00
Time for 1-mismatch full-index search:
00:06:34
# reads processed: 41393135
# reads with at least one reported alignment:
680324 (1.64%)
# reads that failed to align: 40712811
(98.36%)
Reported 680324 alignments to 1 output
stream(s)
Time searching: 00:06:35
Overall time: 00:06:35
```

(*see* **Note 10**).

The output files "MOCK_Rfam_maps.csv" and "TRV_Rfam_maps.csv" contain information of sRNA reads that aligned to Rfam sequences in tab-delimited format and can be removed to save disk space. "MOCK8noRfam.fq" and "TRV8noRfam.fq" contain sRNA sequences without matching to Rfam database in fastq format.

The results of this pipeline are summarized in Table 1.

3.3 Pipeline 2: Sample Infected with Known Viruses

This pipeline is devoted to classify and characterize vsRNAs derived from viruses for which a reference genome is available.

3.3.1 Mapping Against Viral Reference Genomes

To find the occurrences of a sRNA read along the viral reference genome, if no mismatches are allowed, reads are aligned against the genome sequence and mapping results are saved in a SAM format file. For viruses whose genomes are not fully resolved, EST or partial sequences can be used. In our test we confront our filtered sRNA reads against the TRV1 genome by typing:

```
/At_TRV$ bowtie virus_index/TRV1 -q
MOCK8noRfam.fq -k 1 -v 0 --best -t --al
MOCK8mapTRV1.fq -S MOCK8mapTRV1.sam
```

Table 1
Summary of raw data analysis and mapping against viral and plant genomes

	MOCK8			TRV8		
	Total	**%**	**Unique**	**Total**	**%**	**Unique**
Sequence tags	43,279,723			44,951,890		
Size filtering (20–26 nts)	36,331,025	83.94%		41,393,135	92.08%	
Mapping with Rfam						
Mapped	2,429,303	6.69%		680,324	1.64%	
Mo mapped	33,901,722	93.31%		40,712,811	98.36%	
Mapping with TRV1						
Mapped	6871	0.02%		17,450,229	42.86%	41,670
No mapped	33,894,851	99.98%		23,262,582	57.14%	
Mapping with TAIR10						
Mapped	19,055,755	56.21%	151,497	5,212,900	12.80%	144,491
No mapped	14,845,967	43.79%		35,499,911	87.20%	

Options:

```
-q <file.fq>       input file in fastq format
-k <integer>       reports <integer> good alignment per read
--best             hits guaranteed best stratum; ties broken by
                   quality
-v <integer>       report end-to-end hits with <integer>
                   mismatches
-t                 print wall-clock time taken by search phases
--al <file.fq>     write aligned reads to a new file
-S <file.sam>      write hits in SAM format
```

The standard output should be as follows:

```
Time loading forward index: 00:00:00
Time for 0-mismatch search: 00:01:37
# reads processed: 33901722
# reads with at least one reported alignment:
6871 (0.02%)
# reads that failed to align: 33894851
(99.98%)
Reported 6871 alignments to 1 output stream(s)
Time searching: 00:01:37
Overall time: 00:01:37
```

or

```
/At_TRV$ bowtie ./virus_index/TRV1 -q
TRV8noRfam.fq -k 1 -v 0 --best -t --al
```

```
TRV8mapTRV1.fq --un TRV8noTRV1.fq -S
TRV8mapTRV1.sam
```

The standard output should be as follows:

```
Time loading forward index: 00:00:00
Time for 0-mismatch search: 00:04:23
# reads processed: 40712811
# rbeads with at least one reported alignment:
17450229 (42.86%)
# reads that failed to align: 23262582
(57.14%)
Reported 17450229 alignments to 1 output
stream(s)
Time searching: 00:04:23
Overall time: 00:04:24
```

(*see* **Note 10** and Table 1).

The output files "MOCK8mapTRV1.fq" and "TRV8mapTRV1.fq" contain vsRNA sequences derived from TRV1 in fastq format. "MOCK8mapTRV1.sam" and "TRV8mapTRV1.sam" contain the alignment results in SAM file format.

3.3.2 Graphical Representation of vsRNAs Along Viral Genomes

The spatial distribution of vsRNAs of different size and strand orientation across the viral reference genome is used to visualize vsRNA diversity. MISIS [12] shows sRNA reads along the viral genome as an histogram where the *x*-axis represents the nucleotide position of the 5′ end (for sense vsRNAs) or 3′ end (for antisense vsRNAs) and *y*-axis represents the number of reads starting at each genomic position.

To display MISIS (Subheading 2.2.2, **item 10**) type

```
/At_TRV$ java -jar appMISIS.jar
```

Click "No" to reject loading the demo data.

First, create a folder in the working directory called "misis_TRV" to save the project. To load the vsRNA alignment file, click on "File" in the toolbar menu and then click on "Load bam/sam". In the emerging window called "Files selection", select the path to the SAM file created in the previous step (/At_TRV/TRV8mapTRV1.sam). Save the project in "misis_TRV" and select the path to the fasta file with the virus reference genome (At_TRV/virus_index/TRV1.fasta). Finally, select the sizes of the vsRNAs (minimal size 20 and maximal size 26) and press "run" to start the analysis.

The output of this analysis comprises eight files. The four files with the "pm" annotation contain data of vsRNA mappings with no mismatches to the reference TRV1 genome and are the following:

"TRV1_table_pm_20_26.txt" is a table with the counting of mapping reads at each nucleotide position and for each specific size class and strand orientation.

"TRV1_project_pm_20_26.b2btf" contains the whole project information.

"TRV1_pm_table_logo_20_26.txt" contains information to create the logos of all mapped nucleotides.

"TRV1_pm_table_logo_first_nt_20_26.txt" contains information to create the logos of the first mapped nucleotides.

The remaining four files contain the same information to find the mappings of a vsRNA read, but up to two mismatches are allowed. In our example, these files should be identical to the corresponding "pm" files because we only allowed perfect matches to the reference viral genome in the Bowtie alignment.

To visualize the results click on "File" in the toolbar menu, then "Load Project" and select the file "TRV1_project_pm_20_26.b2btf". The resulting histogram is shown in Fig. 1.

To compare vsRNA distribution between different samples, an additional table/project can be loaded in the same MISIS window. MISIS allows a broad range of visualization and analysis options. It is thus possible to magnify a certain area of the viral genome or modify the *Y*-axis scale if required, create a bar graph summary, display and export the reference viral sequence, load the logos (i.e., the nucleotide frequency at each genomic position), load and export the consensus reference viral sequence, display only specific vsRNA size classes or create graphical representation of the frequency of each nucleotide at the 5′ end. Furthermore, the count-

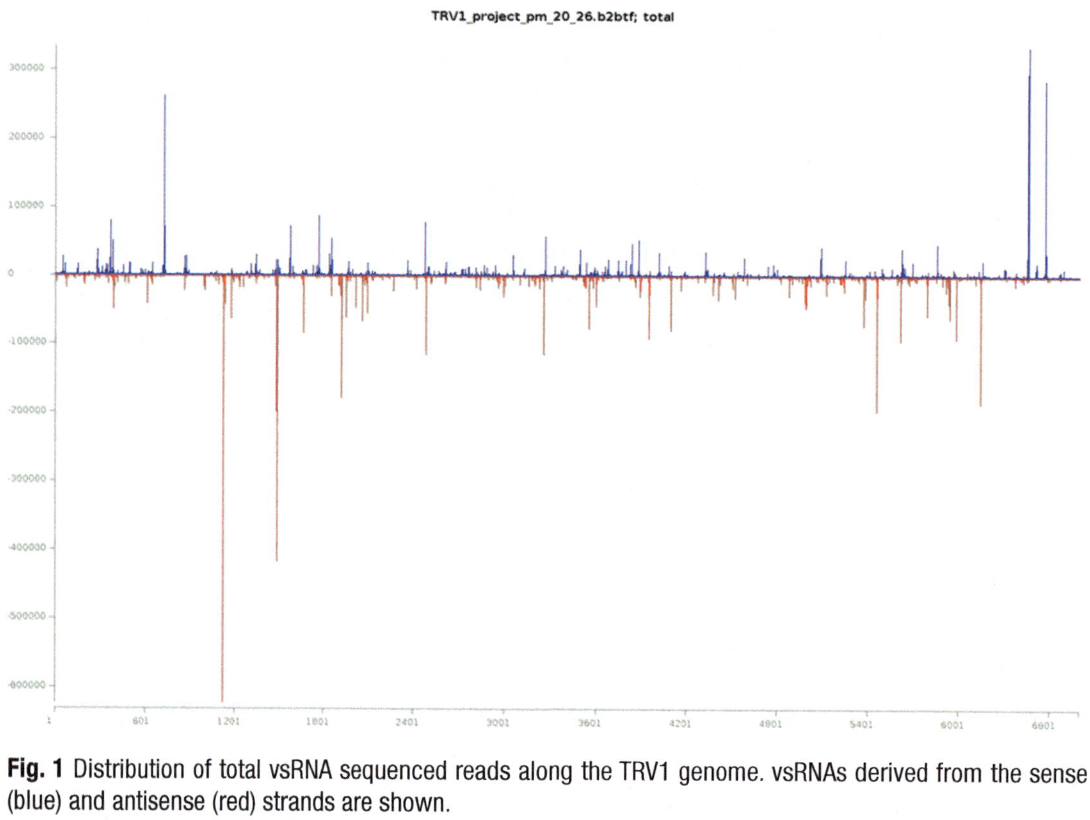

Fig. 1 Distribution of total vsRNA sequenced reads along the TRV1 genome. vsRNAs derived from the sense (blue) and antisense (red) strands are shown.

ing table file generated by MISIS ("TRV1_table_pm_20_26.txt") can be opened using a spreadsheet application such as Excel or Apache OpenOffice Calc for a thorough quantitative analysis.

3.3.3 Analysis of Specific Characteristics of the vsRNA Dataset

The study of distinctive attributes of the vsRNA dataset provides useful information to understand the origin and function of these molecules during the infection. For instance, the size distribution of total and unique sequence reads revealed the contribution of different DCL protein to vsRNA production. The distribution in strand polarity reflects the origin of vsRNAs from positive or negative viral strands and lets us infer the secondary structure of the potential RNA precursors. And the identity of the 5′ nucleotide position of the vsRNAs is used to predict association with multiple effector silencing complexes that mediate different effects on RNA and chromatin [17].

In this section, we applied a number of analyses that can be used to illustrate specific features of the vsRNAs population.

3.3.4 Size Distribution of Total vsRNAs

To obtain the frequency of total vsRNAs of different size classes, execute the Perl script "countfq_20to26.pl" (Subheading 2.2.2, **item 6**):

```
/At_TRV$ perl perl_scripts/countfq_20to26.pl
TRV8mapTRV1.fq
```

The standard output should be as follows:

```
Parsing fastq file, please wait…
Total sequences analyzed = 17450229
length      reads         percentage
20 nts      1072073       6.14360418995074
21 nts      11379993      65.2140037818415
22 nts      3827160       21.9318611807329
23 nts      407482        2.33510975701236
24 nts      508438        2.9136465773601
25 nts      147687        0.846332732940066
26 nts      107396        0.615441780162312
```

This result can be copied to a spreadsheet application for graphical representation as shown in Fig. 2a.

3.3.5 Collapsing Reads and Obtaining Size Distribution of Unique vsRNAs

To collapse reads and obtain a fasta file containing unique vsRNA sequences, execute the Perl script "fqtofaunique.pl" (Subheading 2.2.2, **item 6**):

```
/At_TRV$ perl perl_scripts/fqtofaunique.pl
TRV8mapTRV1.fq
```

The standard output should be as follows:

```
Parsing sequences, please wait…
17450229 total sequences evaluated
41670 unique sequences
```

The resulting file "TRV8mapTRV1-unique.fa" contains 41,670 unique sequences in fasta format and in alphabetical order where the header of each sequence is a random name followed by "_" and an integer that indicates the number of reads found (counts) for this specific vsRNA sequence.

For example, the first two fasta sequences of this file are as follows:

```
>SRR6156925.419430_9
AAAAACTATGTGTATGTTT
>SRR6156925.3321940_4
AAAAAAGAATCAACCGTCTTC
```

These two vsRNAs have been sequenced nine and four times, respectively.

To obtain the size distribution of unique vsRNA sequences from this fasta file use the Perl script "countfa_20to26.pl" (Subheading 2.2.2, **item 6**):

```
/At_TRV$ perl perl_scripts/countfa_20to26nts.
pl TRV8mapTRV1-unique.fa
```

The standard output should be as follows:

```
Parsing fasta file, please wait...
Total sequences analyzed = 41670
length      reads       percentage
20 nts      6694        16.0643148548116
21 nts      10161       24.3844492440605
22 nts      8215        19.7144228461723
23 nts      4715        11.3150947924166
24 nts      5481        13.1533477321814
25 nts      3318        7.9625629949604
26 nts      3086        7.40580753539717
```

This result can be copied to a spreadsheet application for graphical representation as shown in Fig. 2a.

3.3.6 Polarity Distribution of Total and Unique vsRNAs

The dataset generated by MISIS in **step 2** is then used to examine the sense or antisense polarity distribution of total and unique vsRNAs of each size. To extract data from this file execute the Perl script "misis_table_analysis.pl" (Subheading 2.2.2, **item 6**):

```
/At_TRV$ perl perl_scripts/misis_table_analy-
sis.pl misis_TRV/TRV1_table_pm_20_26.txt
```

The standard output should be as follows:

```
#### sense vsRNAs ####
length   total     percentage          unique   percentage
20 nts   568554    6.62549577249614    3439     16.0505927377952
21 nts   5203849   60.6416972709862    5115     23.8728647437693
22 nts   2204903   25.6942621198058    4155     19.3923270792495
23 nts   191034    2.22616490149226    2476     11.5560533930738
24 nts   273287    3.18467878720078    2850     13.301596191543
25 nts   84392     0.983440164403899   1758     8.20498459815178
26 nts   55286     0.644260983614963   1633     7.62158125641744
Sum      8581305                       21426
```

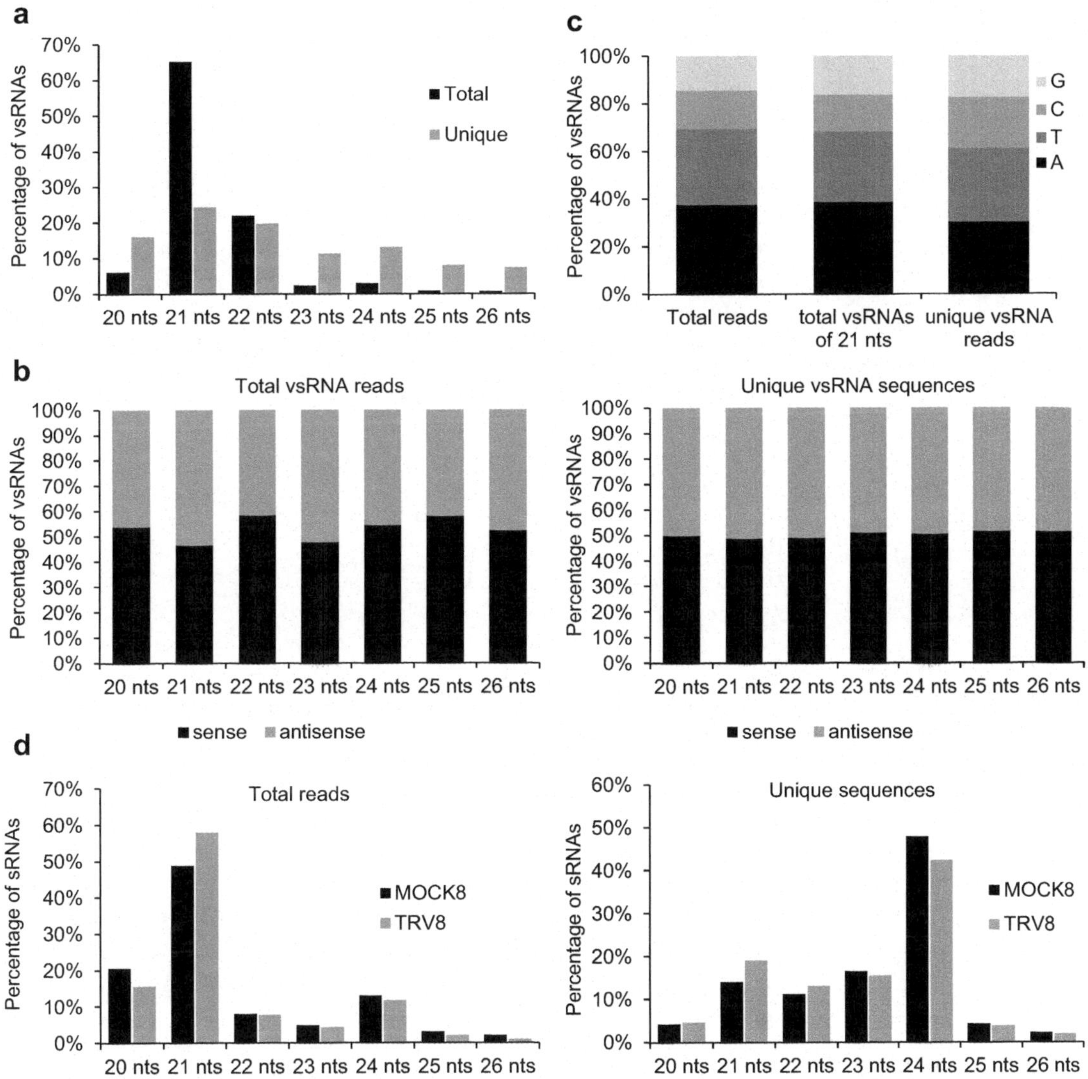

Fig. 2 (**a**) Size distribution of total and unique vsRNA reads. (**b**) Distribution of total and unique vsRNA reads of each size class derived from positive and negative viral strands. (**c**) 5′ end nucleotide composition of vsRNAs. (**d**) Size distribution of total and unique endogenous sRNA reads from mock-inoculated (MOCK8) and TRV-infected (TRV8) plants

```
#### antisense vsRNAs ####
length   total     percentage          unique   percentage
20 nts   503519    5.67734034027127    3255     16.0788381742739
21 nts   6176144   69.6380304984009    5046     24.9259039715471
22 nts   1622257   18.2914748170128    4060     20.0553250345781
23 nts   216448    2.44052153339007    2239     11.0600671803991
24 nts   235151    2.65140393581003    2631     12.9964433906343
25 nts   63295     0.713671692304501   1560     7.70598695909899
26 nts   52110     0.587557182810451   1453     7.17743528946848
sum      8868924                       20244
both     17450229                      41670
```

This result can be copied to a spreadsheet application for graphical representation as shown in Fig. 2b.

3.3.7 5′ End Nucleotide Composition of Total and Unique vsRNAs

MISIS generates bar graphs in which the relative frequency of each 5′ terminal nucleotide of different categories of vsRNAs is represented. This analysis can be done using full-length reference or in user-defined specific viral regions, including sense or antisense viral strands. However, it could be interesting to have the raw data of the 5′ end nucleotide composition of vsRNAs if further statistical analyses are required or for a custom graphical representation with a user-defined format using a spreadsheet application. This data can be extracted with the Perl script "nucomp_fq.pl" (Subheading 2.2.2, **item 6**):

```
/At_TRV$ perl perl_scripts/nucomp_fq.pl
TRV8mapTRV1.fq
```

The standard output should be as follows:

```
Checking sequences…
Total sequences analyzed: 17450229
5'_end_nt    counts     percentage
A            6542630    37.4930896322335
T            5561653    31.8715187061442
C            2817851    16.1479313537948
G            2528095    14.4874603078275
N            0          0
```

This result can be copied to a spreadsheet application for graphical representation as shown in Fig. 2c.

Moreover, it is possible to report the nucleotide composition of vsRNAs of each size class by combining the output of the Perl script used to filter sRNAs by length in Pipeline 1 (Subheading 3.2, **step 4**) ("trim.fastq.length.pl") and the above Perl script. For example, type the following command to obtain the 5′ nucleotide composition of total vsRNAs of 21 nts:

```
/At_TRV$ perl perl_scripts/trim.fastq.length.pl
-i TRV8mapTRV1.fq -o TRV8_21nt.fq -m 21 -x 21 |
perl perl_scripts/nucomp_fq.pl TRV8_21nt.fq
```

The output file "TRV8_21nt.fq" is a fastq file containing TRV1-derived vsRNAs of 21 nts in length.

The 5′ nucleotide composition of vsRNAs of 21 nts should be (Fig. 2c) as follows:

```
Checking sequences…
Total sequences analyzed: 11379993
5'_end_nt    counts     percentage
A            4392230    38.5960694351921
T            3383110    29.7285771616907
C            1747991    15.3602115572479
G            1856662    16.3151418458693
N            0          0
```

To obtain the nucleotide composition of unique vsRNA reads use the Perl script "nucomp_fa.pl" (Subheading 2.2.2, **item 6**):

```
/At_TRV$ perl ./perl_scripts/nucomp_fa.pl
TRV8mapTRV1-unique.fa
```

The result, as shown in Fig. 2c, should be as follows:

```
Checking sequences...
Total sequences analyzed: 41670
5'_end_nt    counts       percentage
A            12602        30.2423806095512
T            12886        30.9239260859131
C            8955         21.4902807775378
G            7227         17.3434125269978
N            0            0
```

3.4 Pipeline 3: Sample Infected with Unknown Viruses

This pipeline is focused on the characterization of vsRNAs derived from viruses for which a previous reference genome is not available or from unknown viruses present in our plant sample. Given that vsRNAs produced in plants upon viral infections are overlapping in sequence, they can be assembled into long contiguous fragments of the viral genome and can be used for virus identification.

3.4.1 Detection of Viruses in a Plant Sample

VirusDetect is a software package that analyzed large-scale sRNA datasets to perform reference-guided assembly as well as de novo assembly using the sRNA assembly program Velvet [18]. The VirusDetect online tool can be used by following the next simple steps (*see* **Note 11**):

1. Register a user account in VirusDetect web page: http://virus-detect.feilab.net/cgi-bin/virusdetect/vd_help.cgi.
2. Login at VirusDetect Online tool.
3. Upload the input fastq file. Use as input file the fastq file generated after removing noncoding RNA contaminants in Pipeline 1 (Subheading 3.2, **step 5** "TRV8noRfam.fq").

   Alternatively, fastq files can be uploaded via ftp:

```
/At_TRV$ ftp virusdetect.feilab.net
Name: your_username
Password: your_password
ftp> put TRV8noRfam.fq
```

4. Upload a fasta file containing sequences of your plant reference transcriptome/genome in case it is not already loaded (named the file with the prefix "host_"). In our example, Arabidopsis is already in the list of predefined hosts, so select it.
5. Analyze the data using the plant virus database at either 95 or 97 or 100% of sequence identity at the nucleotide level with respect to the reference viral sequences (95% in this example).

6. Once the analysis is completed, download the resulting folder "result_TRV8noRfam.fq" and unpack it.

Output files:

"blastn.html" lists the genome reference viruses, identified by BLASTN against the National Center for Biotechnology Information (NCBI) database, that align with virus contigs generated upon assembly of vsRNA reads. In our test, the presence of TRV was confirmed in the TRV-infected sample named TRV8 with normalized sequencing depths (reads per million, RPM) of 162.6 and 152.4 for TRV1 and TRV2, respectively (Table 2). In addition, BLASTN identified several groups of contigs that were identical or highly homologous to the nucleotide genome sequence of an unexpected virus named *Turnip mosaic virus* (TuMV). Deep sequencing of the vsRNA library revealed TuMV-derived vsRNAs but at a much lower normalized depth (0.7) (*see* **Note 12** and Table 2).

"blastx.html" lists the reference viruses, identified by BLASTX, that align with viral contigs generated upon assembly of the vsRNA sequences. In our example, viruses other than TRV or TuMV were not identified by BLASTX, and consequently, this file was not created.

"undetermined.html" lists contigs with no homologies to viral reference genome entries of the non-redundant databases of the GenBank. In this file, clusters of potential vsRNA contigs are highlighted in green. In our example, three contigs were selected as potential viral contigs with 21- and 22-nt vsRNA enrichment higher than 53% (data not shown).

"undetermined_blast.html" lists assembled contigs with hits in the virus reference database but that don't fully satisfy the default filter options. This includes a coverage cutoff of a virus contig with respect to the reference virus sequences higher than 0.75, a coverage cutoff of a reported virus reference sequence with respect to the assembled virus contigs higher than 0.1 or a depth cutoff of a reported virus reference lower than 5. Potential novel virus contigs formed by high percentage of mapping vsRNAs of 21 and 22 nts

Table 2
VirusDetect online tool results summary

Reference	Length	Coverage (%)	Depth (Norm)	% identity	Description
AF169561	9835	7617 (77.4)	0.7	99.93	*Turnip mosaic virus*
AF166084	6791	6771 (99.7)	162.6	99.59	*Tobacco rattle virus* RNA1
Z36974	3855	1642 (42.6)	152.4	99.43	*Tobacco rattle virus* RNA2

are highlighted in green. In our analysis, three contigs with sizes ranging from 84 nts to 604 nts are found, of which two of them share sequence identity the RNA2 of TRV and *Pea early browning tobravirus*, respectively (data not shown).

"blastn.reference.fa", "blastx.reference.fa", "contig_sequences.blastn.fa", "contig_sequences.blastx.fa", "contig_sequences.fa", and "contig_sequences.undetermined.fa" are fasta files of the reference virus sequences identified in the analysis, contigs of assembled reads or contigs with hits to virus reference sequences, respectively.

"blastn.sam", "blastx.sam", "blastn.xls", and "blastx.xls" are sam format files and Excel files containing the alignment information of each contig.

Folders "blastn_references" or "blastx_references" contains html files with detailed alignment information of the virus contig to the reference virus sequences.

3.4.2 Analysis of vsRNA Populations

Once all viruses infecting any given sample have been identified, proceed as described in Pipeline 2 to characterize the vsRNA population.

3.5 Pipeline 4: Analysis of Virus-Associated Endogenous Small RNAs

A significant fraction of the sRNA transcriptome is altered in plants exposed to viral attacks. This includes the massive accumulation of virus-associated endogenous siRNAs that reflects widespread silencing of host genes during viral infections [19]. Pipeline 4 performs a comparative analysis of the populations of host-specific endogenous sRNAs between infected and noninfected plants.

3.5.1 Mapping sRNA Reads Against the Plant Reference Transcriptome

Use TAIR (version 10) as cDNA reference sequence to search for Arabidopsis sRNAs that map to protein coding genes using Bowtie (*see* Subheading 2.2.2, **item 7** and Subheading 2.2.3, **item 3**):

```
/At_TRV$ bowtie plant_index/TAIR10cdna -q
MOCK8noRfam.fq -k 1 -v 0 --best -t --al
MOCK8mapAt.fq -S MOCK8mapAt.sam
```

Options:

```
-q <file.fq>    input file in fastq format
-k <integer>    reports <integer> good alignment per read
--best          hits guaranteed best stratum; ties broken by
                quality
-v <integer>    report end-to-end hits with <integer> mismatches
-t              print wall-clock time taken by search phases
--al <file.fq>  write aligned reads to a new file
-S <file.sam>   write hits in SAM format
```

The standard output should be as follows:

```
Time loading forward index: 00:00:00
Time for 0-mismatch search: 00:05:18
# reads processed: 33901722
# reads with at least one reported alignment:
19055755 (56.21%)
# reads that failed to align: 14845967
(43.79%)
Reported 19055755 alignments to 1 output
stream(s)
Time searching: 00:05:18
Overall time: 00:05:18
```

or

```
/At_TRV$ bowtie plant_index/TAIR10cdna -q
TRV8noRfam.fq -k 1 -v 0 --best -t --al
TRV8mapAt.fq -S TRV8mapAt.sam
```

The standard output should be as follows:

```
Time loading forward index: 00:00:00
Time for 0-mismatch search: 00:05:44
# reads processed: 40712811
# reads with at least one reported alignment:
5212900 (12.80%)
# reads that failed to align: 35499911
(87.20%)
Reported 5212900 alignments to 1 output
stream(s)
Time searching: 00:05:44
Overall time: 00:05:44
```

See **Note 10**.

The mapping results are summarized in Table 1.

3.5.2 Collapsing Reads to Unique sRNA Sequences

To collapse reads and obtain fasta files containing unique sRNA sequences use the Perl script "fqtofaunique.pl" (Subheading 2.2.2, **item 5**) as follows:

```
/At_TRV$ perl perl_scripts/fqtofaunique.pl
MOCK8mapAt.fq
```

The standard output should be as follows:

```
Parsing sequences, please wait…
19055755 total sequences evaluated
151497 unique sequences
```

or

```
/At_TRV$ perl perl_scripts/fqtofaunique.pl
TRV8mapAt.fq
```

The standard output should be as follows:

```
Parsing sequences, please wait…
5212900 total sequences evaluated
144491 unique sequences
```

The results are summarized in Table 1.

3.5.3 Size Distribution of Total and Unique sRNA Sequences

Use the Perl scripts described in Pipeline 2 to obtain the size distribution of total reads or unique sequences (Subheading 3.2, **steps 1** and **3**, respectively):

```
/At_TRV$ perl perl_scripts/countfq_20to26nts.
pl MOCK8mapAt.fq
```

The standard output should be as follows:

```
Parsing fastq file, please wait...
Total sequences analyzed = 19055755
length     reads      percentage
20 nts     3915193    20.5459872883546
21 nts     9296099    48.7836824098547
22 nts     1526113    8.00867244567324
23 nts     897877     4.71184164573904
24 nts     2463915    12.9300308489483
25 nts     567734     2.97933091604085
26 nts     388824     2.04045444538933
```

or

```
/At_TRV$ perl perl_scripts/count_fq_20to26nts.
pl TRV8mapAt.fq
```

The standard output should be as follows:

```
Parsing fastq file, please wait...
Total sequences analyzed = 5212900
length       reads          percentage
20 nts       810496         15.5478908093384
21 nts       3014825        57.8339312091159
22 nts       403928         7.74862360682154
23 nts       220534         4.23054345949472
24 nts       604162         11.5897485085077
25 nts       106014         2.03368566440945
26 nts       52941          1.01557674231234
```

or

```
/At_TRV$ perl perl_scripts/countfa_20to26nts.
pl MOCK8mapAt-unique.fa
```

The standard output should be as follows:

```
Parsing fasta file, please wait...
Total sequences analyzed = 151497
length       reads          percentage
20 nts       6363           4.20008316996376
21 nts       21260          14.0332811870862
22 nts       16779          11.0754668409275
23 nts       24990          16.4953761460623
24 nts       72465          47.8326303491158
25 nts       6303           4.16047842531535
26 nts       3337           2.20268388152901
```

or

```
/At_TRV$ perl perl_scripts/countfa_20to26nts.
pl TRV8mapAt-unique.fa
```

The standard output should be as follows:

```
Parsing fasta file, please wait...
Total sequences analyzed = 144491
length      reads       percentage
20 nts      6672        4.61758863873875
21 nts      27449       18.9970309569454
22 nts      18872       13.0610211016603
23 nts      22307       15.4383317992124
24 nts      61062       42.2600715615505
25 nts      5386        3.72756780699144
26 nts      2743        1.89838813490114
```

Data regarding size distribution of total reads and unique plant-derived sRNA sequences can be copied to a spreadsheet application for graphical representation as shown in Fig. 2d.

3.5.4 Graphical Representation of sRNAs Mapping to the Reference Transcriptome

Use the IGV [20] (*see* Subheading 2.2.2, **item 11**) desktop application to visualize sRNA mappings against the reference transcriptome by typing

```
At_TRV/$ java -Xmx750m -jar IGV_2.3.97/igv.jar
```

Options:

```
-Xmx(integer)m Start IGV with (integer) MB of memory.
```

In the "Genomes" tab of the menu tool bar, click on "Load Genome from file" and search for the Arabidopsis TAIR10 cDNA fasta file used to create the Bowtie index (*see* Subheading 2.2.3, **item 3**) ("/At_TRV/plant_index/TAIR10_cdna_modif.fasta"). Then, open it.

IGV requires sRNA alignment files in BAM format. Prior to loading the BAM files into IGV it is necessary to sort them by creating indexed files using SAMtools [11] (Subheading 2.2.2 **item 9**) as follows:

To transform SAM files to BAM format, type

```
At_TRV/$ samtools view -b -h -S MOCK8mapAt.sam -o MOCK8mapAt.bam
```

or

```
At_TRV/$ samtools view -b -h -S TRV8mapAt.sam -o TRV8mapAt.bam
```

Options:

```
-b output file in BAM format
-h print header for the SAM file
-S input is SAM format
-o output file name
```

To sort the BAM files, type

```
/At_TRV$ samtools sort MOCK8mapAt.bam MOCK8mapAt.sort
```

or

```
/At_TRV$ samtools sort TRV8mapAt.bam TRV8mapAt.sort
```

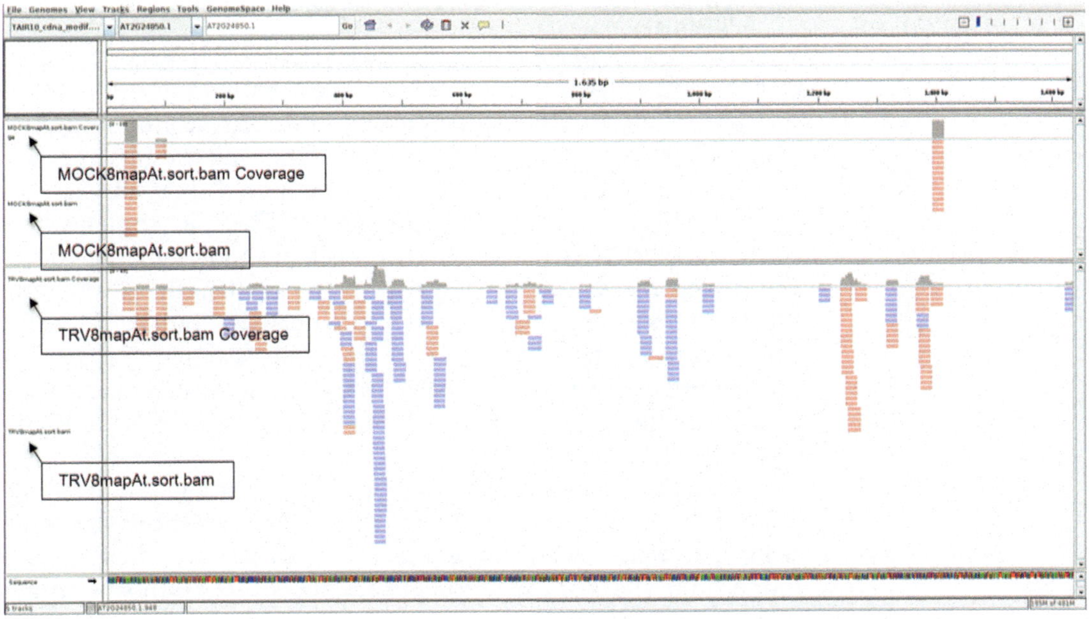

Fig. 3 sRNA mapping from mock-inoculated (MOCK8) and TRV-infected (TRV8) plants along the Arabidopsis transcript locus AT4G24690.1 using the IGV snapshot viewer. sRNAs derived from the sense (red) and antisense (blue) strands are shown

To create an index of the BAM files, type

```
/At_TRV$ samtools index MOCK8mapAt.sort.bam
```

or

```
/At_TRV$ samtools index TRV8mapAt.sort.bam
```

Finally, to upload the sRNA-mapping BAM files select "File" in the toolbar menu and then "Load from file", search for the sorted bam files generated previously ("/At_TRV/MOCK8mapAt.sort.bam" or "/At_TRV/TRV8mapAt.sort.bam") and open them. Loading both samples at the same time is useful to compare sRNA abundance between noninfected and infected samples for a specific Arabidopsis transcript. As an example, Fig. 3 illustrates the output alignment of sRNAs from both MOCK8 and TRV8 samples along the Arabidopsis locus AT2G24850.1.

3.5.5 Quantification of Endogenous sRNA Abundance for Each Locus

To count the number of total sRNAs produced at any given gene locus, use the eXpress program [14] (*see* **Note 13**) (Subheading 2.2.2, **item 12**) as follows:

```
/At_TRV$ express-1.5.1-linux_x86_64/ex-
press plant_index/TAIR10_cdna_modif.fasta
MOCK8mapAt.sam -o express_MOCK8
```

or

```
/At_TRV$ express-1.5.1-linux_x86_64/express
plant_index/TAIR10_cdna_modif.fasta TRV8mapAt.
sam -o express_TRV8
```

Options:

```
-o Sets the name of the directory in which eX-
press will write all of its output
```

The following files will be created in the directories "express_MOCK" and "express_TRV": "params.xprs" and "results.xprs".

The file "results.xprs" has a tab-delimited format and can be opened using a spreadsheet application. This file contains the counting of sRNAs mapping for each locus, among other data. Data from the second ("target_id"), third ("length"), and fifth ("tot_counts") columns are used in this analysis. "target_id" is the ID given to the transcript locus in the input fasta file, "length" is the length in base pairs of each transcript sequence, and "tot_counts" is the number of sRNA mapping (uniquely or ambiguously) to each transcript.

Next, the Perl script "srna_count_comparison.pl" (Subheading 2.2.2, **item 6**) is used to integrate the sRNA abundance from both samples in a new tab-delimited file. This script also calculates the fold change (FC) or the ratio between the sRNA counts per transcript in infected versus non-infected samples and the $\log_2$FC (FCs lower than 1 become negative, while FCs higher than 1 become positives for graphical representation).

To execute the script, type the following in the command line:

```
/At_TRV$ perl perl_scripts/srna_count_com-
parison.pl express_MOCK8/results.xprs express_
TRV8/results.xprs MOCK TRV
```

The standard output should be as follows:

```
################################################
########################
Number of transcripts analyzed: 41671
Number of transcripts without sRNAs mapping
any sample, MOCK and TRV: 17863
Number of transcripts with sRNAs mapping only
to sample MOCK: 4402
Number of transcripts with sRNAs mapping only
to sample TRV: 5636
Number of transcripts with sRNAs mapping to
both samples: 13770
################################################
######################
The job finished without any problem, congratu-
lations!
```

The output file "MOCK&TRV_comparison.csv" has a tab-delimited format and can be opened with a spreadsheet application. It contains the following columns: "transcript_id", "transcript_length", "#_of_sRNAs_mappings_in_MOCK", "#_of_sRNAs_mappings_in_TRV", "Fold_change:TRV:MOCK", and "log2_FC".

In our dataset, there are 186 Arabidopsis transcripts for which the abundance of sRNAs in the infected sample is higher ($log_2FC > 2$) than in the noninfected sample, whereas 5636 transcripts produce sRNAs in virus-infected plants but not in mock-inoculated ones (*see* **Note 14**).

4 Notes

1. All the analyses described in this chapter have been done using Ubuntu version 16.04. However, operating systems different to Linux can also be used by installing and executing a Virtual Machine (i.e., Virtualbox) in a Windows PC or MacOS X environment. Alternatively, MacOS versions are available for most of the programs used here, although the execution procedure and program installation may differ.
2. Generally, programs or applications are installed and executed using text-based commands in a Terminal (also called command line or command prompt). Press Ctrl+Alt+t to open a Terminal in Ubuntu 16.04. The Linux user must have root (or superuser) access to run administrative tasks (like installing programs). Here, "`$`" or "`working_directory/$`" are used to indicate what has to be written in the command line in order to run a program or execute a Unix command. Knowledge of Linux/Unix command usage is not required, albeit advisable, to perform these analyses.
3. If you get any bug during compilation you may need to install the zlib and bzlib2 libraries by typing

```
$sudo apt-get install libz-dev
$sudo apt-get install libbz2-dev
$sudo apt-get install zlib1g-dev
```

4. The system requires the complete location (or path) of the executable files to run the corresponding programs. To save time and circumvent potential setbacks during the execution of a program, it is highly recommended to install programs in the so-called working directory. Alternative, it is possible to add a specific directory to $PATH variable, which is an environment variable where the system search for the executable, by typing

```
$ export PATH=/path_to_your_executable_
file:$PATH
```

5. Noncoding RNA sequences from different organisms can be downloaded from other publicly available databases such as RNAcentral (http://rnacentral.org/), lncrnadb (), tRNADB-CE (http://trna.ie.niigata-u.ac.jp/cgi-bin/trnadb/index.cgi), or ncRNA database (http://biobases.ibch.poznan.pl/ncRNA/), among others.

6. From Rfam version 12 onward the file "Rfam.fasta.gz" has been replaced with a folder called "fasta_files" containing fasta sequence files. To download all files of the current version (version 13, accessed in October 2017), type

```
/working_directory$ wget ftp://ftp.ebi.ac.uk/
pub/databases/Rfam/CURRENT/fasta_files/RF*.fa.gz
```

To merge all into one unique file, type

```
/working_directory$ gunzip RF*.fa.gz
/working_directory$ rm RF*.fa.gz
/working_directory$ cat RF*.fa > Rfam-CURRENT.
fasta
/working_directory$ rm RF*.fa
```

7. NGS QC toolkit can be used for data obtained from any other platform [21].
8. Illumina adapter sequences are available at https://support.illumina.com/content/dam/illumina-support/documents/documentation/chemistry_documentation/experiment-design/illumina-adapter-sequences-1000000002694-10.pdf. The 3′ adapters RPI1 to RPI48 used by TruSeq Small RNA Library Prep Kit (Illumina) for library preparation are listed in the "RNA PCR Index Primers" section. The 3′ adapters D701 to D712 used by SMARTer smRNA-Seq Kit for Illumina (Clontech Laboratories) are listed in the "Illumina TruSeq HT index" section. To remove the 3′ adapter sequence using AdapterRemoval (Subheading 2.2.2, **item 5**) type the following in the command prompt:

```
$ AdapterRemoval --file1 fastq_file.fq.gz
--adapter1 "your_adapter_sequence"
```

The output of this program includes three files:

"your_output.discarded" contains sRNA sequences smaller than 14 nts in length.

"your_output.settings" contains log data, trimming statistics and sRNA length distribution.

"your_output.truncated" contains sRNA sequences larger than 15 nts in length.

Change the name of the files with another fastq file name before executing the program again in order to avoid overwriting.

Other software tools such as Quadapt [22] or Trimmomatic [23] can be used to trim adaptor sequences or to discard low-quality reads.

9. The Standard output (or stdout) in Unix refers to the standardized data flow that is produced by command line programs. The default destination is the displayed screen of the computer terminal that executes the program. Here, the

standard output of the programs used gives some information about the current status of the running program or directly the result of a specific command line program. The standard output can be easily redirected to a file by adding to the command line options "> file_name".

10. The execution time may be different according to the computer capacity. Here we used a desktop computer with Core i7-6800K CPU, 32 GB of RAM and 6 threads.
11. Detailed information of the workflow followed by VirusDetect online tool is shown here: http://virusdetect.feilab.net/cgi-bin/virusdetect/vdo_help.cgi. Alternatively, VirusDetect standalone program can be used to analyze your samples. See, for information of system requirements, dependences, VirusDetect installation and usage. Other tools such as VirFind [24] or ViRome [25] are also designed for virus discovery through vsRNA assembly or alignment against virus reference sequences, respectively.
12. Our sample referred to as TRV8 was harvested from plants that had been inoculated in our lab with an infectious TRV clone and maintained in the growth chamber until collection. We envisioned two possible scenarios to explain the presence of sRNAs with sequence homology to the TuMV genome in this sample: (1) wrong assignment to the TRV-infected library during sorting of multiplexed samples [26], (2) contamination during plant manipulation including RNA extraction and/or library preparation.
13. The eXpress software is available online, albeit updated versions are not expected in the future. Other tools such as Kallisto [27]can be used for quantification of sRNAs mapping to target transcripts. However, the output file of Kallisto should be modified to have exactly the same format as the output file of eXpress when using the Perl script "sRNA_count_comparison.pl".
14. Introducing sample replicates is strongly recommended to minimize variability in the sRNA counts due to technical manipulation or biological variation and to increase reproducibility and robustness of the statistical tests [4].

Acknowledgments

This work was supported by National Research Grants BIO2012-39973, PCIN-2013-064, and BIO2015-70752 from Ministerio de Economía y Competitividad, Spain.

References

1. Ding SW, Voinnet O (2007) Antiviral immunity directed by small RNAs. Cell 130:413–426
2. Llave C (2010) Virus-derived small interfering RNAs at the core of plant-virus interactions. Trends Plant Sci 15:701–707
3. Head SR, Kiyomi Komori H, LaMere SA et al (2014) Library construction for next-generation sequencing: Overviews and challenges. BioTechniques 56:61–77
4. McCormick KP, Willmann MR, Meyers BC (2011) Experimental design, preprocessing, normalization and differential expression analysis of small RNA sequencing experiments. Silence 2:2
5. Fahlgren N, Sullivan CM, Kasschau KD et al (2009) Computational and analytical framework for small RNA profiling by high-throughput sequencing. RNA 15:992–1002
6. Farazi TA, Brown M, Morozov P et al (2012) Bioinformatic analysis of barcoded cDNA libraries for small RNA profiling by next-generation sequencing. Methods 58:171–187
7. Leinonen R, Sugawara H, Shumway M (2011) The sequence read archive. Nucleic Acids Res 39:2010–2012
8. Andrews S (2017) FastQC: a quality control tool for high throughput sequence data. https://www.bioinformatics.babraham.ac.uk/projects/fastqc/. Accessed 30 Oct 2017
9. Lindgreen S (2012) AdapterRemoval: easy cleaning of next generation sequencing reads. BMC Res Notes 5:337–343
10. Langmead B, Trapnell C, Pop M et al (2009) Ultrafast and memory-efficient alignment of short DNA sequences to the human genome. Genome Biol 10:R25
11. Li H, Handsaker B, Wysoker A et al (2009) The sequence alignment/Map format and SAMtools. Bioinformatics 25:2078–2079
12. Seguin J, Otten P, Baerlocher L et al (2014) MISIS: a bioinformatics tool to view and analyze maps of small RNAs derived from viruses and genomic loci generating multiple small RNAs. J Virol Methods 195:120–122
13. Roberts A, Pachter L (2017) eXpress: streaming quantification for high-throughput sequencing. https://pachterlab.github.io/eXpress/overview.html. Accessed 30 Oct 2017
14. Donovan WP, Zhang Y, Howell MD (2011) Large-scale sequencing of plant small RNAs. Methods Mol Biol 744:159–173
15. Hafner M, Renwick N, Farazi TA et al (2012) Barcoded cDNA library preparation for small RNA profiling by next-generation sequencing. Methods 58:164–170
16. Babu CVS, Gassmann M (2016) Assessing integrity of plant RNA with the Agilent 2100 Bioanalyzer System. Agil Appl Note, Waldbronn, pp 5990–8850E
17. Donaire L, Wang Y, Gonzalez-Ibeas D et al (2009) Deep-sequencing of plant viral small RNAs reveals effective and widespread targeting of viral genomes. Virology 392:203–214
18. Zheng Y, Gao S, Padmanabhan C et al (2017) VirusDetect: an automated pipeline for efficient virus discovery using deep sequencing of small RNAs. Virology 500:130–138
19. Cao M, Du P, Wang X et al (2014) Virus infection triggers widespread silencing of host genes by a distinct class of endogenous siRNAs in Arabidopsis. Proc Natl Acad Sci U S A 111:14613–14618
20. Robinson JT, Thorvaldsdóttir H, Winckler W, Guttman M, Lander ES, Getz G, Mesirov JP (2011) Integrative genomics viewer. Nat Biotechnol 29(1):24–26
21. Patel RK, Jain M (2012) NGS QC toolkit: a toolkit for quality control of next generation sequencing data. PLoS One 7:e30619
22. Martin M (2011) Cutadapt removes adapter sequences from high-throughput sequencing reads. EMBnet J 17:10–12
23. Bolger AM, Lohse M, Usadel B (2014) Trimmomatic: a flexible trimmer for Illumina sequence data. Bioinformatics 30:2114–2120
24. Ho T, Tzanetakis IE (2014) Development of a virus detection and discovery pipeline using next generation sequencing. Virology 471–473:54–60
25. Watson M, Schnettler E, Kohl A (2013) ViRome: an R package for the visualization and analysis of viral small RNA sequence datasets. Bioinformatics 29:1902–1903
26. Illumina, Inc (2017) Effects of index misassignment on multiplexing and downstream analysis. https://www.illumina.com/content/dam/illumina-marketing/documents/products/whitepapers/index-hopping-white-paper-770-2017-004.pdf. Accessed 30 Oct 2017.
27. Bray NL, Pimentel H, Melsted P et al (2016) Near-optimal probabilistic RNA-seq quantification. Nat Biotechnol 34:525–527

Chapter 12

A Sensitized Genetic Screen to Identify Novel Components and Regulators of the Host Antiviral RNA Interference Pathway

Zhongxin Guo, Xian-Bing Wang, Wan-Xiang Li, and Shou-Wei Ding

Abstract

RNA interference (RNAi) acts as a natural defense mechanism against virus infection in plants and animals. Much is known about the antiviral function of the core RNAi pathway components identified mostly by genetic screens based on specific RNAi of cellular mRNAs. Here we describe a sensitized genetic screening system for the identification of novel components and regulators in the antiviral RNAi pathway established in the model plant species *Arabidopsis thaliana*. Our genetic screen identifies antiviral RNAi (*avi*)-defective *Arabidopsis* mutants that develop visible disease symptoms after infection with CMV-Δ2b, a Cucumber mosaic virus mutant deficient in the expression of its viral suppressor of RNAi. Loss of RNAi suppression renders CMV-Δ2b highly susceptible to antiviral RNAi so that it replicates to high levels and induces disease development only in *avi* mutants. This chapter provides the methods for the propagation of CMV-Δ2b, preparation of the mutant plants for virus inoculation, identification and characterization of *avi* mutants, and cloning of the genes responsible for the mutant phenotype by either the genetic linkage to T-DNA insertion or a mapping-by-sequencing approach.

Key words Antiviral RNAi, siRNA, Innate immunity, *avi* mutant, Causal gene identification

1 Introduction

RNA interference (RNAi) by Dicer-dependent small interfering RNAs (siRNAs) and microRNAs (miRNAs) is a highly conserved gene regulatory mechanism in eukaryotic organisms [1]. An important function of RNAi in fungi, plants and invertebrates is to act as a natural antiviral defense mechanism [2]. Recent studies have also demonstrated a similar antiviral function of RNAi in mammals [3–6]. In antiviral RNAi, the virus-specific double-strand RNA (dsRNA) produced during RNA and DNA virus infection is processed by Dicer into 21- to 24-nucleotide virus-derived siRNAs [7, 8]. Similar to cellular miRNAs and siRNAs, these viral siRNAs are loaded into the effector protein Argonaute (AGO) in RNA-induced silencing complex (RISC) to guide specific virus clearance by RNAi [2].

Kappei Kobayashi and Masamichi Nishiguchi (eds.), *Antiviral Resistance in Plants: Methods and Protocols*, Methods in Molecular Biology, vol. 2028, https://doi.org/10.1007/978-1-4939-9635-3_12,

Antiviral RNAi has been extensively characterized in plants, insects, and nematodes [2]. In plants and nematodes, the viral primary siRNAs processed from the viral dsRNA replicative intermediates trigger further siRNA amplification by host RNA-dependent RNA polymerase (RDR), producing a population of the viral secondary siRNAs that are much more abundant than the viral primary siRNAs and essential for antiviral RNAi [9, 10]. In contrast, fruit flies and mosquitoes may depend on the host-mediated reverse transcription of the viral RNAs into viral DNA for the amplification of viral siRNAs [11–13].

Much is known about the antiviral function and mechanism of the core RNAi pathway components, several of which were identified by the unbiased genetic screens based on specific RNAi of cellular mRNAs [14–18]. We have recently reported the identification of two families of Arabidopsis genes required for RNAi by viral siRNAs in the infected plants but not cellular miRNAs or siRNAs in heathy plants [19, 20]. *Antiviral RNAi gene 1* (*AVI1*) and *AVI2* were identified by a sensitized genetic screen system developed in *Arabidopsis thaliana* ecotype Columbia-0 (Col-0). Genetic screens have not been used widely for the genetic dissection of antiviral RNAi in the Arabidopsis model probably because of two main roadblocks. First, highly redundant or overlapping siRNA pathways control antiviral RNAi in Arabidopsis so that genetic inactivation of a single pathway may not phenotypically alter virus susceptibility. For example, antiviral RNAi against *Cucumber mosaic virus* (CMV) and other RNA viruses is mediated by at least two members in the Dicer [Dicer-like 2 (DCL2) and DCL4], Argonaute (Ago1 and Ago2) and RDR (RDR1 and RDR6) families. Second, plant viruses each encode at least one viral suppressor of RNAi (VSR), such as VSR-2b of CMV, the expression of which is often sufficient to erase the difference in virus susceptibility between wild-type and antiviral RNAi-defective mutant Arabidopsis plants.

We largely bypass these roadblocks by screening Arabidopsis mutants with CMV-Δ2b, a previous characterized mutant of CMV deficient in the expression of its VSR-2b (Fig. 1) and known to be targeted for clearance by a single RDR6-dependent pathway of antiviral RNAi [21]. CMV-Δ2b spreads systemically in wild-type Col-0 plants, but replicates to low levels and causes no visible disease symptoms because of the increased susceptibility to antiviral RNAi. Importantly, CMV-Δ2b becomes highly virulent and replicates to high levels in single gene mutants defective in the RDR6-dependent biogenesis of secondary siRNAs (Fig. 2) [21], including those carrying a loss-of-function mutation in either *RDR6* or *Suppressor of Gene Silencing 3* (*SGS3*) [15, 16]. Our studies have shown that use of CMV-Δ2b to screen a population of Arabidopsis mutants in Col-0 background such as those in Salk homozygous T-DNA insertion mutant pools can identify specific components of the antiviral RNAi pathway [19, 20]. In this chapter, we describe the key protocols used in the sensitized genetic screening system.

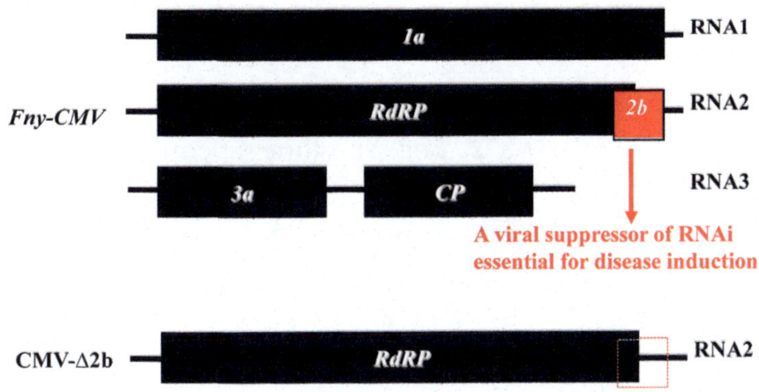

Fig. 1 The genome structure of CMV and CMV-Δ2b. CMV contains 3 positive-strand genomic RNAs that encode five proteins in total. Its VSR-2b is encoded in the +1 reading frame of the open reading frame (ORF) for the viral RNA-dependent RNA polymerase (RdRP, protein 2a). In the genomic RNA2 of CMV-Δ2b, the three 5′-terminal A*T*G codons, including the start codon, of VSR-2b ORF is mutated to A*C*G. These substitutions abolish the translation of VSR-2b, but do not alter the amino acids encoded by viral RdRP-2a in the overlapping reading frame since they are at the third codon positions

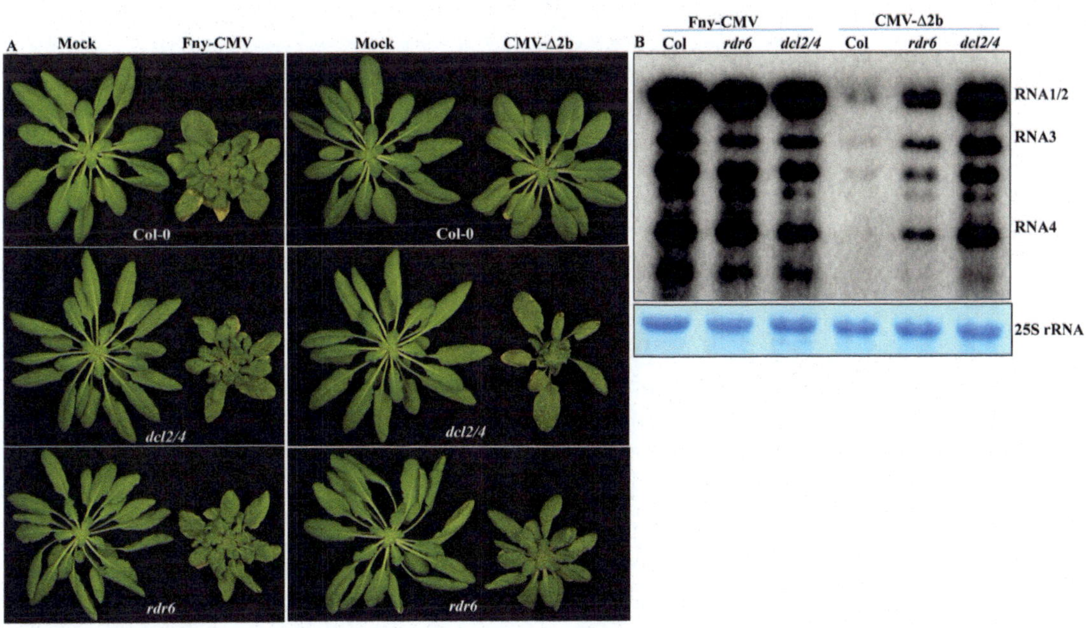

Fig. 2 The principle of the sensitized genetic screen to identify ***avi*** mutants. (**a**) The wild-type (Col-0) and mutant plants were photographed 2 weeks after infection with the wild-type CMV (strain Fny) or its VSR-deficient mutant, CMV-Δ2b. (**b**) The accumulation of the viral RNAs in wild-type and mutant plants 2 weeks after infection with CMV or CMV-Δ2b detected by Northern blot hybridization. Infection with CMV-Δ2b induces an enhanced disease susceptibility phenotype in the single (***rdr6***) and double (***dcl2 dcl4***) mutant plants compared to Col-0 plants. In contrast, knockout of ***RDR6*** alone or ***DCL2*** and ***DCL4*** together does not significantly enhance either the symptom severity or virus accumulation levels during wild-type CMV infection

2 Materials

2.1 Virus Propagation

1. *Nicotiana clevelandii*.
2. Soil: Sunshine supermix 360 (Sun Gro Horticulture).
3. Steam soil sterilizer.
4. Pots: for seedlings.
5. Growth room.
6. CMV (Fny strain), a tripartite positive-strand (+) RNA virus (Fig. 1).
7. CMV-Δ2b, CMV mutant defective in its VSR protein, 2b (Fig. 1).
8. Silicon carbide:400 mesh.
9. Model 119 Powder Blower (Blowout Medical LLC).
10. Plastic gloves.
11. Solution A: 0.5 M sodium citrate, 5 mM EDTA (pH 9.0), 0.5% thioglycolic acid (add just before use).
12. Solution B: 5 mM sodium borate, 0.5 mM EDTA (pH 9.0); 2% Triton X 100.
13. Solution C: 5 mM sodium borate, 0.5 mM EDTA (pH 9.0).
14. Chloroform.
15. 100% glycerol.
16. 40% (w/v) polyethylene glycol (PEG) solution (PEG6000 in Solution C).
17. Sucrose cushion: 10% sucrose in Solution C.
18. Blender.
19. High-speed centrifuge.
20. Ultracentrifuge.
21. Spectrophotometer.

2.2 Growth of Homozygous Salk Line Plants

1. *Arabidopsis thaliana* Salk homozygous T-DNA insertion mutant pools CS27941 and CS27951, CS27942 and CS27952, CS27943 and CS27953.
2. Soil: Sunshine supermix 360 (Sun Gro Horticulture).
3. Steam soil sterilizer.
4. Pots: for seedlings.
5. Growth room.

2.3 Identification of Arabidopsis Mutants Defective in Antiviral RNAi

1. Seedlings of Salk homozygous T-DNA insertion mutant lines from Subheading 3.2.
2. CMV-Δ2b: Purified virus particles of CMV mutant not expressing in its VSR protein, 2b from Subheading 3.1.

3. Silicon carbide: 400 mesh.
4. Model 119 Powder Blower (Blowout Medical LLC).
5. Plastic gloves.

2.4 Identification of the Gene Responsible for the Mutant Phenotype

1. *avi* mutants identified in Subheading 3.3.
2. Wild type *A. thaliana* Col-0.
3. Computer system with internet connection.
4. Materials for genotyping Salk lines: Check for the detailed protocol for genotyping Salk lines on the website http://signal.salk.edu/tdnaprimers.2.html.
5. mutCAN: Computational pipeline available at https://github.com/JinfengLu/mutCAN.

2.5 Total RNA Extraction

1. Plant materials: *A. thaliana* mutants of interest, and Col-0, *rdr6-15*, and *dcl2-1/dcl4-2* for the control experiments.
2. TRIzol reagent (Thermo Fisher Scientific).
3. Chloroform.
4. Isopropanol.
5. Ethanol.
6. DEPC-treated ddH_2O: Add 0.1% diethylpyrocarbonate (DEPC) to double distilled H_2O (ddH_2O), mix thoroughly and incubate at 37 °C for 12 h, and autoclave for 15 min to remove DEPC.
7. High-speed Centrifuge.

2.6 Northern Blot Detection of Viral Genomic RNAs

1. Agarose, electrophoresis grade.
2. 10× MOPS buffer: 0.2 M 3-(*N*-morpholino)propanesulfonic acid (MOPS), 0.08 M sodium acetate, 0.01 M EDTA (pH 8.0). Adjust pH to 7.0 with NaOH. Sterilize the solution by filtration through a 0.2-micron Millipore filter. Store in the dark.
3. 37% (w/w) formaldehyde.
4. DEPC-treated ddH_2O.
5. 2× RNA loading buffer: 95% formamide, 0.025% (w/v) SDS, 0.025% (w/v) bromophenol blue, 0.025% (w/v) xylene cyanol FF, 0.5 mM EDTA (pH 8.0).
6. Hybridization buffer: PerfectHyb™ Plus hybridization buffer (Sigma).
7. 20× SSC: 0.3 M trisodium citrate, 3 M NaCl, adjust pH at 7.0 with diluted HCl.
8. 10% SDS: Dissolve sodium dodecyl sulfate (SDS) in ddH_2O. Do not autoclave. Store at room temperature.
9. Methylene blue staining solution: 0.5% methylene blue in ddH_2O.

10. α-^{32}P- CTP: 3000 Ci/mmol 10 mCi/mL, 250 μCi.
11. DecaLabel DNA Labeling Kit (Thermo Scientific).
12. Nylon membrane: Amersham Hybond™ -N+ Membrane (GE Healthcare).
13. Saran Wrap.
14. X-ray film: Kodak® BioMax® MS film (Sigma).
15. Whatman paper.
16. Electrophoresis set and power supply.
17. Fume hood.
18. Vacu-Blot vacuum blotting system.
19. UV cross-linker: UV Stratalinker 1800 (Stratagene).
20. Hybridization oven (Thomas Scientific).

2.7 Northern Blot Detection of Viral siRNAs

1. 40% acrylamide solution (19:1): 38% acrylamide, 2% bis-acrylamide aqueous solution.
2. Urea, electrophoresis grade.
3. 5× TBE: Dissolve 54 g tris(hydroxymethyl)aminomethane and 27.5 g boric acid, and add 20 mL of 0.5 M EDTA.
4. 10% ammonium persulfate: Dissolved 10% (w/v) ammonium persulfate in ddH_2O.
5. TEMED: Tetramethylethylenediamine.
6. DEPC-treated ddH_2O.
7. 2× RNA loading buffer: 95% formamide, 0.025% (w/v) SDS, 0.025% (w/v) bromophenol blue, 0.025% (w/v) xylene cyanol FF, 0.5 mM EDTA (pH 8.0).
8. Oligonucleotides: A mixture of seven 40-nt oligo DNAs are synthesized as probes to detect the negative-strand viral siRNAs that target the genomic RNA3 of CMV [22].
9. γ-^{32}P-ATP: 3000 Ci/mmol, 10 mCi/mL, Lead, 100 μCi.
10. KinaseMax™ 5′ End-Labeling Kit (Thermo Scientific).
11. PerfectHyb™ Plus hybridization buffer (Sigma).
12. Semidry transfer cell: Trans-Blot SD DNA/RNA blotting kit (Bio-Rad).
13. Amersham Hybond™ -NX membrane (GE Healthcare).
14. Electrophoresis set and power supply.
15. UV cross-linker: UV Stratalinker 1800 (Stratagene).
16. Hybridization oven (Thomas Scientific).
17. BAS Storage Phosphor Screen (GE Healthcare).
18. FLA7000 Typhoon Storage Phosphorimager.

3 Methods

3.1 Virus Propagation

Viruses are propagated in *N. clevelandii* and purified using the published protocol [23, 24] with modification.

1. Steam soil for 4 h and prepare pots with the sterilized soil.
2. Germinate and grow tobacco *N. clevelandii* in soil at room temperature with 10 h light and 14 h darkness daily (*see* **Note 1**).
3. After about 3 weeks, *N. clevelandii* seedlings with about six expanded true leaves are infected with wild-type Fny-CMV or its mutant CMV-Δ2b, which does not express its VSR-2b due to the introduction of point nucleotide substitutions (Fig. 1). Dilute virus particle stock solution to the final concentration of 20 ng/μL. Three expanded leaves of each seedling are dusted with silicon carbide and mechanically inoculated with Fny-CMV or CMV-Δ2b by gentle rubbing with gloved finger.
4. Two weeks after viral infection, 50 g upper uninoculated systemically infected leaves are collected for virus purification.
5. Homogenize the infected leaves in 1 mL/g leaf materials of ice-cold Solution A with freshly added 0.5% thioglycolic acid and an equal volume of chloroform using a blender. After centrifuging at 15,000 × *g*, 4 °C, for 10 min, filter the supernatant through a dampened miracloth, add PEG from the stock solution to a final concentration of 10% and keep it at 4 °C for half an hour with gentle stirring.
6. Centrifuge at 15,000 × *g*, 4 °C, for 15 min, remove supernatant, add 20 mL Solution B to the pellet, and incubate 30 min at 4 °C to resuspend the pellet.
7. Centrifuge at 10,000 × *g*, 4 °C, for 10 min.
8. The supernatant is overlaid on the Sucrose Cushion (~2 mL) in ultracentrifuge tubes. Ultracentrifuge at 140,000 × *g*, 4 °C, for 2 h.
9. Remove supernatant, add 3 mL Solution C to the pellet, incubate overnight at 4 °C to resuspend the virus in the pellet, centrifuge at 10,000 × *g*, 4 °C, for 5 min and keep the virus-containing supernatant.
10. Measure the adsorption at 260 nm to quantify the purified virions. The A_{260} of 1 mg/mL suspension of CMV is 5.0.
11. The final virus suspension can be used immediately or mixed with equal volume of glycerol for storage at −80 °C (*see* **Note 2**).

3.2 Growth of Homozygous Salk Line Plants

Each Salk homozygous T-DNA insertion mutant pool comprises thousands of confirmed mutant lines, which need to be grown separately.

1. Steam soil for 4 h and prepare pots with the sterilized soil.
2. Germinate *Arabidopsis* seeds in soil and keep the pots with plastic cover at 4 °C in a cold room for 48 h before transferring the pots to a growth room programmed for 9 h in light and 15 h in dark at about 24 °C (*see* **Note 3**).
3. Seeds will germinate after about 4 days. Ten days after germination in the growth room, transfer and plant 12 seedlings of each Salk line into a new pot (*see* **Note 4**).
4. After 2 weeks, seedlings with about ten expanded true leaves are ready for virus inoculation.

3.3 Identification of Arabidopsis Mutants Defective in Antiviral RNAi

Arabidopsis mutant lines are subjected to the screening of antiviral RNAi-defective mutants through mechanical inoculation of CMV-Δ2b.

1. Dilute CMV-Δ2b particle stock solution to the final concentration of 20 μg/mL. Three expanded leaves of each seedling are dusted with silicon powder and mechanically inoculated with CMV-Δ2b by gentle rubbing. For each Salk mutant line, ten seedlings are infected with CMV-Δ2b with additional two seedlings uninoculated to serve as healthy mutant plant controls.
2. Five to 10 min after mechanical inoculation, spray water to remove the silicon powder and the inoculum from the inoculated plants. The inoculated plants are placed in a growth room programmed for 9 h in light and 15 h in dark at about 24 °C (*see* **Note 4**).
3. From 2 weeks post inoculation with CMV-Δ2b, visually inspect the inoculated plants of each Salk mutant line for the development of disease symptoms by comparison to the uninoculated healthy control plants of the same mutant and to the wild-type Col-0 plants infected with CMV-Δ2b (*see* **Note 5**). A candidate *avi* mutant is identified if the majority of the inoculated plants from a Salk line show clearly visible disease symptoms, such as leaf mosaic or whole plant stunting after the infection with CMV-Δ2b.

3.4 Identification of the Gene Responsible for the Mutant Phenotype

1. The T-DNA insertion site is known for each of the homozygous Salk lines. The T-DNA insertion can be confirmed by genotyping according to the protocol published on the website (*see* **Note 6**).
2. Individual candidate *avi* mutant is backcrossed to the parental Col-0 plants followed by self-crosses to generate a final

segregating F2 population. Inoculate approximately 80 seedlings from the F2 population of each *avi* mutant with CMV-Δ2b and determine the ratio of the inoculated F2 plants that do or do not develop the disease symptoms of the parental *avi* mutant, defined as the mutant and wild-type phenotypes, respectively. A 3:1 segregation ratio of the inoculated plants exhibiting the wild-type and mutant phenotype would indicate a recessive causal mutation in a single gene responsible for the mutant phenotype. The recessive mutation in a given *avi* mutant (e.g., *avi2*) is caused by T-DNA insertion in a specific gene when all of the F2 plants with the mutant phenotype are homozygous for the T-DNA insertion [20].

3. A recessive causal mutation is not caused by T-DNA insertion as annotated for the Salk mutant line when the mutant phenotype of a given *avi* mutant (e.g., *avi1*) does not cosegregate with the T-DNA insertion [19]. For those *avi* mutants, a mapping-by-sequencing approach is necessary for the identification of the causal mutation as described [19]. Briefly, identify approximately 80 plants each with the wild-type or mutant phenotype in the F2 population of each *avi* mutant to generate the wild-type and mutant pools for DNA extraction and whole genome re-sequencing. The causal mutation is identified by a computational pipeline such as mutCAN [19] for mutation calling by coverage depth and nucleotide frequency (*see* **Note** 7).
4. In a final step to verify the causal mutation by the complementation analysis, a binary plasmid vector for the expression of the wild-type candidate gene is constructed and transformed into the homozygous *avi* mutant plants. The resulting transgenic plants are infected with CMV-Δ2b to determine whether expression of the wild-type gene rescues the resistance phenotype lost in the *avi* mutant plants.

3.5 Northern Blot Detection of the Viral Venomic RNAs

The analysis of virus replication is critical for the characterization of *avi* mutants. Northern blot hybridization to analyze the accumulation of CMV genomic RNAs is performed according to the published protocol [10] with minor modifications.

1. About 2 g upper uninoculated systemic leaves from ten plants are pooled for extracting total RNA. Total RNA is extracted using TRIzol according to the manufacturer's protocol.
2. Prepare 1.2% denaturing formaldehyde–MOPS agarose gel. Add 1.2 g agarose to 87 mL DEPC-treated ddH_2O and boil. After the agarose solution is cooled down to 50 °C, add 10 mL 10× MOPS buffer and 3 mL formaldehyde (37%), mix well, and pour into the gel tray in a fume hood. Let stand under room temperature until the gel solidifies.

3. Mix the 5 μg total RNA of each sample in the same volume (adjusting the sample volume with DEPC-treated ddH_2O) with an equal volume of the 2× RNA loading buffer. Before loading, heat the sample mixture for 5 min at 75 °C and cool on ice for at least 5 min.
4. Run the agarose gel electrophoresis in a fume hood with 100 V for 2 to 3 h until the dye is close to the bottom of the gel.
5. Transfer the RNA in the agarose gel to nylon membrane using Vacu-Blot vacuum blotting system for 2–3 h. Wash the membrane briefly in 2× SSC before UV cross-linking by placing the membrane on wet Whatman paper inside UV cross-linker.
6. Stain the membrane with methylene blue to visualize rRNA bands. Wash the membrane with ddH_2O thoroughly after taking a picture as a record for equal loading.
7. Pre-hybridization at 65 °C for 2 h in 5–10 mL hybridization buffer.
8. Prepare radioactive probe (*see* **Note 8**) with DecaLabel DNA Labeling Kit according to the manufacturer's manual, and after adding the probe, incubate at 65 °C overnight with rotation at 20 rpm.
9. Wash the membrane once with 2× SSC, 0.1% SDS at 65 °C for 30 min, once with 1× SSC, 0.1% SDS at 65 °C for 20 min, and once with 0.1× SSC, 0.1% SDS at 65 °C for 20 min.
10. Expose the membrane under Saran Wrap to X-ray film.

3.6 Northern Blot Detection of the Viral siRNAs

The analysis of virus-derived siRNA is critical for the characterization of *avi* mutants. Northern blot hybridization to analyze the accumulation of viral siRNAs is performed according to the published protocol [21] with minor modifications.

1. Extract total RNA from each sample using TRIzol as described above.
2. Prepare 15% denaturing PAGE gel. For a gel of 16.5 cm wide, 22 cm long, and 1.5 mm thick, mix 15 mL 40% acrylamide solution (19:1), 19.2 g urea, 4 mL 5× TBE, make it to 40 mL by adding DEPC-treated ddH_2O. Mix to dissolve the urea completely. Add 400 μL 10% ammonium persulfate and 30 μL TEMED, mix well briefly and pour the gel to the level 1 cm below the comb.
3. After gel polymerization, prepare the loading gel by mixing 1.87 mL 40% acrylamide solution (19:1), 0.5 mL 5× TBE, make it to 5 mL by adding DEPC-treated ddH_2O. Mix well and add 50 μL 10% ammonium persulfate and 4 μL TEMED, mix well briefly and pour the loading gel on the top of the denaturing gel, and set the 1.5 mm comb.

4. After gel polymerization, remove the comb, and set up PAGE apparatus. Briefly rinse the wells with 1× TBE to remove excess acrylamide. Pre-run for 10 min at 200 V.
5. Equally load 50~100 μg total RNA (which includes small RNAs) from each sample per lane, concentrate total RNA solution and add water as needed to make it to 20 μL equal volume. Add equal volume of 2× RNA loading buffer.
6. Denature RNA samples at 95 °C for 5 min and quickly move the denatured samples to ice for 5 min. Load the RNA samples into the wells and run the gel electrophoresis at ~400 V for 3–4 h until the bromophenol blue near to the bottom of the gel.
7. Disassemble the gel apparatus and cut the gel into lower and upper parts along the line of the xylene cyanol dye. The upper and lower parts of the gel should contain U6 RNAs (for loading controls) and small RNAs (siRNAs and miRNAs), respectively.
8. Transfer RNA to the nylon membrane for 30 min with a semidry transfer cell according to the manufacturer's manual. Disassemble the semidry transfer apparatus and cross-link the RNA to the membrane by UV. The membrane can be immediately used for hybridization.
9. Prehybridize with 5–10 mL of hybridization buffer at 42 °C for at least 1 h in a hybridization oven.
10. Prepare radioactive probe by labeling oligonucleotides with KinaseMax™ 5′ End-Labeling Kit according to the manufacturer's manual and after adding the probe, incubate at 42 °C overnight with rotation at 18 rpm.
11. Wash the membrane once with 2× SSC, 0.1% SDS at 45 °C for 30 min, once with 1× SSC, 0.1% SDS at 45 °C for 20 min, once with 0.1× SSC, 0.1% SDS at 45 °C for 20 min.
12. Expose the membrane under Saran wrap to X-ray film or scan the radioactivity signal with a phosphor-imager.

4 Perspective

It is known that expression of VSR-2b during wild-type CMV infection potently suppresses antiviral RNAi, especially the amplification of vsiRNAs by RDR1 and RDR6 [19, 22, 25–27]. As a result, knockout of RDR6 alone or DCL2 and DCL4 together does not significantly enhance either the symptom severity or virus accumulation levels in CMV-infected plants (Fig. 2). Thus, it is unlikely that use of wild-type viruses in genetic screens would identify mutants with specific defects in antiviral RNAi.

Loss of VSR-2b expression dramatically increases the susceptibility of CMV-Δ2b to antiviral RNAi. Consistently, the VSR-deficient mutant virus replicates to significantly higher levels and induces clearly visible disease symptoms in both the single (*rdr6*) and double (*dcl2 dcl4*) mutant plants compared to Col-0 plants, which support only low levels of virus replication and are symptomless (Fig. 2). These unique properties of CMV-Δ2b explain why screening Arabidopsis mutants for the enhanced disease susceptibility phenotype to CMV-Δ2b has led to the successful cloning of *AVI1* (renamed as *ALA2*) and *AVI2* genes from the antiviral RNAi pathway [19, 20]. Interestingly, both *AVI1* and *AVI2* promote antiviral RNAi by enhancing the biogenesis of viral secondary siRNAs and of the endogenous virus-activated siRNAs (vasiRNAs) in the infected plants [19, 20]. Notably, neither *AVI1* nor *AVI2* is necessary for the biogenesis of endogenous siRNAs and miRNAs, suggesting identification of specific antiviral factors by the sensitized genetic screen.

Daoxin Xie and colleagues have recently developed a forward genetic screen to identify the enhancers of *rdr6* in antiviral RNAi in collaboration with us [28, 29]. In this system, the M2 population of EMS-mutagenized *rdr6* progeny plants is screened for the enhanced disease susceptibility phenotype in the infection by CMV-2aTΔ2b, a mutant of Fny-CMV containing 295-nucleotide deletion in VSR-2b coding region of the genomic RNA2 that also leads to early termination of RdRP-2a protein [30]. Unlike CMV-Δ2b, CMV-2aTΔ2b is cleared similarly efficiently in Col-0 plants by either RDR1- or RDR6-dependent pathway so that it replicates to high levels and induces clearly visible disease symptoms only when both pathways are defective [10, 21]. Although designed originally to identify specific factors in RDR1-dependent pathway, both *AVI1* and *AVI2* homologs have been identified by the mutant enhancer screen, which nevertheless independently confirms the critical function of these new components in antiviral RNAi.

Use of a similar strategy has also facilitated the development of a forward genetic screen to identify genes required for antiviral RNAi in the nematode *Caenorhabditis elegans* [31]. In this system, self-replication of Flock house virus genomic RNA1 in absence of its VSR-B2 expression becomes a sensitive reporter of antiviral RNAi in the transgenic worms. Interestingly, cloning EMS-induced causal mutations with the mapping-by-sequencing approach [19] has uncovered four loss-of-function alleles of *drh-1* [31], known to encode Dicer-related helicase 1 (DRH-1) essential for antiviral RNAi but dispensable for RNAi targeting cellular mRNAs [32, 33].

In summary, these recent studies have established a successful genetic screening strategy for the identification of specific components and regulators of antiviral RNAi in both plants and animals.

It is known that VSR-deficient plant, insect and mammalian viruses are highly susceptible to the antiviral response [3–6, 8, 34–37], suggesting the feasibility to develop similar sensitized genetic screens in distinct host systems. It is likely that further exploration of these genetic screens will lead to better mechanistic understanding of the antiviral RNAi in plants and animals.

5 Notes

1. *N. clevelandii* plants are selected for the propagation of CMV-Δ2b because they develop visible chlorosis in the systemically infected leaves and support the replication of the mutant virus at high levels.
2. Purified CMV and CMV-Δ2b stock solutions containing 50% glycerol can be stored at −80 °C for at least 10 years.
3. 4 °C cold treatment of seeds for 48 h ensures even germination.
4. Seedlings are grown under short-day conditions to extend the vegetative phase of Arabidopsis life cycle for virus infection and symptom development.
5. Wild-type Col-0 plants infected with CMV-Δ2b support low levels of virus replication and do not develop visible disease symptoms. Thus, a candidate *avi* mutant is identified if the majority of the inoculated plants from a Salk line show clearly visible disease symptoms, such as leaf mosaic or whole plant stunting after the infection with CMV-Δ2b.
6. The detailed protocol for genotyping Salk lines can be found on the website http://signal.salk.edu/tdnaprimers.2.html.
7. For the pipeline of computational programs for mapping the causal mutation, please refer to [19].
8. The 3′-terminal untranslated regions of all the genomic and subgenomic RNAs from a particular CMV isolate are highly homologous (>90% identity). This region of the genomic RNA2 is routinely amplified as the single probe to detect all CMV RNAs [10].

Acknowledgments

This project was supported by grants from the US Department of Agriculture Research Service (6659-22000-025), the US-Israel Binational Agricultural Research and Development Fund, National Institutes of Health (R01AI52447 and GM94396), by the Agricultural Experimental Station of the University of California, Riverside (to S.-W.D.)

References

1. Matranga C, Zamore PD (2007) Small silencing RNAs. Curr Biol 17:R789–R793
2. Ding SW (2010) RNA-based antiviral immunity. Nat Rev Immunol 10:632–644
3. Li Y, Basavappa M, Lu J, Dong S, Cronkite DA, Prior JT, Reinecker HC, Hertzog P, Han Y, Li WX, Cheloufi S, Karginov FV, Ding SW, Jeffrey KL (2016) Induction and suppression of antiviral RNA interference by influenza a virus in mammalian cells. Nat Microbiol 2:16250
4. Li Y, Lu J, Han Y, Fan X, Ding SW (2013) RNA interference functions as an antiviral immunity mechanism in mammals. Science 342:231–234
5. Maillard PV, Ciaudo C, Marchais A, Li Y, Jay F, Ding SW, Voinnet O (2013) Antiviral RNA interference in mammalian cells. Science 342:235–238
6. Qiu Y, Xu Y, Zhang Y, Zhou H, Deng YQ, Li XF, Miao M, Zhang Q, Zhong B, Hu Y, Zhang FC, Wu L, Qin CF, Zhou X (2017) Human virus-derived small RNAs can confer antiviral immunity in mammals. Immunity 46:992–1004
7. Hamilton AJ, Baulcombe DC (1999) A species of small antisense RNA in posttranscriptional gene silencing in plants. Science 286:950–952
8. Li HW, Li WX, Ding SW (2002) Induction and suppression of RNA silencing by an animal virus. Science 296:1319–1321
9. Garcia-Ruiz H, Takeda A, Chapman EJ, Sullivan CM, Fahlgren N, Brempelis KJ, Carrington JC (2010) Arabidopsis RNA-dependent RNA polymerases and dicer-like proteins in antiviral defense and small interfering RNA biogenesis during turnip mosaic virus infection. Plant Cell 22:481–496
10. Wang XB, Wu Q, Ito T, Cillo F, Li WX, Chen X, Yu JL, Ding SW (2010) RNAi-mediated viral immunity requires amplification of virus-derived siRNAs in *Arabidopsis thaliana*. Proc Natl Acad Sci U S A 107:484–489
11. Goic B, Stapleford KA, Frangeul L, Doucet AJ, Gausson V, Blanc H, Schemmel-Jofre N, Cristofari G, Lambrechts L, Vignuzzi M, Saleh MC (2016) Virus-derived DNA drives mosquito vector tolerance to arboviral infection. Nat Commun 7:12410
12. Goic B, Vodovar N, Mondotte JA, Monot C, Frangeul L, Blanc H, Gausson V, Vera-Otarola J, Cristofari G, Saleh MC (2013) RNA-mediated interference and reverse transcription control the persistence of RNA viruses in the insect model Drosophila. Nat Immunol 14:396–403
13. Tassetto M, Kunitomi M, Andino R (2017) Circulating immune cells mediate a systemic RNAi-based adaptive antiviral response in Drosophila. Cell 169:314–325.e313
14. Cogoni C, Macino G (1999) Gene silencing in Neurospora crassa requires a protein homologous to RNA-dependent RNA polymerase. Nature 399:166–169
15. Dalmay T, Hamilton A, Rudd S, Angell S, Baulcombe DC (2000) An RNA-dependent RNA polymerase gene in Arabidopsis is required for posttranscriptional gene silencing mediated by a transgene but not by a virus. Cell 101:543–553
16. Mourrain P, Beclin C, Elmayan T, Feuerbach F, Godon C, Morel JB, Jouette D, Lacombe AM, Nikic S, Picault N, Remoue K, Sanial M, Vo TA, Vaucheret H (2000) Arabidopsis SGS2 and SGS3 genes are required for posttranscriptional gene silencing and natural virus resistance. Cell 101:533–542
17. Park W, Li J, Song R, Messing J, Chen X (2002) CARPEL FACTORY, a dicer homolog, and HEN1, a novel protein, act in microRNA metabolism in *Arabidopsis thaliana*. Curr Biol 12:1484–1495
18. Tabara H, Sarkissian M, Kelly WG, Fleenor J, Grishok A, Timmons L, Fire A, Mello CC (1999) The rde-1 gene, RNA interference, and transposon silencing in C-elegans. Cell 99:123–132
19. Guo Z, Lu J, Wang X, Zhan B, Li W, Ding SW (2017a) Lipid flippases promote antiviral silencing and the biogenesis of viral and host siRNAs in Arabidopsis. Proc Natl Acad Sci U S A 114:1377–1382
20. Guo Z, Wang XB, Wang Y, Li WX, Gal-On A, Ding SW (2017b) Identification of a new host factor required for antiviral RNAi and amplification of viral siRNAs. Plant Physiol 176:1587–1597
21. Wang XB, Jovel J, Udomporn P, Wang Y, Wu Q, Li WX, Gasciolli V, Vaucheret H, Ding SW (2011) The 21-nucleotide, but not 22-nucleotide, viral secondary small interfering RNAs direct potent antiviral defense by two cooperative argonautes in *Arabidopsis thaliana*. Plant Cell 23:1625–1638
22. Diaz-Pendon JA, Li F, Li WX, Ding SW (2007) Suppression of antiviral silencing by cucumber mosaic virus 2b protein in Arabidopsis is associated with drastically reduced accumulation of three classes of viral small interfering RNAs. Plant Cell 19:2053–2063
23. Scott H (1963) Purification of cucumber mosaic virus. Virology 20:103–106

24. Peden KW, Symons RH (1973) Cucumber mosaic virus contains a functionally divided genome. Virology 53:487–492
25. Duan CG, Fang YY, Zhou BJ, Zhao JH, Hou WN, Zhu H, Ding SW, Guo HS (2012) Suppression of Arabidopsis ARGONAUTE1-mediated slicing, transgene-induced RNA silencing, and DNA methylation by distinct domains of the cucumber mosaic virus 2b protein. Plant Cell 24:259–274
26. Fang YY, Zhao JH, Liu SW, Wang S, Duan CG, Guo HS (2016) CMV2b-AGO interaction is required for the suppression of RDR-dependent antiviral silencing in Arabidopsis. Front Microbiol 7:1329
27. Li HW, Lucy AP, Guo HS, Li WX, Ji LH, Wong SM, Ding SW (1999) Strong host resistance targeted against a viral suppressor of the plant gene silencing defence mechanism. EMBO J 18:2683–2691
28. Gao H, Yang M, Yang H, Qin Y, Zhu B, Xu G, Xie C, Wu D, Zhang X, Li WX, Yan J, Song S, Qi T, Ding SW, Xie D (2017) Arabidopsis ENOR3 regulates RNAi-mediated antiviral defense. J Genet Genomics 45:33–40
29. Zhu B, Gao H, Xu G, Wu D, Song S, Jiang H, Zhu S, Qi T, Xie D (2017) Arabidopsis ALA1 and ALA2 mediate RNAi-based antiviral immunity. Front Plant Sci 8:422
30. Soards AJ, Murphy AM, Palukaitis P, Carr JP (2002) Virulence and differential local and systemic spread of cucumber mosaic virus in tobacco are affected by the CMV 2b protein. Mol Plant-Microbe Interact 15:647–653
31. Coffman SR, Lu J, Guo X, Zhong J, Jiang H, Broitman-Maduro G, Li WX, Lu R, Maduro M, Ding SW (2017) *Caenorhabditis elegans* RIG-I homolog mediates antiviral RNA interference downstream of dicer-dependent biogenesis of viral small interfering RNAs. MBio 8:e00264–e00217
32. Lu R, Yigit E, Li WX, Ding SW (2009) An RIG-I-like RNA helicase mediates antiviral RNAi downstream of viral siRNA biogenesis in *Caenorhabditis elegans*. PLoS Pathog 5:e1000286
33. Guo X, Zhang R, Wang J, Ding SW, Lu R (2013) Homologous RIG-I-like helicase proteins direct RNAi-mediated antiviral immunity in *C. elegans* by distinct mechanisms. Proc Natl Acad Sci U S A 110:16085–16090
34. Ding SW, Li WX, Symons RH (1995) A novel naturally-occurring hybrid gene encoded by a plant RNA virus facilitates long-distance virus movement. EMBO J 14:5762–5772
35. Scholthof HB, Scholthof KB, Jackson AO (1995) Identification of tomato bushy stunt virus host-specific symptom determinants by expression of individual genes from a potato virus X vector. Plant Cell 7:1157–1172
36. Han YH, Luo YJ, Wu Q, Jovel J, Wang XH, Aliyari R, Han C, Li WX, Ding SW (2011) RNA-based immunity terminates viral infection in adult Drosophila in the absence of viral suppression of RNA interference: characterization of viral small interfering RNA populations in wild-type and mutant flies. J Virol 85:13153–13163
37. Guo Z, Li Y, Ding SW (2019) Small RNA-based antimicrobial immunity. Nat Rev Immunol 19, 31–44

Chapter 13

Design, Synthesis, and Functional Analysis of Highly Specific Artificial Small RNAs with Antiviral Activity in Plants

Alberto Carbonell and José-Antonio Daròs

Abstract

Artificial microRNAs (amiRNAs) and synthetic trans-acting small interfering RNAs (syn-tasiRNAs) are two classes of artificial small RNAs (sRNAs) that have been broadly used to confer antiviral resistance in plants. However, methods for designing, synthesizing and functionally analyzing antiviral artificial sRNAs have not been optimized for time and cost-effectiveness and high-throughput applicability since recently. Here we present a systematic methodology for the simple and fast-forward design, generation, and functional analysis of large numbers of artificial sRNA constructs engineered to induce high levels of antiviral resistance in plants. Artificial sRNA constructs are transiently expressed in *Nicotiana benthamiana* plants, which are subsequently inoculated with the virus of interest. The antiviral activity of each artificial sRNA construct is assessed by monitoring viral symptom appearance, and through molecular analysis of virus accumulation in plant tissues. This approach is aimed to easily identify artificial sRNAs with high antiviral activity that could be expressed in transgenic plants for highly durable antiviral resistance.

Key words Plant virus, Antiviral resistance, Artificial small RNAs, AmiRNAs, Syn-tasiRNAs

1 Introduction

Artificial microRNAs (amiRNAs) and synthetic trans-acting small interfering RNAs (syn-tasiRNAs) are two classes of artificial small RNAs (sRNAs) engineered to selectively silence transcripts in plants, including viral RNAs [1]. AmiRNAs and syn-tasiRNAs are produced in planta by expressing a functional miRNA or tasiRNA precursor with modified miRNA or tasiRNA sequences, respectively. AmiRNAs are generated from DICER-LIKE1 (DCL1) cleavage of precursors with foldback structures, while syn-tasiRNAs are produced after cleavage of a *TAS* precursor by an miRNA/ARGONAUTE (AGO) complex, RNA-DEPENDENT RNA POLYMERASE 6 conversion of one of the cleavage products to

Kappei Kobayashi and Masamichi Nishiguchi (eds.), *Antiviral Resistance in Plants: Methods and Protocols*, Methods in Molecular Biology, vol. 2028, https://doi.org/10.1007/978-1-4939-9635-3_13,

double-stranded RNA, and DLC4-mediated processing of the dsRNA into 21-nucleotide phased syn-tasiRNAs in register with the miRNA-guided cleavage site. Despite their biogenesis pathways are different, both classes of artificial sRNAs associate with an AGO protein, generally AGO1, to target and cleave highly sequence complementary transcripts [2].

AmiRNAs have been extensively used to selectively confer antiviral resistance in transgenic plants [3]. However, because antiviral amiRNAs typically target a single viral RNA site, this resistance can be overcome if virus variants accumulate mutations in the amiRNA target site [4, 5]. To avoid these limitations, multiple amiRNAs have been expressed from different precursors or from a single polycistronic precursor to target several regions within a single viral RNA [6–9]. Syn-tasiRNAs have recently emerged as an alternative to amiRNA to induce antiviral resistance in plants. This is another possibility of expressing multiple artificial sRNAs from a single precursor, which should also allow the targeting of multiple sites within a viral RNA or of multiple sequence unrelated viruses, and therefore could induce more durable antiviral resistance [10]. Indeed, syn-tasiRNAs have already been reported to induce simultaneous resistance to two different viruses [11], and to interfere with viroid infections as well [12].

Despite both classes of artificial sRNAs have been used to confer antiviral resistance, methods for designing, synthesizing and functionally analyzing artificial sRNAs have not been optimized for time and cost-effectiveness and high-throughput applicability since recently [13–15]. In this chapter, we describe a detailed methodology for the automated design, high-throughput generation and functional analysis of large number of artificial sRNA constructs with antiviral activity in plants. First, we detail the design process of highly specific antiviral amiRNAs with the P-SAMS ("Plant Small RNA Maker Suite") web tool [15] and the molecular methods for their cloning into *Bsa*I/*ccd*B "B/c" vectors, a recently reported new generation of amiRNA vectors for efficient gene silencing in plants [13, 14]. Next, the experimental plan to analyze the antiviral activity of large numbers of amiRNA constructs is described. AmiRNA constructs are transiently expressed in *Nicotiana benthamiana* plants, which are subsequently inoculated with the virus of interest. The antiviral activity of each amiRNA construct is assessed by monitoring viral symptoms, and through molecular analysis of virus accumulation in plant tissues. Finally, we describe the design, generation, and functional analysis of syn-tasiRNA constructs that include several artificial sRNA sequences with high antiviral activity, which were previously selected from the initial amiRNA screening. This methodology has been successfully applied to identify highly active artificial sRNAs against *Potato spindle tuber viroid* [12] and *Tomato spotted wilt virus* [16].

2 Materials

2.1 Antiviral AmiRNA Design

1. Computer with Internet connection.
2. Web browser (e.g., Google Chrome, Safari, Mozilla Firefox, Internet Explorer).

2.2 Antiviral AmiRNA Cloning

1. Oligo Annealing buffer: 60 mM Tris–HCl (pH 7.5), 500 mM NaCl, 60 mM $MgCl_2$, 10 mM DTT (*see* **Note 1**).
2. Thermocycler.
3. Water bath.
4. Sterile H_2O.
5. T4 DNA ligase (5 U/μL, Thermo Fisher Scientific).
6. *Bsa*I (10 U/μL, New England Biolabs).
7. Competent cells of *Escherichia coli* (*E. coli*) *ccd*B-sensitive strain (e.g., DH5α, DH10B, TOP10).
8. LB agar plate with kanamycin: 10 g/L tryptone, 5 g/L yeast extract, 10 g/L NaCl, 15 g/L of Bacto agar, and 50 mg/L of kanamycin.
9. Liquid LB with kanamycin: 10 g/L tryptone, 5 g/L yeast extract, 10 g/L NaCl, and 50 mg/L of kanamycin.
10. GeneJET Plasmid Miniprep Kit (Thermo Fisher Scientific).
11. B/c vectors: Each vector contains a unique combination of bacterial and plant antibiotic resistance genes. Eudicot amiRNA B/c vectors are *pENTR-AtMIR390a-B/c* (Addgene plasmid 51,778), *pFK210B-AtMIR390a-B/c* (Addgene plasmid 51,777), *pMDC123SB-AtMIR390a-B/c* (Addgene plasmid 51,775) and *pMDC32B-AtMIR390a-B/c* (Addgene plasmid 51,778). Monocot amiRNA vectors are *pENTR-OsMIR390-B/c* (Addgene plasmid 61,468), *pH7WG2B-OsMIR390-B/c* (Addgene plasmid 61,465), *pMDC123SB-OsMIR390-B/c* (Addgene plasmid 61,466), and *pMDC32B-OsMIR390-B/c* (Addgene plasmid 61,467) [13, 14] (Table 1) (*see* **Note 2**).
12. Oligonucleotides

M13-F:	CCCAGTCACGACGTTGTAAAACGACGG
M13-R:	CAGAGCTGCCAGGAAACAGCTATGACC
attB1:	ACAAGTTTGTACAAAAAAGCAGGCT
attB2:	ACCACTTTGTACAAGAAAGCTGGGT

13. Competent cells of *Agrobacterium tumefaciens* GV3101 strain.

Table 1
***BsaI/ccdB*-based ("B/c") vectors for direct cloning of amiRNAs and syn-tasiRNAs**

Vector	Small RNA class	Bacterial antibiotic resistance	Plant antibiotic resistance	GATEWAY use	Backbone	Promoter	Terminator	Plant species
pENTR-AtMIR390a-B/c	amiRNA	Kanamycin	–	Donor	*pENTR*	–	–	–
pFK210B-AtMIR390a-B/c	amiRNA	Spectomycin	BASTA	–	*pGreen III*	*CaMV* 35S	*rbcS*	Eudicots
pMDC123SB-AtMIR390a-B/c	amiRNA	Kanamycin	BASTA	–	*pMDC123*	*CaMV* 2×35S	–	Eudicots
pMDC32B-AtMIR390a-B/c	amiRNA	Kanamycin Hygromycin	Hygromycin	–	*pMDC32*	*CaMV* 2×35S	*nos*	Eudicots
pENTR-OsMIR390-B/c	amiRNA	Kanamycin	–	Donor	*pENTR*			–
pMDC123SB-OsMIR390-B/c	amiRNA	Kanamycin	BASTA	–	*pMDC123*			Monocots
pMDC32B-OsMIR390-B/c	amiRNA	Kanamycin Hygromycin	Hygromycin	–	*pMDC32*			Monocots
pH7WG2B-OsMIR390-B/c	amiRNA	Spectinomycin	Hygromycin	–	*pH7WG2*			Monocots
pENTR-AtTAS1c-B/c	syn-tasiRNA	Kanamycin	–	Donor	*pENTR*	–	–	–
pMDC123SB-AtTAS1c-B/c	syn-tasiRNA	Kanamycin Hygromycin	BASTA	–	*pMDC123*	*CaMV* 2×35S	*nos*	*A. thaliana*
pMDC32B-AtTAS1c-B/c	syn-tasiRNA	Kanamycin Hygromycin	Hygromycin	–	*pMDC32*	*CaMV* 2×35S	*nos*	*A. thaliana*

2.3 Functional Analysis of Antiviral AmiRNAs

2.3.1 Agroinfiltration in Nicotiana benthamiana

1. *Nicotiana benthamiana*: 3–4-week-old *N. benthamiana* plants.
2. LB with antibiotics: 10 g/L tryptone, 5 g/L yeast extract, 10 g/L NaCl, 50 mg/L rifampicin, and 50 mg/L kanamycin.
3. Sterilized stock solutions: 1 M $MgSO_4$, 1 M $CaCl_2$, 20% glucose, 1 M MES buffer (pH 5.2), and 0.1 M acetosyringone.
4. M9 solution: 6 g/L of Na_2HPO_4, 3 g/L of KH_2PO_4, 0.5 g/L NaCl, and 1 g/L NH_4Cl. Adjust pH to 5.2 with KOH and autoclave. Add 2 mL 1 M $MgSO_4$, 0.1 mL 1 M $CaCl_2$, and 10 mL 20% glucose in sterile hood.
5. Vir Induction medium: Add 5 mL 20% glucose, 5 mL 1 M MES buffer (pH 5.2), 500 μL 0.1 M acetosyringone, 50 μL 1 M $CaCl_2$, and 1 mL 1 M $MgSO_4$ to 500 mL of sterile M9 solution.
6. Infiltration solution: Add 5 mL 1 M $MgCl_2$, 5 mL 1 M MES buffer (pH 5.2), and 750 μL 0.1 M acetosyringone to 500 mL of sterile H_2O.
7. 1 mL syringes without needles.

2.3.2 Analysis of Antiviral AmiRNA Accumulation

Total RNA Extraction

1. TRIzol reagent (Invitrogen).
2. Chloroform.
3. Isopropanol.
4. Cold 75% ethanol.
5. NanoDrop (Thermo Fisher Scientific).

Northern Blot Analysis of AmiRNAs

1. 5× Tris–borate–EDTA (TBE): 54 g/L Tris base, 27.5 g/L boric acid, 20 mL 0.5 M EDTA (pH 8.0).
2. 30% polyacrylamide (37.5:1): Dissolve 29.2 g acrylamide and 0.8 g bisacrylamide in 100 mL H_2O.
3. Urea, electrophoresis grade.
4. TEMED: *N,N,N,N′*-tetramethylethylenediamine.
5. Ammonium persulfate: 10% aqueous solution.
6. Formamide loading buffer: 95% formamide, 0.1% (w/v) bromophenol blue, 0.1% (w/v) xylene cyanol FF, 5 mM EDTA (pH 8.0).
7. Gel plate, an electrophoresis apparatus, and a power supply.
8. Blotting paper.
9. Whatman Nytran SuperCharge nylon membrane (Sigma-Aldrich).
10. Semidry electroblotting apparatus.
11. UV cross-linker.

12. Probe oligonucleotide: DNA or LNA oligonucleotide complementary to the amiRNA to be analyzed.
13. T4 polynucleotide kinase.
14. [^{32}P] γ-ATP, 6000 Ci/mmol, 10 mCi/mL
15. Mini Quick Spin Oligo Columns (Roche Life Science).
16. PerfectHyb™Plus buffer (Sigma-Aldrich) or equivalent hybridization solution.
17. 20× SSC (saline sodium citrate): 3.0 M NaCl, 0.3 M sodium citrate, pH 7.0.
18. 10% SDS: 10% (w/v) sodium dodecyl sulfate in sterile H_2O.
19. Wash buffer 1: 2× SSC, 0.1% SDS. Prepare with 20× SSC, 10% SDS solution, and sterile H_2O.
20. Wash buffer 2: 0.1× SSC, 0.1% SDS. Prepare with 20× SSC, 10% SDS solution, and sterile H_2O.
21. 3× SSC: Dilute 20× SSC with sterile H_2O.

2.3.3 Virus Inoculation

1. Inoculation buffer: 50 mM potassium phosphate pH 8.0, 1% PVP (polyvinylpyrrolidone) 10, 1% PEG (polyethylene glycol) 6000 and 10 mM 2-mercaptoethanol.
2. Carborundum powder (0.013 mm).
3. Cotton swabs.

2.4 Antiviral Syn-tasiRNA Design

All of the materials required are the same as Subheading 2.1.

2.5 Antiviral Syn-tasiRNA Cloning

1. Syn-tasiRNA vectors: *pENTR-AtTAS1c-B/c* (Addgene plasmid 51774), *pMDC123SB-AtTAS1c-B/c* (Addgene plasmid 51772) and *pMDC32B-AtTAS1c-B/c* (Addgene plasmid 51773) (Table 1).
2. Other reagents listed in Subheading 2.2.

2.6 Functional Analysis of Antiviral Syn-tasiRNA

All of the materials required are the same as Subheading 2.3.

3 Methods

The main steps regarding the design, synthesis, and analysis of antiviral artificial sRNAs are summarized in Fig. 1.

3.1 Antiviral AmiRNA Design

1. Download the virus sequences (e.g., from NCBI).
2. Go to P-SAMS amiRNA Designer website (http://p-sams.carringtonlab.org/amirna/designer).

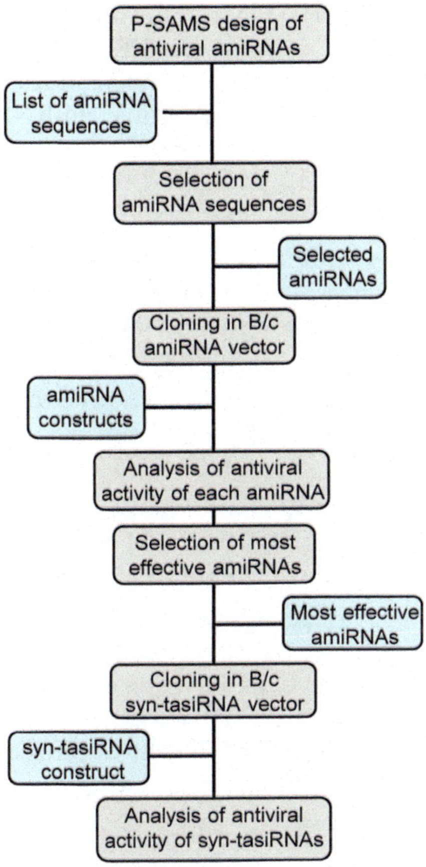

Fig. 1 Diagram of the steps for the design and synthesis of antiviral amiRNAs and syn-tasiRNAs, and for the analysis of their antiviral activity. Each step is described in light grey boxes. The product of each step is shown in light blue boxes

3. Click "Get Started" and then "Design an amiRNA" to start the design process.
4. Select the transcriptome of the species of interest and click "Yes."
5. Click "Option 2" to target an exogenous transcript such as a viral RNA.
6. Enter or paste the FASTA sequence(s) of target viral RNA(s) and click "Next" (*see* **Note 3**).
7. Click "Yes" to have the results automatically filtered based on target specificity (*see* **Note 4**).
8. Click "Submit" to submit the job.
9. Select "Click to see Results" to display the results. AmiRNAs predicted to target uniquely viral RNA(s) are output as "Optimal Results"; amiRNAs predicted to target viral RNA(s) and endogenous transcripts are output as "Sub-optimal Results."

10. Click "Download" to download the result summary including a list of amiRNA sequences together with the sequence of the two oligonucleotides required for cloning each amiRNA insert in compatible amiRNA B/c vectors (Table 1).

3.2 Antiviral AmiRNA Cloning

1. Resuspend the two oligonucleotides in sterile H_2O to a final concentration of 100 μM.
2. Assemble the oligonucleotide annealing reaction in a PCR tube as follows: 2 μL of 100 μM forward oligonucleotide, 2 μL of 100 μM reverse oligonucleotide and 46 μL of oligo annealing buffer.
3. Transfer the tube to a thermocycler set to heat the annealing reaction 5 min at 94 °C and then cool down to 20 °C at a rate of 0.05 °C/s (*see* **Note 5**).
4. Dilute the annealed oligonucleotides to a final concentration of 0.15 μM by mixing 3 μL annealed oligonucleotides and 37 μL H_2O (*see* **Note 6**).
5. Assemble the digestion–ligation reaction as follows: 50 ng B/c vector, 1 μL diluted annealed oligonucleotides, 1 μL 10× T4 DNA ligase buffer, 1 μL of 5 U/μL T4 DNA ligase, 1 μL of 10 U/μL *Bsa*I and add sterile H_2O to 10 μL. Use the amiRNA B/c vector of your choice (Table 1) (*see* **Note 7**).
6. Mix the reactions by pipetting and incubate for 5 min at 37 °C.
7. Transform 1-5 μL of the digestion–ligation product into a *ccd*B-sensitive *E. coli* strain and plate on an LB agar plate with kanamycin.
8. Pick two colonies per construct, grow in 4 mL of liquid LB with kanamycin and purify plasmids with a miniprep kit.
9. Sequence two clones per construct with appropriate oligonucleotides: M13-F and M13-R for *pENTR*-based vectors, attB1and attB2 for *pMDC32B*-, *pMDC123SB*-, *pFK210B*-, or *pH7WG2B*-based vectors.
10. Transform 0.5 μL of the purified plasmid into *A. tumefaciens* GV3101. Plate 1/10 of the culture in a LB agar plate including rifampicin and kanamycin.
11. Incubate the plate at 28 °C during 48 h.
12. Store the plate with grown colonies at 4 °C until use.

3.3 Functional Analysis of Antiviral AmiRNAs

3.3.1 Experimental Setup

The main steps of the experimental plan for the functional analysis of antiviral amiRNAs are summarized in Fig. 2 and described below. Typically, three blocks (A, B, and C) of three plants are required for the functional analysis of each antiviral amiRNA construct. It is highly recommended to compare the activity of

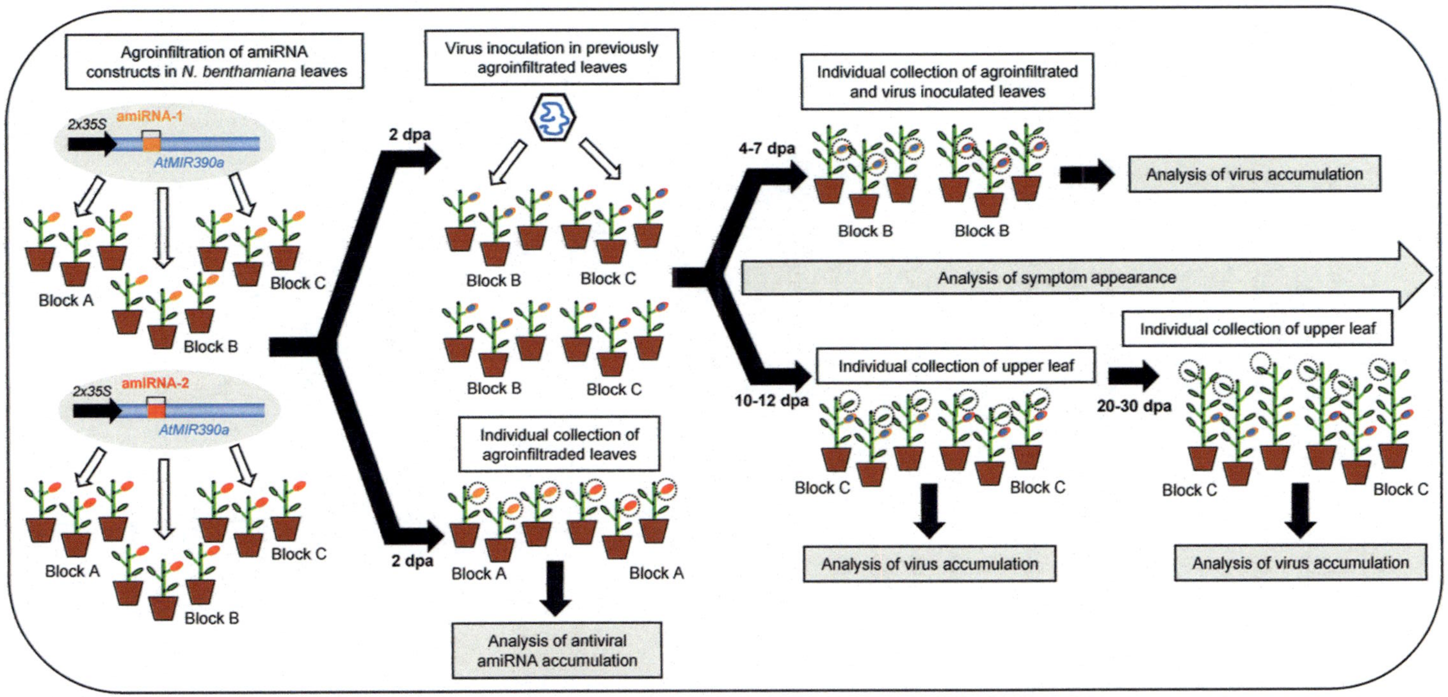

Fig. 2 Experimental plan to functionally analyze antiviral amiRNAs. Main experimental steps are described in white boxes. Main analytical steps are described in light grey boxes. "dpa" refers to days postagroinfiltration. Due to space limitations, the functional analysis plan is shown for only two amiRNA constructs, but this setup can be applied for the functional analysis of multiple amiRNA constructs in large-scale amiRNA screenings

antiviral amiRNAs with a control amiRNA targeting an exogenous sequence other than the viral RNA (*E. coli* β-glucuronidase (GUS), green fluorescent protein (GFP), etc.) (*see* **Note 8**).

3.3.2 Agroinfiltration of N. benthamiana Leaves with Antiviral AmiRNA Constructs

1. For each amiRNA construct to be analyzed, pick a single colony of *A. tumefaciens* transformed with the corresponding construct using a toothpick.
2. Transfer to a culture tube that contains 5 mL of LB with rifampicin and kanamycin.
3. Incubate the starter culture at 28 °C for 24 h on a shaking incubator.
4. Transfer 1–5 mL of the starter culture to 50 mL of LB containing kanamycin in a 250 mL flask.
5. Shake for 4–6 h at 28 °C until OD_{600} ~0.55.
6. Centrifuge 45 mL of culture for 15 min at 5000 × *g* in a 50 mL conical tube.
7. Remove the supernatant and resuspend in equal volume of Vir Induction medium.
8. Transfer back to flasks and shake overnight (~14 h) at 28 °C.
9. Measure OD_{600} of the overnight culture.
10. Calculate the volume of Infiltration solution to resuspend the pellet to an OD_{600} ~1.1.
11. Centrifuge the culture for 15 min at 5000 × *g* in a 50 mL conical tube.
12. Remove the supernatant and resuspend pellet in calculated volume of Infiltration solution.
13. Measure OD_{600}, and normalize the culture to a final OD_{600} of 1.0.
14. Mix the amiRNA culture with an equal volume of a *A. tumefaciens* culture including an empty vector or control construct such as *pMDC32-GUS* [17].
15. Infiltrate the whole abaxial surface of the fourth emerged leaf of nine 3–4-week-old *N. benthamiana* plants (Fig. 2).
16. Two days postagroinfiltration (dpa), collect independently the infiltrated leaf of each of the three plants of block A (Fig. 2). Store tissue at −80 °C until needed.

3.3.3 Analysis of Antiviral AmiRNA Accumulation

This step is recommended to confirm that amiRNAs are being expressed in planta.

Total RNA Extraction

1. Place frozen leaf in a labeled mortar and pour 30–50 mL liquid nitrogen.
2. Lightly crush leaves after liquid nitrogen has completely evaporated; grind vigorously until tissue is powdered.

3. Add 5 mL of TRIzol and continue homogenization to mix partially frozen sample.
4. Transfer homogenized sample to a 15 mL conical tube.
5. Centrifuge at 7000 × *g* for 10 min at 4 °C to pellet cell debris.
6. Transfer supernatant to a new 15 mL conical tube.
7. Add 3 mL of chloroform and shake vigorously for 20 s.
8. Incubate at room temperature for 3 min.
9. Centrifuge at 7000 × *g* for 10 min at 4 °C to separate phases.
10. Transfer aqueous phase to new 15 mL conical tube.
11. Add 5 mL cold isopropanol.
12. Invert several times and incubate for 10 min at room temperature.
13. Centrifuge at 10,000 × *g* for 10 min at 4 °C to pellet RNA.
14. Wash the pellet with cold 75% ethanol by adding the ethanol, inverting twice and pouring off.
15. Air-dry for 2 min and remove residual ethanol by pipetting.
16. Air-dry for 2 min and resuspend the pellet in 100–200 μL sterile H_2O.
17. Transfer to a 1.5 mL microcentrifuge tube.
18. Quantify RNA concentration with a NanoDrop.

Northern Blot Analysis of AmiRNAs

Confirm the accumulation of amiRNA in planta by Northern blot hybridization following standard protocols. The following are some specifications.

1. Prepare a 17% polyacrylamide gel containing 7 M Urea in 0.5× TBE. Mix 17 mL of 30% polyacrylamide (acrylamide–bisacrylamide, 37.5:1), 12.6 g of urea, 1.5 mL of 5× TBE, and 2 mL H_2O. Mix thoroughly by inversion. Do not shake or vortex as this incorporates oxygen to the solution which inhibits polymerization. Heat to 65 °C for 10 min to dissolve the urea. Stir for additional 5 min on the bench to allow final resuspension of urea, and keep 20 min on ice to chill the solution. Add 25 μL TEMED and mix by inversion. Add 150 μL of 10% ammonium persulfate and mix quickly by inversion. Pour the gel and allow it to polymerize for at least 30 min.
2. Prerun the gel at 180 V in 0.5× TBE for 1 h. Rinse wells before loading samples as urea accumulates at the bottom of the wells.
3. Add an equal amount of formamide loading buffer to the total RNA. Heat the samples for 10 min at 65 °C and immediately quench on ice briefly.
4. Load 10–20 μg of total RNA. Run at 180 V in 0.5× TBE until the bromophenol blue reaches the bottom of the gel (~4 h).

5. Assemble the blot sandwich with blotting paper (three sheets, prewetted with 0.5× TBE), gel, a sheet of positively charged nylon membrane, and another three sheets of blotting paper prewetted with 0.5× TBE in a semidry chamber. Transfer for 30 min at 1 mA per cm^2 of membrane. Cross-link RNA to the membrane at 0.12 J/cm^2 using a UV cross-linker. Store membranes between two sheets of filter paper until use.
6. Prepare the following reaction mix to radiolabel the probe oligonucleotide (*see* **Note 9**): 1 μL 10 μM probe oligonucleotide, 1 μL 10× polynucleotide kinase buffer, 3 μL sterile H_2O, 4 μL [^{32}P] γ-ATP (6000 Ci/mmol; 10 mCi/mL), and 1 μL T4 polynucleotide kinase.
7. Incubate for 60 min at 37 °C. Purify probe on spin columns according to manufacturer's instructions. Quantify the incorporated radioactivity (CPM/μL).
8. Place the membrane (RNA-side-up) in a hybridization tube and prehybridize with rotation for at least 5 min at 38–42 °C (*see* **Note 10**) in 5 mL of hybridization buffer.
9. Mix 1,000,000–2,000,000 CPM of probe with 200 μL of PerfectHyb™Plus buffer, incubate for 2 min at 95 °C, and transfer immediately to ice briefly.
10. Replace the hybridization solution, add the probe to the hybridization tube and incubate for 12–16 h at 38–42 °C (*see* **Note 10**).
11. Remove hybridization solution and wash the membrane five times with preheated wash solutions as follows: wash buffer 1 thorough rinse, wash buffer 1 for 5 min with rotation at 38–42 °C, wash buffer 1 for 20 min with rotation at 55 °C, wash buffer 1 for 20 min with rotation at 55 °C, wash buffer 2 for 20 min with rotation at 55 °C, and wash buffer 2 for 30 min with rotation at 55 °C (*see* **Note 11**).
12. Rinse membrane briefly in 3× SSC, then air-dry briefly and cover in transparent plastic wrap.
13. Autoradiograph.

3.3.4 Virus Inoculation

The mechanical inoculation of plants from blocks B and C is described next.

1. From a previously obtained virus infected tissue, prepare a crude extract in 20 volumes of Inoculation buffer.
2. Deposit several drops (5 μL each) of 10% carborundum in Inoculation buffer on leaf surface. Note that the number of drops will depend on the leaf size, although one or two drops per leave should be enough.
3. Dip a cotton swab in the tube containing the infected-plant extract.

4. Apply the extract to the whole surface of the leaf that was previously agroinfiltrated 2 days before with the artificial sRNA construct (Fig. 2) (*see* **Note 12**).

3.3.5 Analysis of Symptom Appearance

After virus inoculation, plants must be monitored for viral symptoms appearance (Fig. 2). Depending on the virus, symptoms can appear on inoculated and/or upper noninoculated tissue. The day of symptom appearance in either tissue must be annotated.

3.3.6 Analysis of Virus Accumulation

Virus presence or absence can be determined through different types of assays including ELISA, Western blot, Northern blot, or quantitative RT-PCR (*see* **Note 13**). Two time points are recommended for virus detection: (1) an early time point a few days after the controls show viral symptoms in upper noninoculated tissues (e.g., 10–12 dpa if symptoms appear 7–9 dpa in controls), and (2) a late time point (20–30 dpa) to analyze if the plant is definitely free of virus close to the end of its life cycle (Fig. 2). Note that dpa time is always 2 days ahead of virus inoculation.

3.4 Antiviral Syn-tasiRNA Design

1. Select 2–5 amiRNA sequences with high antiviral activity based on the amiRNA functional analysis described above (*see* Subheading 3.3). These sequences will be expressed as syn-tasiRNAs in a single construct.
2. Go to P-SAMS syn-tasiRNA Designer website (http://p-sams.carringtonlab.org/syntasi/designer) [15].
3. Click "Get Started" to start the design process.
4. Click "Generate Oligos."
5. Enter 2–5 syn-tasiRNA sequences.
6. Click "Submit" to submit the job.
7. Select "Click to see Results" to display results.
8. Click "Download" to download the sequence of the syn-tasiRNAs and the sequence of the two oligonucleotides required for cloning the syn-tasiRNA cassette in compatible syn-tasiRNA B/c vectors (Table 1).

3.5 Antiviral Syn-tasiRNA Cloning

Follow the steps described in Subheading 3.2 using the syn-tasiRNA vector of your choice (Table 1).

3.6 Functional Analysis of Antiviral Syn-tasiRNA

Follow the experimental plan described in Fig. 2 to functionally analyze the antiviral activity of the antiviral syn-tasiRNA construct. As noted for the antiviral amiRNA functional analysis (*see* Subheading 3.3.1), it is highly recommended to compare the activity of the antiviral syn-tasiRNA construct with a control syn-tasiRNA including several syn-tasiRNA sequences targeting an artificial transcript (e.g. GUS, GFP, etc.) (*see* **Note 14**).

4 Notes

1. Prepare Oligo Annealing Buffer, make 1 mL aliquots and store at −20 °C.
2. B/c vectors must be ordered at Addgene website (http://www.addgene.org/).
3. Multiple viral sequences can be entered, for example, when several sequence variants of the same virus are required to be targeted simultaneously.
4. This option allows the design of specific amiRNAs that will exclusively target viral RNA(s) without affecting endogenous transcripts.
5. Alternatively, the annealing reaction can be done in a water bath or thermoblock by heating during 5 min at 94 °C and then turning off the apparatus. Let the reaction to cool down until it reaches room temperature.
6. Do not store the diluted oligonucleotides.
7. If amiRNA inserts are cloned in *pENTR*-based B/c vectors, then the amiRNA cassette can be transfer through GATEWAY recombination to a destination vector such as *pMDC32* or *pMDC99*.
8. As for antiviral amiRNAs, use P-SAMS amiRNA Designer to design a highly specific amiRNA targeting an exogenous sequence other than the viral RNA. Confirm that the selected amiRNA does not extensively base pair with viral RNA(s) to avoid undesired silencing of the virus.
9. Oligonucleotide sequence is antisense to the artificial sRNA that is analyzed. Start by using a DNA probe for artificial sRNA detection as DNA oligonucleotides are cheap and work well most of the times. However, if sRNA detection with a DNA probe fails then use an LNA probe. Order the LNA oligonucleotide with every other three nucleotides locked, including the first one (e.g., an LNA probe to detect *A. thaliana* miR172 is A + TGC + AGC + ATC + ATC + AAG + ATT + CT, where the + indicates the locked nucleotide).
10. If using an LNA probe prehybridize and hybridize at ~20 °C below the calculated dissociation temperature (T_d) [T_d (°C) = 4(G + C) + 2(A + T)] for the corresponding ^{32}P-labeled oligonucleotide.
11. This works in the majority of cases. If there is still a background problem, proceed to an additional incubation of the membrane with 0.1× SSC/0.1% SDS for 60 min at 50 °C, or with 0.1× SSC/1% SDS for 60 min at 50 °C.
12. Use the cotton swab as a pencil to gently "paint" the whole surface of the leaf.

13. The choice of a particular assay will depend on the virus to be detected. For instance, ELISA has been optimized for *Tomato spotted wilt virus* detection [18].
14. The control syn-tasiRNA construct can be designed using P-SAMS syn-tasiRNA Designer (http://p-sams.carrington-lab.org/syntasi/designer).

Acknowledgments

This work was supported by grants from Ministerio de Ciencia, Innovación y Universidades (MCIU, Spain), Agencia Estatal de Investigación (AEI, Spain) and Fondo Europeo de Desarrollo Regional (FEDER, European Union) [RTI2018-095118-A-100 and RYC-2017-21648 to A.C.; BIO2017-83184-R to J.-A.D.].

References

1. Carbonell A (2017) Artificial small RNA-based strategies for effective and specific gene silencing in plants. In: Dalmay T (ed) Plant gene silencing: mechanisms and applications. CABI Publishing, Boston, MA, pp 110–127
2. Carbonell A (2017) Plant ARGONAUTEs: features, functions, and unknowns. Methods Mol Biol 1640:1–21. https://doi.org/10.1007/978-1-4939-7165-7_1
3. Niu QW, Lin SS, Reyes JL, Chen KC, Wu HW, Yeh SD, Chua NH (2006) Expression of artificial microRNAs in transgenic *Arabidopsis thaliana* confers virus resistance. Nat Biotechnol 24(11):1420–1428. https://doi.org/10.1038/nbt1255
4. Lin SS, Wu HW, Elena SF, Chen KC, Niu QW, Yeh SD, Chen CC, Chua NH (2009) Molecular evolution of a viral non-coding sequence under the selective pressure of amiRNA-mediated silencing. PLoS Pathog 5(2):e1000312. https://doi.org/10.1371/journal.ppat.1000312
5. Lafforgue G, Martinez F, Sardanyes J, de la Iglesia F, Niu QW, Lin SS, Sole RV, Chua NH, Daros JA, Elena SF (2011) Tempo and mode of plant RNA virus escape from RNA interference-mediated resistance. J Virol 85(19):9686–9695. https://doi.org/10.1128/JVI.05326-11
6. Fahim M, Millar AA, Wood CC, Larkin PJ (2012) Resistance to *Wheat streak mosaic virus* generated by expression of an artificial polycistronic microRNA in wheat. Plant Biotechnol J 10(2):150–163. https://doi.org/10.1111/j.1467-7652.2011.00647.x
7. Kung YJ, Lin SS, Huang YL, Chen TC, Harish SS, Chua NH, Yeh SD (2012) Multiple artificial microRNAs targeting conserved motifs of the replicase gene confer robust transgenic resistance to negative-sense single-stranded RNA plant virus. Mol Plant Pathol 13(3):303–317. https://doi.org/10.1111/j.1364-3703.2011.00747.x
8. Lafforgue G, Martinez F, Niu QW, Chua NH, Daros JA, Elena SF (2013) Improving the effectiveness of artificial microRNA (amiR)-mediated resistance against *Turnip mosaic virus* by combining two amiRs or by targeting highly conserved viral genomic regions. J Virol 87(14):8254–8256. https://doi.org/10.1128/JVI.00914-13
9. Kis A, Tholt G, Ivanics M, Varallyay E, Jenes B, Havelda Z (2016) Polycistronic artificial miRNA-mediated resistance to *Wheat dwarf virus* in barley is highly efficient at low temperature. Mol Plant Pathol 17(3):427–437. https://doi.org/10.1111/mpp.12291
10. Carbonell A, Carrington JC, Daros JA (2016) Fast-forward generation of effective artificial small RNAs for enhanced antiviral defense in plants. RNA Dis 3(1)
11. Chen L, Cheng X, Cai J, Zhan L, Wu X, Liu Q, Wu X (2016) Multiple virus resistance using artificial trans-acting siRNAs. J Virol Methods 228:16–20. https://doi.org/10.1016/j.jviromet.2015.11.004
12. Carbonell A, Daros JA (2017) Artificial microRNAs and synthetic trans-acting small interfering RNAs interfere with viroid infection. Mol Plant Pathol 18(5):746–753. https://doi.org/10.1111/mpp.12529

13. Carbonell A, Takeda A, Fahlgren N, Johnson SC, Cuperus JT, Carrington JC (2014) New generation of artificial MicroRNA and synthetic trans-acting small interfering RNA vectors for efficient gene silencing in Arabidopsis. Plant Physiol 165(1):15–29. https://doi.org/10.1104/pp.113.234989
14. Carbonell A, Fahlgren N, Mitchell S, Cox KL Jr, Reilly KC, Mockler TC, Carrington JC (2015) Highly specific gene silencing in a monocot species by artificial microRNAs derived from chimeric miRNA precursors. Plant J 82(6):1061–1075. https://doi.org/10.1111/tpj.12835
15. Fahlgren N, Hill ST, Carrington JC, Carbonell A (2016) P-SAMS: a web site for plant artificial microRNA and synthetic trans-acting small interfering RNA design. Bioinformatics 32(1):157–158. https://doi.org/10.1093/bioinformatics/btv534
16. Carbonell A, López C, Daròs JA (2019) Fast-forward identification of highly effective artificial small RNAs against different Tomato spotted wilt virus isolates. Mol Plant-Microbe Interact 32(2):142–156. https://doi.org/10.1094/MPMI-05-18-0117-TA
17. Montgomery TA, Howell MD, Cuperus JT, Li D, Hansen JE, Alexander AL, Chapman EJ, Fahlgren N, Allen E, Carrington JC (2008) Specificity of ARGONAUTE7-miR390 interaction and dual functionality in TAS3 trans-acting siRNA formation. Cell 133(1):128–141. https://doi.org/10.1016/j.cell.2008.02.033
18. Mitter N, Zhai Y, Bai AX, Chua K, Eid S, Constantin M, Mitchell R, Pappu HR (2016) Evaluation and identification of candidate genes for artificial microRNA-mediated resistance to tomato spotted wilt virus. Virus Res 211:151–158. https://doi.org/10.1016/j.virusres.2015.10.003

Chapter 14

Resistance Breeding Through RNA Silencing of Host Factors Involved in Virus Replication

Masamichi Nishiguchi, Emran Md. Ali, Hui Chen, Masayuki Ishikawa, and Kappei Kobayashi

Abstract

RNA silencing is a sequence-specific suppression of gene expression conserved in eukaryotes including fungi, plants, and animals. Based on this mechanism, crop improvements have been made to confer pathogen resistance and abiotic stress tolerance. Here we have applied this technique to produce virus resistant tomato plants using host genes involved in viral replication. Tomato homologs of *Arabidopsis TOM1* involved in tobamovirus replication has been isolated and used to construct the plasmids that carried inverted repeats of the genes for induction of RNA silencing. Tomato plants were transformed by the plasmids via *Agrobacterium*, and tested for virus resistance. Actually, the T2 and T3 plants showed resistance to tomato mosaic virus. Here we describe the method to construct RNA silencing-inducing plasmids, to transform tomato plants and to check the introduction of transgenes and virus resistance.

Key words Tomato, Host gene, Plasmid, *TOM1* homologs, Transformation, *LeTH1*, Tomato mosaic virus, Virus resistance, RNA silencing

1 Introduction

There have been many host genes reported, which are involved in the virus infection process [1]. If the expression of these host genes is inhibited, the accumulation level of the virus might be decreased. This kind of resistance is referred to as recessive virus resistance [2]. One of these host genes is *TOM1* in *Arabidopsis* [3]. A homolog of *TOM1*, *TOM3*, is also known [3]. This is involved in tobamovirus replication (*see* review [4]). *Arabidopsis tom1 tom3* double mutant plants showed the complete abolishment of virus accumulation [5]. These homologs of *TOM1* have also been reported in tobacco, tomato, melon, and rice [6]. Tobacco homologs,

Kappei Kobayashi and Masamichi Nishiguchi (eds.), *Antiviral Resistance in Plants: Methods and Protocols*, Methods in Molecular Biology, vol. 2028, https://doi.org/10.1007/978-1-4939-9635-3_14,

NtTOM1 and *NtTOM3*, were engineered to produce double-stranded RNA and used for generating transgenic tobacco plants, which were posttranscriptionally silenced in those genes and showed high resistance to tobamoviruses [6].

RNA silencing, a general term used to describe the posttranscriptional gene silencing in plants, also called quelling in fungi and RNA interference in animals, is a highly conserved process of sequence-specific RNA degradation, translational inhibition, and DNA methylation. This process is induced by double-stranded RNA (dsRNA) [7, 8]. The dsRNA is cleaved by a family of Dicer-like endonucleases into 21–25 bp of small interfering RNA (siRNA) [9]. The negative strand of siRNA is then incorporated into the RNA-induced silencing complex (RISC) and acts as a guide to recognize mRNA for degradation [10]. This mechanism can also be induced by viruses and acts as a host defense mechanism against them [11–14]. More recently we showed the virus resistance in transgenic tomato plants which were RNA-silenced with the endogenous tomato homolog of *LeTH1*, *LeTH2*, or *LeTH3* [15]. Here we present the protocol for producing transgenic tomatoes with an inverted repeat of a tomato gene to see any inhibitory effect on the multiplication of ToMV.

2 Materials

2.1 General Equipment and Consumables

1. Thermal cycler.
2. Microtubes: 1.5 mL, 2.0 mL, 200 μL.
3. Electroporation equipment.
4. Autoclave.
5. Growth chamber.
6. Laminar flow cabinet.
7. Autoclavable flask (100, 300 and 500 mL).
8. Sterilized plastic petri dish (9 cm in diameter).
9. Plant box: Transparent, autoclavable plastic box of 300 mL.
10. Scalpel or razor blade.
11. Forceps or tweezers.
12. Adhesive tape.
13. Miracloth: Pore size of 22–25 μm (Calbiochem, CA, USA).
14. Sterilized paper towel.
15. Plastic pots: 100, 250, and 500 mL.
16. Vermiculite.
17. Clay pots: 1 and 2 L.
18. Commercial potting soils.
19. ELISA reader.

2.2 Plasmid Construction for *TOM1* Silencing

2.2.1 Amplification of *TOM1* Gene

1. Seedlings of tomato: *Solanum lycopersicum* L. (formerly *Lycopersicon esculentum*) cv. Micro-Tom.
2. RNA extraction kit: RNeasy Mini Kit (Qiagen, Hilden, Germany) or equivalent.
3. RNase-free DNase: Recombinant DNase I (RNase-free) (TakaraBio, Kusatsu, Japan) or equivalent.
4. Reverse transcriptase: PrimeScript Reverse Transcriptase (TakaraBio) or equivalent.
5. RT Buffer: It is usually supplied with reverse transcriptase.
6. 10 mM dNTPs: A mixture of cDNA synthesis grade dATP, dCTP, dGTP, and dTTP at 10 mM each.
7. RNase inhibitor: Recombinant RNase Inhibitor (TakaraBio) or equivalent.
8. Oligo(dT) primer: 1 μM 15-mer oligo(dT).
9. BmLeTH1F: CGAGGATCCGATGCACGCTGTTG (BamHI site underlined).
10. ScLeTH1R: CTGGAGCTCATTTCATCATCCG (SacI site underlined).
11. Mg^{+} free 10× PCR buffer: 100 mM Tris–HCl, pH 8.3, 500 mM KCl.
12. 2 mM dNTPs: A mixture of dATP, dCTP, dGTP, and dTTP at 2 mM each.
13. 25 mM $MgCl_2$.
14. Taq DNA Polymerase.
15. 50× TAE stock: 2 M Tris base, 1 M acetate, 50 mM Na_2 EDTA (pH 8.0)
16. Agarose gel electrophoresis.
17. PCR purification kit.

2.2.2 Cloning of TOM1 Gene and Construction of Hairpin RNA Expression Vector

1. pBI121: Binary vector expressing *Escherichia coli* β-glucuronidase (GUS) under the control of *Cauliflower mosaic virus* 35S RNA promoter (35Spro).
2. BamHI, SacI.
3. QIAquick Gel Extraction Kit (Qiagen) or equivalent.
4. Ligation kit: Ligation high ver.2 (Toyobo, Osaka, Japan) or equivalent.
5. *E. coli* competent cells: Chemical competent cells of *Escherichia coli* HB101 or other commonly used strains.
6. SOC medium: 2% tryptone, 0.5% yeast extract, 10 mM NaCl, 20 mM glucose, 10 mM $MgCl_2$, 10 mM $MgSO_4$, 2.5 mM KCl. Adjust pH to 7.0 with 1 N NaOH.
7. LB medium: 1% (w/v) tryptone, 0.5% (w/v) yeast extract, 1% (w/v) NaCl. Adjust pH to 7.0 with 1 N NaOH. Sterilize by autoclave.

8. LB-K100 plates: LB medium supplemented with 1.5% (w/v) agar is sterilize by autoclave and added with 100 μg/mL kanamycin before pouring into petri dishes.
9. Miniprep kit: QIAprep Spin (Qiagen) or equivalent.
10. PCR reagents: Subheading 2.2.1, **items 9–12**.
11. PCR primers.

LeTH1R-GUSF:	TTCAACCCGACCCTATCTACCCGCTTCGCGTCG
BmGUSR:	CGAGGATCCATTGTTTGCCTCCCTGC (BamHI site is underlined)
GUSR-LeTH1F:	CGGGTAGATAGGGTCGGGTTGAAACAGCG
BmLeTH1R:	CGAGGATCCGATGCACGCTGTTGTATCAG (BamHI site is underlined)

2.3 Tomato Transformation

1. Agrobacterium competent cells: Electrocompetent cells of *Agrobacterium tumefaciens* LBA4404 strain.
2. LB-RK plates: LB medium supplemented with 1.5% (w/v) agar should be sterilize by autoclave and added with 50 μg/mL rifampicin and 25 μg/mL kanamycin before pouring into petri dishes.
3. Seeds of tomato (*Solanum lycopersicum* L.) cv Micro-Tom.
4. 70% ethanol.
5. Seed sterilization solution: 10% (v/v; 1% active chlorine) sodium hypochlorite, 0.1% Tween 20.
6. MS stock solution A: 95 g/L KNO_3, 82.5 g/L NH_4NO_3, 8.5 g/L KH_2PO_4, 1115 mg/L $MnSO_4{\cdot}4H_2O$, 430 mg/L $ZnSO_4{\cdot}7H_2O$, 310 mg/L H_3BO_3.
7. MS stock solution A1: 250 mg/L $Na_2MoO_4{\cdot}2H_2O$, 25 mg/L $CuSO_4{\cdot}5H_2O$, 25 mg/L $CoCl_2{\cdot}6H_2O$, 830 mg/L KI.
8. MS stock solution B: 44 g/L $CaCl_2{\cdot}2H_2O$.
9. MS stock solution C: 37 g/L $MgSO_4{\cdot}7H_2O$.
10. MS stock solution D: 5560 mg/L $FeSO_4{\cdot}7H_2O$, 7460 mg/L $Na_2{\cdot}$EDTA.
11. MS liquid medium: To 800 mL double distilled water, add 20 mL MS stock solution A, 1 mL MS stock solution A1, 10 mL MS stock solution B, 10 mL MS stock solution C, and 5 mL MS stock solution D, 15 g sucrose and adjust pH to 5.8 with 0.1 N NaOH and volume to 1 L.
12. MS agar plates: To MS liquid medium (**item 11**), add 0.1% (v/v) Plant Preservative Mix (PPM; Plant Cell Technology) and 0.8% (w/v) agar (plant tissue culture grade), and autoclave. Cool down to 55 °C and pour into plastic dishes.
13. KCMS medium: Dilute MS stock solutions as in **item 11**, add 100 mg myoinositol, 1.3 mg thiamine–HCl, 0.2 mg 2, 4-D,

0.1 mg kinetin, 200 mg KH_2PO_4, 30 g sucrose, 488 mg 2-(*N*-morpholino)ethanesulfonic acid (MES), adjust pH to 5.6 with 0.1 N NaOH and volume to 1 L. Add 0.8% (w/v) agar (plant tissue culture grade), autoclave, cool down to 55 °C, and pour into plastic dishes.

14. Coculture medium: MS liquid medium added with 0.5% (v/v) of 20 mg/mL acetosyringone stock solution in dimethyl sulfoxide (final concentration, 100 μg/mL) after autoclave.
15. Wash medium: MS liquid medium containing 25 g/mL meropenem and 0.1% PPM.
16. KCMS/AS medium: Dilute MS stock solutions as in **item 11**, add 100 mg myoinositol, 1.3 mg thiamine–HCl, 0.2 mg 2, 4-D, 0.1 mg kinetin, 200 mg KH_2PO_4, 30 g sucrose, 10 g glucose, 488 mg MES, adjust pH to 5.4 with 0.1 N NaOH and volume to 1 L. Add 0.8% (w/v) agar (plant tissue culture grade), autoclave, cool down to 55 °C, add 10 mg/L acetosyringone, and pour into plastic dishes.
17. 2Z medium: Dilute MS stock solutions as in **item 11**, add 100 mg myoinositol, 2 mg *trans*-zeatin, 1 mL Gamborg's vitamin solution (Sigma-Aldrich), 20 g sucrose, 488 mg MES, 25 mg meropenem. Adjust pH to 5.8 with 0.1 N NaOH and volume to 1 L, and add 0.8% (w/v) agar. After autoclaving, cool down to 55 °C, add 100 mg/L kanamycin, and pour into plastic dishes.
18. 1Z medium: Dilute MS stock solutions as in **item 11**, add 100 mg myoinositol, 1 mg thiamine–HCl, 1 mg *trans*-zeatin, 20 g sucrose, 488 mg MES, 25 mg meropenem. Adjust pH to 5.8 with 0.1 N NaOH and volume to 1 L, and add 0.8% (w/v) agar. After autoclaving, cool down to 55 °C, add 100 mg/L kanamycin, and pour into plastic dishes.
19. MSSV medium: Dilute MS stock solutions as in **item 11**, add 1 mL Gamborg's vitamin solution, 2 mg indole 3 butyric acid, 30 g sucrose, 488 mg MES, 25 mg meropenem. Adjust pH to 5.8 with 0.1 N NaOH and volume to 1 L, and add 0.8% (w/v) agar. After autoclaving, cool down to 55 °C, add 100 mg/L kanamycin, and pour into plant boxes.

2.4 Analysis of Transgenic Tomato

2.4.1 Verification of the Transgene, Collection of T2 Seeds, and Expression Analysis of the Target Gene

1. DNA extraction kit: DNeasy Plant Mini Kit (Qiagen) or equivalent.
2. PCR primers.

GUS-linker-F:	CATGAAGATGCGGACTTACG
LeTH1-769F:	ATGTTACGGCGTTTCCCTG
LeTH1-1159R:	TCCAGAGTCCTGATGAACC
LeActin-F:	TGGCATCATACTTTCTACAATG
LeActin-R:	CTAATATCCACCTCACATTTCAT

3. Reagents for PCR and RT-PCR analyses: *see* Subheading 2.2.1, **items 2–7** and **10–15**.
4. Seed selection medium: MS agar plates containing 100 mg/L kanamycin.

2.4.2 Virus Infection and Enzyme-Linked Immunosorbent Assay (ELISA)

1. Seedlings of transgenic or wild-type plants.
2. ToMV-L: *Tomato mosaic virus* (ToMV) L strain [16]. Other strains of ToMV or other Tobamovirus spp. can also be used.
3. 10 mM sodium phosphate buffer, pH 7.0.
4. Carborundum: Carborundum powder (600 mesh).
5. Inoculation rods: Glass rods with flattened paddle on an end.
6. Plastic boxes: Plastic boxes with transparent lids. For example, 33 cm long, 24 cm wide and 5 cm high. No need to be tightly sealed.
7. Coating buffer: 1.59 g Na_2CO_3, and 2.93 g $NaHCO_3$ dissolved in 900 mL H_2O, adjust pH to 9.6 with HCl and make up to 1 L.
8. Polyethylene bag: 13 cm long and 7.5 cm wide, 0.1 mm thickness.
9. Microtiter plate.
10. 10× PBST: Concentrated stock for phosphate buffered saline with 0.05% Tween 20. Dissolve 80 g NaCl, 2 g KCl, 14.4 g Na_2HPO_4, and 2.4 g KH_2PO_4 in distilled water, adjust pH to 7.4 and volume to 1 L. Dilute tenfold to prepare PBST working solution.
11. Primary antibody: Anti-ToMV antibody from Japan Plant Protection Association (Tokyo, Japan) is diluted 1000-fold in PBST. Alternatively, one may use the antibodies or antisera raised against the virus to be tested. The optimal concentration of the primary antibody should be empirically determined.
12. Secondary antibody: Goat anti-rabbit IgG antibody alkaline phosphatase conjugate (Sigma-Aldrich or other manufacturer) is diluted 1000-fold in PBST.
13. Substrate buffer: 1 M diethanolamine buffer, pH 9.8, 0.5 mM $MgCl_2$. Add 97 mL of diethanolamine and 100 mg of $MgCl_2$ to 800 mL of water, adjust pH to 9.8 with 10 M HCl solution and volume to 1 L with water.
14. Phosphatase substrate: A phosphatase substrate 5 mg tablet (Sigma-Aldrich) is dissolved in 5 mL substrate buffer. Alternatively, 1 mg/mL *p*-nitrophenyl phosphate in the substrate buffer.
15. Stop solution: 3 N NaOH solution.

3 Methods

3.1 Plasmid Construction for TOM1 Silencing

3.1.1 Amplification of LeTH1 Gene by RT-PCR

1. Young tomato seedlings of 2–3 weeks after sowing are subjected to RNA extraction using an RNA extraction kit.
2. About 10 μg of total RNA preparation is treated with the RNase-free DNaseI in the presence of RNase inhibitor in a 50 μL reaction mixture. After DNase-treatment, RNA is purified using an RNA extraction kit.
3. The cDNA is synthesized from the DNase-treated RNA using oligo(dT) primer with a reverse transcriptase. The reaction is set in thin walled tubes as follows: Mix 1 μL 10 mM dNTP, 1 μL 10 μM oligo(dT) primer, and 1 μL 1 μg/μL total RNA, and add H_2O up to 10 μL.
4. This mixture is denatured at 65 °C for 5 min, quickly chilled in ice-water, and then added with 4 μL 5× RT buffer, 0.5 μL 20 U/μL RNase Inhibitor, 1 μL 200 U/μL reverse transcriptase and 4.5 μL H_2O. Total reaction volume is 20 μL.
5. The reaction is performed in a thermal cycler as follows: 30 °C for 10 min; 42 °C for 60 min; 95 °C for 5 min.
6. The cDNA is used as a template for PCR to amplify *LeTH1* gene (a tomato homolog of *Arabidopsis TOM1*). PCR reaction is set in thin walled tubes as follows: 5 μL Mg^{++} free 10× PCR buffer, 6 μL 25 mM $MgCl_2$, 5 μL 2 mM dNTPs, 1 μL 10 μM BmLeTH1F, 1 μL 10 μM ScLeTH1R, 1 μL cDNA from the previous step (*see* **Note 1**), 2.5 μL Taq DNA polymerase and H_2O up to 50 μL. The primers were designed for cloning entire ORF of *LeTH1* with BamHI and SacI recognition sites at 5′ and 3′ ends, respectively. PCR tubes are set in a thermal cycler at 2 min of 94 °C preheating, followed by a 25-cycle amplification program of 1 min at 94 °C, 1 min at 55 °C, and 1 min at 72 °C, and a final extension at 72 °C for 5 min.
7. The PCR reaction is analyzed on a 1% agarose gel.
8. The amplified DNA fragment is purified with a PCR purification kit.

3.1.2 Cloning of TOM1 Gene and Construction of Hairpin RNA Expression Vector

1. The purified PCR products from Subheading 3.1.1, **step 8** is digested with BamHI and SacI for more than 8 h to ensure complete digestion.
2. The reaction is separated on a 1% agarose gel and the *LeTH1* gene fragment is gel-purified.
3. The pBI121 plasmid is digested with BamHI and SacI to remove GUS gene coding sequence.

4. The product is subjected to a 1% agarose gel electrophoresis to confirm the generation of two DNA fragments, the vector and GUS gene fragments.
5. The vector fragment is gel-purified and ligated with a gel-purified, PCR-amplified fragment from **step 2** using a ligation kit according to the manufacturer's protocol.
6. Ligated solution is used for transformation of *E. coli*, and the resultant plasmid pBI-LeTH1 is obtained. The transformant colonies formed on LB-K100 plates are grown in LB medium containing 100 μg/mL kanamycin. Plasmid DNA is extracted using a miniprep kit and analyzed on a 1% agarose gel electrophoresis after digestion with BamHI and SacI. The pBI-LeTH1 is further used to construct a plasmid carrying an *LeTH1* cDNA inverted repeat with an intervening GUS sequence.
7. Two PCR mixtures are prepared separately as in Subheading 3.1.1, **step 6** except for primers and template: one with primers LeTH1R-GUSF and BmGUSR with pBI121 as a template for the amplification of a GUS fragment, and the other with primers GUSR-LeTH1F and BmLeTH1R with pBI-LeTH1 as a template for the amplification of LeTH1 cDNA. The reaction mixtures including 1 ng/μL template are incubated at 94 °C for 4 min, and subjected to 12 PCR cycles of 94 °C for 30 s-50 °C for 1 min-72 °C for 3 min, followed by incubation at 72 °C for 7 min.
8. Amplified DNAs are gel-purified, combined and further used as templates for overlap extension PCR (OE-PCR) amplification with primers BmLeTH1R and BmGUSR to obtain LeTH1 cDNA fused with a GUS sequence. Amplified DNA is purified using a PCR purification kit, digested by BamHI, and gel-purified using a gel purification kit.
9. The pBI-LeTH1 digested with BamHI is ligated using a ligation kit with the gel-purified LeTH1-GUS DNA fragment and used to transform *E. coli.*
10. Plasmid DNA is extracted as in **step 6** and analyzed on a 1% agarose gel electrophoresis after digestion with ScaI to select the plasmid with expected structure (carrying an inverted copy of LeTH1 sequence, GUS spacer sequence, and LeTH1sequence in the downstream of 35Spro).

3.2 Tomato Transformation

1. Transform *Agrobacterium* with the plasmid constructed in Subheading 3.1.2, **step 10**. The bacterial transformation is carried out with a standard protocol of electroporation and the bacteria are spread on LB-RK plates.
2. Sterilize about 100 tomato seeds in a laminar flow cabinet by immersing in 70% ethanol for 2 min, followed by 15 min in Seed sterilization solution, and then rinse the seeds three times with sterilized distilled water for 9 min (*see* **Note 2**).

3. Sow about 10 seeds per plate on MS agar plates. Incubate the plates in a growth chamber at 24 °C with 16 h light period and 8 h dark period with 60% humidity for 5–7 days.
4. Spread *Agrobacterium* carrying the binary vector on three LB-RK plates 2–3 days before transformation and culture at 28 °C until use.
5. Cut cotyledons into three equal pieces and excise hypocotyls into 7–10 mm segments just 0.5 cm below the cotyledonary nodes with a sterilized scalpel or razor blade (*see* **Note 3**).
6. Transfer the explants to KCMS medium and culture at 25 °C overnight.
7. Scrape off and resuspend the transformed *Agrobacterium* colonies from **step 4** above in 5 mL of coculture medium, dip the explants in the bacterial suspension for 2 min, remove the suspension on sheets of sterilized paper towel, and then place the explants on KCMS/AS medium at 25 °C for 2–3 days.
8. Wash the explants twice with 20–30 mL MS liquid medium (*see* **Note 4**), and with 15–20 mL wash medium, blot-dry the explants on the sterilized paper towel to remove the medium, lay the explants on 2Z medium at 25 °C for 2 weeks.
9. Transfer the explants to a new 2Z medium, culture further for 2–3 weeks at 25 °C.
10. Transplant kanamycin-resistant calluses to 1Z medium and culture at 25 °C for 2–3 weeks. Transfer calluses with kanamycin-resistant shoots to a new 1Z medium and culture further for 1–2 weeks.
11. Cut the shoot about 3 cm long at the base and transfer onto MSSV agar medium in a plant box to induce root formation.
12. Transfer the well-rooting plant to vermiculite in a plastic pot and let it acclimatize to atmosphere.
13. Transplant the acclimatized plant to a clay pot containing soil. The plant is grown to obtain T1 seeds.

3.3 Analysis of Transgenic Tomato

3.3.1 Verification of the Transgene and Collection of T2 Seeds

1. Extract Genomic DNA from the T0 transformant tomato plants using a plant DNA extraction kit according to the manufacturer's protocol.
2. To verify the transgene, set up PCR reaction in thin-walled tubes as in Subheading 3.1.1, **step 6**, but with primers GUS-linker-F and LeTH1-1159R and the genomic DNA as a template.
3. PCR scheme and gel electrophoresis conditions are the same as those in Subheading 3.1.1, **steps 6** and 7 (*see* **Note 5**).
4. Surface-sterilize the T1 seeds obtained from the transgenic lines verified to have the transgene, and sow them on the seed selection

medium. Record the ratio of kanamycin-resistant seedlings to verify the single-copy transgene with the segregation ratio of 3:1.

5. Transfer kanamycin-resistant seedlings to vermiculite and then to soil as in Subheading 3.2, **steps 11** and **12**. Grow at least 10 individuals per transgenic line to collect T2 seeds.
6. Test for the kanamycin resistance of T2 population as in **step 4** above. T2 population with 100% kanamycin resistance is regarded to be homozygous for the transgene. Such population can be used for the expression analysis of target gene or the production of T3 seeds.

3.3.2 Expression Analysis of the Target Gene

1. Grow three or more T2 or T3 plants per transgenic line and the control nontransformant plants for 2–3 weeks and extract total RNA as in Subheading 3.1.1, **steps 1** and **2**.
2. Perform RT-PCR analysis as in Subheading 3.1.1, **steps 3–7** but with primers LeTH1-769F and LeTH1-1159R (*see* **Note 6**). In parallel with the expression analysis of the target gene, internal control should be analyzed, for example, using LeActin-F and LeActin-R primers (*see* **Note 7**).
3. Select the transgenic lines showing reduced levels of LeTH1 mRNA accumulation for further analysis on virus resistance.

3.3.3 Virus Resistance Assay with Enzyme-Linked Immunosorbent Assay (ELISA)

1. The assay is performed based on the method reported by Koenig [17].
2. Grow three or more T2 plants per LeTH1-silenced transgenic lines and the control nontransformant plants for a month.
3. Detach third and fourth true leaves from each individual plant, dust with carborundum, and inoculate with ToMV-L at 10 μg/mL in 10 mM sodium phosphate buffer, pH 7.0 by placing a 10 μL droplet of virus suspension on the leaf surface and rub with an inoculation rod (*see* **Note 8**).
4. Wrap the cut end of petiole of the inoculated leaves with wet tissue paper and place them in a plastic box. Incubate in a growth chamber at 24 °C with a 16 h light period and 8 h dark period. Alternatively, whole plants can be infected with the virus and tested for virus resistance (*see* **Note 9**).
5. Put the inoculated leaves after 15 days postinoculation in a polyethylene bag containing 400 μL coating buffer and mash the leaf tissue by rubbing the outer surface of polyethylene bag with a round end of plastic rod such as a marker pen or a ball-point pen.
6. Transfer the 200 μL leaf sap into a well of the microtiter plate and incubate the plate overnight at 4 °C or for 2 h at room temperature to coat the well surface with antigens.
7. After antigen coating, wash the wells of the plate with PBST for three times.

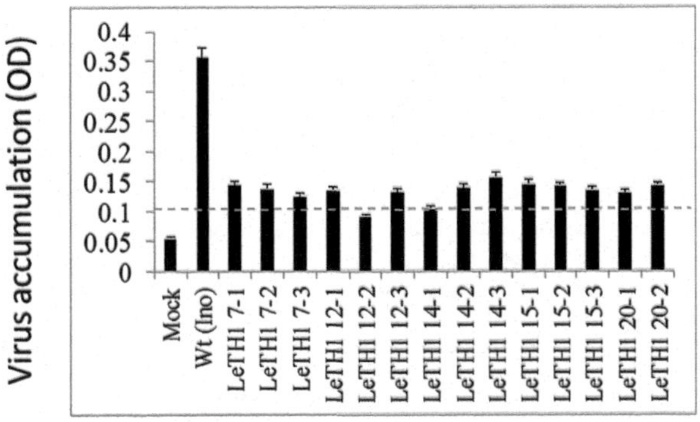

Fig. 1 ELISA assay of transgenic LeTH1-silenced lines inoculated with ToMV Detached leaves of wild-type and LeTH1-silenced lines were inoculated with ToMV-L (10 μg/mL) and analyzed as described in Subheading 3.3.3. Each absorbance values with standard deviations (SD) were analyzed in three independent ELISA assays. Dashed line shows the negative/positive threshold (reprinted from [15] with permission from Springer)

8. Add 200 μL primary antibody to each well of the plate (*see* **Note 10**), incubate at RT for 1 h and wash the wells of the plate three times with PBST.
9. Add 200 μL secondary antibody to each well of the, and incubated at RT for 1 h and wash the wells three times with PBST.
10. Add 200 μL phosphatase substrate to each well and incubate 15–30 min at room temperature until the control samples produce clearly visible yellow color.
11. Stop the reaction by adding 50 μL stop solution to each well and measure the absorbance at 405 nm using an ELISA reader. Representative results are shown in Fig. 1. Infection is usually considered positive when the absorbance value is at least twice larger than the mean of those of the uninoculated healthy control samples.

4 Notes

1. Optimization of the amount of DNA for PCR might be needed in some cases; starting with 0.5 μL up to 5 μL can be used in a 50 μL-PCR reaction. We find that 1.0 μL is often sufficient for successful PCR.
2. It is easier for handling about 100 tomato seeds to put them in a small bag made of Miracloth with stapler.
3. Only cotyledons and hypocotyls are selected as explants. It is important to avoid shoot apical meristem to reduce nontransformed shoots.
4. Wash the explants in a sterilized 50-mL conical tube and drain by using a sterilized funnel with two layers of sterilized Miracloth inserted into a sterilized 50-mL conical tube.

5. Optimization of PCR conditions might be required for different template DNAs. Different annealing temperatures, extension times, cycle numbers, concentrations of Mg^{++} can be examined when necessary.
6. Many parameters must be taken into consideration for the determination of RNA expression levels by RT-PCR. These include (1) quality of total RNA used for cDNA synthesis, (2) purity of cDNA used as a template for PCR, and (3) storage conditions of RNA and cDNA for repeated use.
7. Selection of the housekeeping gene is also critical as an internal control. Its validity relies on the stability of its expression level across the sample panel being analyzed.
8. It is highly recommended to select the same age and size of the leaves for all comparable lines for testing the virus resistance. We have experienced the variation in virus titer among different leaves of a plant.
9. When infecting whole plants, inoculate first and second true leaves of 3–4 leaf stage plants with the virus and test for virus accumulation in third and fourth true leaves as described in Subheading 3.3.3, **steps 4–10** at 15 days postinoculation.
10. A high quality, specific, sensitive, and stable antiserum is very important for the good ELISA results. Make sure to use the right antiserum and store it under optimal conditions. We find that storing under wrong conditions such as at room temperature might decrease its efficacy.

Acknowledgments

We are grateful to Dr. H. Nyunoya for the advice on tomato transformation. This work was supported by the Program for Promotion of Basic and Applied Research in Bio-oriented Industry (BRAIN) and the Ministry of Education, Culture, Sports and Technology of Japan with the Grant-in-Aid for Scientific Research for Scientific Research (C), No. 24580065 to MN.

References

1. Nagy PD (2008) Yeast as a model host to explore plant virus-host interactions. Annu Rev Phytopathol 46:217–242
2. Hashimoto M, Neriy Y, Yamaji Y, Namba S (2016) Recessive resistance to plant viruses: potential resistance genes beyond translation initiation factors. Front Microbiol 7:1695. https://doi.org/10.3389/fmicb.2016.01695
3. Yamanaka T, Ohta T, Takahashi M, Meshi T, Schmidt R, Dean C, Naito S, Ishikawa M (2000) *TOM1*, an *Arabidopsis* gene required for efficient multiplication of a tobamovirus, encodes a putative transmembrane protein. Proc Natl Acad Sci U S A 97:10107–10112. https://doi.org/10.1073/pnas.170295097
4. Ishibashi K, Ishikawa M (2016) Replication of tobamovirus RNA. Annu Rev Phytopathol 54:55–78.https://doi.org/10.1146/annurev-phyto-080615-100217

5. Yamanaka T, Imai T, Satoh R, Kawashima A, Takahashi M, Tomita K, Kubota K, Meshi T, Naito S, Ishikawa M (2002) Complete inhibition of tobamovirus multiplication by simultaneous mutations in two homologous host genes. J Virol 76:2491–2497
6. Asano M, Satoh R, Mochizuki A, Tsuda S, Yamanaka T, Nishiguchi M, Hirai K, Meshi T, Naito S, Ishikawa M (2005) Tobamovirus-resistant tobacco generated by RNA interference directed against host genes. FEBS Lett 579:4479–4484
7. Fire A, Xu S, Montgomery MK, Kostas SA, Driver SE, Mello CC (1998) Potent and specific genetic interference by double-stranded RNA in *Caenorhabditis elegans*. Nature 391:806–811
8. Kennerdell JR, Carthew RW (1998) Use of dsRNA-mediated genetic interference to demonstrate that *frizzled* and *frizzled 2* act in the wingless pathway. Cell 95(7): 1017–1026
9. Bernstein E, Caudy AA, Hammond SM, Hannon GJ (2001) Role for a bidentate ribonuclease in the initiation step of RNA interference. Nature 409:363–366
10. Hammond SM, Bernstein E, Beach D, Hannon GJ (2000) An RNA-directed nuclease mediates post-transcriptional gene silencing in *Drosophila* cells. Nature 404:293–296
11. Baulcombe DC (2004) RNA silencing in plants. Nature 431:356–363
12. Hamilton AJ, Baulcombe DC (1999) A species of small antisense RNA in posttranscriptional gene silencing in plants. Science 286:950–952
13. Voinnet O (2001) RNA silencing as a plant immune system against viruses. Trends Genet 17:449–459
14. Waterhouse PM, Wang MB, Lough T (2001) Gene silencing as an adaptive defence against viruses. Nature 411:834–842
15. Ali ME, Ishii Y, Taniguchi J, Waliullah S, Kobayashi K, Yaeno T, Yamaoka N, Nishiguchi M (2018) Conferring virus resistance in tomato by RNA silencing of each of three tomato homologs of *TOM1* from *Arabidopsis*. Arch Virol 163:1357–1362. https://doi.org/10.1007/s00705-018-3747-4
16. Ohno T, Aoyagi M, Yamanashi Y, Saito H, Ikawa S, Meshi T, Okada Y (1984) Nucleotide sequence of the tobacco mosaic virus (tomato strain) genome and comparison with the common strain genome. J Biochem 96:1915–1923
17. Koenig R (1981) Indirect ELISA methods for the broad specificity detection of plant viruses. J Gen Virol 55:53–62

Chapter 15

Large-Scale Inoculation and Evaluation Methods for Attenuated Plant Viruses

Kenji Kubota and Yasuhiro Tomitaka

Abstract

Cross-protection is a phenomenon in which a plant that is infected with a virus becomes immune to a secondary infection by the same or related viruses. Although molecular mechanisms underlying this phenomenon are not completely understood, cross-protection induced by an attenuated strain with mild symptoms has been successfully used to prevent damage by more severe strains. In the development and selection of an effective attenuated strain among candidate isolates, evaluating their infectivity and efficiency of cross-protection is important. We describe two protocols to check the infection efficiency and distribution in a plant based on immunostaining results. In addition, a practical inoculation method that uses a spray gun to apply attenuated viruses to a large number of seedlings is presented.

Key words Inoculation, Cross-protection, Infectivity, Distribution, Spray gun, Carborundum, Blotting, Filter paper, Toothpick, Hammer, Immunostaining

1 Introduction

Currently, no practical means are available for treating crops that are infected with a virus. However, cross-protection is an effective tool to prevent viral diseases in plants. Cross-protection is a phenomenon in which a plant that is infected with a virus becomes immune to a secondary infection by the same or related viruses; thus, it is effective in preventing viral diseases. Various attenuated strains of viruses have been developed and used to prevent viral diseases in vegetable crops, including tobamoviruses [tomato mosaic virus in tomato, cucumber green mottle mosaic virus (CGMMV) in melon, and pepper mild mottle virus (PMMoV) in green pepper] [1–4], cucumovirus (cucumber mosaic virus in tomato) [5], potexvirus (pepino mosaic virus in tomato) [6], and potyvirus (zucchini yellow mosaic virus in cucumber) [7].

Kappei Kobayashi and Masamichi Nishiguchi (eds.), *Antiviral Resistance in Plants: Methods and Protocols*, Methods in Molecular Biology, vol. 2028, https://doi.org/10.1007/978-1-4939-9635-3_15,

A well-attenuated strain that is suitable for practical use in crop protection may have several important features [8]. First, an attenuated strain itself should cause no or slight symptoms, or at least cause only acceptable damage to the quality or quantity of harvests. To this end, the stability of pathogenicity is important (i.e., an attenuated strain should not or should only rarely revert to wild-type virulent strains during the propagation and maintenance of inoculum and the cultivation of inoculated crops). Second, an attenuated strain should have a high infectivity and should retain it during long storage periods. Attenuated strains generally have lesser titers than virulent strains; hence, it is important to check the infectivity by inoculating plants with an attenuated virus to create and evaluate the practicality of an attenuated strain. Third, an attenuated strain should exhibit sufficient cross-protecting effect against infection with similar virulent strains. Because cross-protection functions only when an attenuated virus infects a plant prior to the invasion of virulent viruses and only where it replicates in a plant, a good attenuated virus should rapidly and widely spread from inoculated tissues to other tissues, including noninoculated upper leaves and roots, which are invasion sites of soil-borne viruses.

While selecting a suitable attenuated virus among several candidate isolates, it is necessary to inoculate a large number of plants with the isolates and rapidly check their infectivity. A convenient, large-scale inoculation technique is desirable for testing the infectivity and for practical applications wherein thousands of seedlings need to be simultaneously inoculated (particularly for annual crops). Also, because plants that are infected with an attenuated virus are difficult to distinguish from uninfected plants on the basis of their symptoms, an evaluation of the infectivity requires molecular diagnostic techniques, such as RT-PCR or DAS-ELISA, which are laborious and expensive when dealing with several samples. Therefore, a more convenient method is required.

To evaluate the ability of a virus to spread in an infected host, genetically engineered viruses expressing fluorescent proteins (such as green fluorescent protein) have been used. However, the pathogenicity and ability of the engineered virus to move from cell-to-cell and over longer distances may differ from those of the original virus. Furthermore, this genetic engineering approach cannot be applied to all plant viruses.

In the following protocols, we describe a large-scale virus inoculation method for use in vegetable seedlings. This method, which utilizes a high-pressure spray gun, was developed for inoculating tobacco plants with tomato spotted wilt virus [9]. The original method has been improved and applied to inoculate >2000 pepper seedlings with an attenuated PMMoV isolate. Inoculation can be performed in as short a period as 1 h, which results in higher infectivity than by manual inoculations.

We also describe two immunoblotting approaches, namely toothpick and hammer blotting. The former approach is suitable for checking the infection rate in a large number of plants. Equipments, their costs, and the labor needed for toothpick blotting are much less than those required for RT-PCR, DAS-ELISA, or dot-blot immunoassays. Hammer blotting is a technique for visualizing the distribution of a virus in plant tissues, including leaf lamina [10]. This method can also be employed to evaluate cross-protection between distinct viruses by immunostaining two blots derived from the same tissue with antibodies specific to each virus. We provide an example of hammer blotting, which was performed using cucumber leaves that were mix-inoculated with two distinct tobamoviruses [CGMMV and kyuri green mottle mosaic virus (KGMMV)], thereby showing a mutual exclusion between the two viruses.

2 Materials

2.1 Large-Scale Virus Inoculation of Pepper Seedlings with an Attenuated PMMoV Strain

1. Seedling plug trays: Plug trays of 128- or 200-cell.
2. Potting soil.
3. Green pepper seeds: *Capsicum annuum* L.
4. Gravity feed spray gun.
5. Air compressor.
6. Seedling fixator: The apparatus made from metal (Fig. 1a) is commercially available from Kyoto Biken Laboratories, Inc., Kyoto, Japan (*see* **Note 1**). The fixator fits to the width and depth of a plug tray and fixes the seedlings into a comb-shaped structure. The spacing of combs is of the same size as the array of cells in the plug tray. Horizontally inserting the comb beneath the cotyledons forms a grating that supports the cotyledons of seedlings from below.
7. Nettings: Plastic netting of 2-cm mesh (Fig. 1b).
8. Inoculum: Sap prepared from *Nicotiana benthamiana* leaves inoculated with the attenuated strain, L3-163 of PMMoV [4].
9. Carborundum powder (600 mesh).
10. 100 mM sodium phosphate buffer (pH 7.0).
11. Water: Distilled water is recommended.
12. Mortar and pestle.
13. High-speed centrifuge.
14. Latex gloves.
15. Goggles.

Fig. 1 Large-scale inoculation of pepper seedlings with attenuated PMMoV. (**a**) Platform fitting a nursery tray. (**b**) Nursery tray with pepper seedlings placed on the platform and covered with a netting. (**c**) Inoculation of attenuated virus by spraying the inoculum solution containing carborundum powder

2.2 Toothpick Blotting

1. Filter paper: Advantec Toyo, No. 7 or Whatman 3 MM).
2. Pencil: 2H.
3. Scale ruler.
4. Toothpicks: Round toothpicks with one blunt end.
5. Latex gloves.
6. Hybridization oven.

2.3 Hammer Blotting

1. Cucumber seedlings with 3–5 true leaves (*see* **Note 2**).
2. Inocula: Purified virion preparations of CGMMV-W [11] and KGMMV-C [12] prepared at 200 ng/μL in 100 mM sodium phosphate buffer (pH 7.0). Used alone or as a mixture.
3. Carborundum powder.

4. Glass rods.
5. Filter paper: Advantec Toyo, No. 7, Whatman, 3 MM, etc.
6. Paper towels.
7. Clear, flat polyethylene bags.
8. Hammer: Steel hammer with a slightly rounded surface.
9. Silicone rubber sheet: 2–4 mm thick.
10. Steel block: 2–4 cm thick.
11. Felt: A sheet of felt larger in area than the steel block.
12. Latex gloves.
13. Hybridization oven.

2.4 Immunostaining

1. Primary antibodies: Rabbit anti-CGMMV immunoglobulin G (IgG) and rabbit anti-KGMMV IgG (Japan Plant Protection Association, Tokyo, Japan).
2. Secondary antibodies: Goat anti-rabbit IgG, conjugated with alkaline phosphatase.
3. Tris-buffered saline (TBS): Prepare 10× stock solution of 1.5 M NaCl, 0.1 M Tris–HCl (pH 7.4). Sterilize by autoclave. Dilute tenfold before use.
4. Chlorophyll removal buffer: 1× TBS, 2% (v/v) Triton X-100 is added dropwise with stirring and stored at room temperature.
5. Blocking buffer: 1× TBS containing 2% skimmed milk, prepared just before use.
6. Primary antibody buffer: Dilute the primary antibody with blocking buffer at 1000~10,000× dilution), prepared immediately before use.
7. Secondary antibody buffer: Dilute the secondary antibody with blocking buffer at 1000~10,000× dilution), prepared immediately before use.
8. Alkaline phosphatase buffer: 100 mM NaCl, 100 mM Tris–HCl (pH 9.5), 50 mM $MgCl_2$, prepared from stock solutions of 5 M NaCl, 1 M Tris–HCl (pH 9.5), and 1 M $MgCl_2$.
9. AP substrate buffer: NBT/BCIP Stock Solution (Sigma-Aldrich, St. Louis, MO) diluted with 50 volumes of alkaline phosphatase buffer (*see* **Note 3**), prepared immediately before use and protected from light.
10. Plastic containers.
11. Electric shaker.
12. Tweezers with flat ends.
13. Pipette.

3 Methods

3.1 Large-Scale Inoculation of Pepper Seedlings with an Attenuated PMMoV Strain

1. Preparation of pepper seedlings: Sow pepper seeds in each cell of 128- or 200-cell plug tray filled with potting soil. Grow normally until cotyledons have completely developed (*see* **Note 4**).
2. Preparation of inoculation equipment: Adjust the height of the seedling fixator stand such that the cotyledon leaves are supported by iron sticks (Fig. 1a). Place the plug tray planting of pepper seedlings on the seedling fixator. Assemble a grating by setting the comb-shaped structure at right angles so that it supports cotyledons from below. Place a netting over the seedlings (Fig. 1b) to prevent the strong air flow from the spray from blowing down the seedlings.
3. Preparation of the spray gun: Adjust the air pressure of the compressor to 0.25~0.3 MPa. Test run the spray with water (*see* **Note 5**).
4. Inoculum preparation: Prepare the inoculum from *N. benthamiana* leaves inoculated with PMMoV L3-163 strain. Leaf sap is obtained by grinding the inoculated leaves with mortar and pestle with three volumes of 0.1 M sodium phosphate buffer (pH 7.0). To prevent the spray nozzle from clogging, the sap is spun in a centrifuge at ~10,000 × *g* for 5 min at 4 °C. The supernatant is used as the inoculum. Add approximately 50 mg of carborundum powder per 1 mL of inoculum solution and mix well. Keep the solution on ice until all settings are completed. Pour the inoculum solution into the cup of the spray gun. Typically, 50 mL of inoculum solution is required for inoculating 128 seedlings.
5. Spray the inoculum at the seedlings from 15 cm above. Move the gun from side to side so that the inoculum is uniformly applied (Fig. 1c).
6. When the inoculation of one tray is completed, move the plug tray to a shady location to prevent damage to the leaves from strong sunlight.
7. Repeat procedures 2–6 for inoculating another tray.
8. To wash off the carborundum and ionic compounds from leaves, shower the plants with tap water immediately after the inoculation, failing which the leaves may become necrotic or wilted.
9. Wash the equipment with water and dry.
10. Keep the inoculated plants in the shade for 1 day.
11. After 1 day, continue growing the plants as usual.

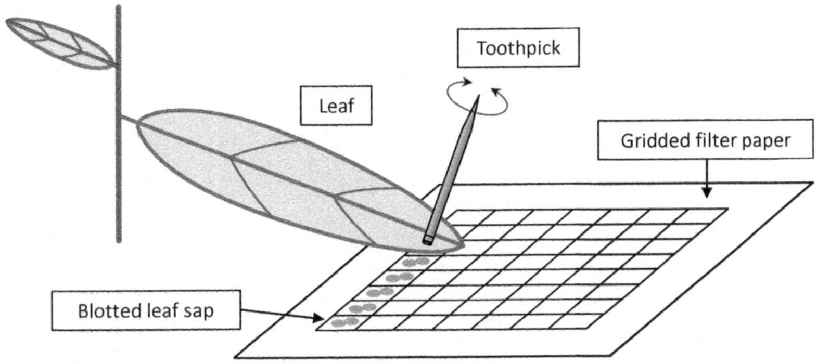

Fig. 2 A scheme of toothpick blotting. Leaf sap of test plants is blotted in each square on a gridded filter paper by pushing the leaf lamina with the round tip of a toothpick. The blotted paper is immunostained with appropriate antibodies

3.2 Toothpick Blotting

1. Cut a filter paper in an appropriate size (e.g., 5 × 10 cm) (*see* **Note 6**). Using a pencil and a ruler, draw grid lines on a filter paper to form 7–10 mm (per side) squares. Alternatively, grid lines may be printed on a printer.
2. Place the filter paper below a sample leaf (not wet). With the round tip of a toothpick, grind the leaf lamina onto a square of the filter paper. Confirm the blotting of the green sap. Repeat twice or thrice (Fig. 2) (*see* **Note 7**).
3. Remove leaf debris on the filter paper with the pointed tip of a toothpick. Repeat **steps 2** and **3** for each sample and grid.
4. Completely air dry the filter papers at room temperature or in a hybridization oven set at 50 °C (*see* **Note 8**).
5. Store the blotted filter papers in a plastic bag with desiccant at room temperature.
6. Immunostain the filter paper following the protocols described in Subheading 3.4.
7. Evaluation: blots from infected tissues yield purple staining, whereas those from uninfected tissues remain unstained. Determine the infection of each blotted tissue by comparing with positive and negative controls.

3.3 Hammer Blotting

1. Inoculation: Mature leaves of cucumber are inoculated with the inocula of purified CGMMV, KGMMV, or a mixture of the two viruses. The leaves are dusted with carborundum powder, and droplets of solution containing virion are placed on the leaves and inoculated by rubbing with a glass rod. After inoculation, the leaves are washed with water and normally grown for 3 weeks.
2. Setting of blotting paper: Spread out a felt on a flat bench and place a steel block on it. Place a silicone rubber sheet on the block (Fig. 3).

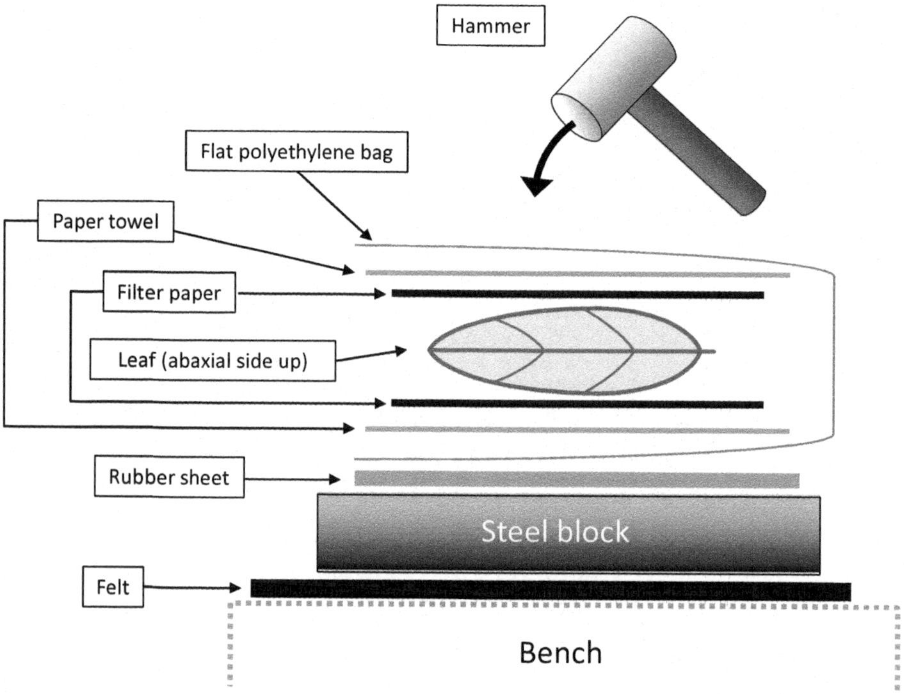

Fig. 3 A scheme of hammer blotting. A leaf is sandwiched with filter papers. Leaf sap is blotted by gently hitting the leaf with a hammer. Paper towels and a flat poly bag prevent contamination with excess leaf sap. Felt, steal block, and rubber sheet help to reduce shock and noise. Each blotted paper can be separately immunostained with different antibodies

3. Sandwich an inoculated leaf between two sheets of filter papers and a paper towel. Cover the filter paper and towel with a clear flat polyethylene bag to avoid contaminating the hammer with sap. Place this preparation on the silicone rubber sheet, with leaf abaxial side up (Fig. 3).
4. Blotting by a hammer: Caution should be taken to not hit your hand with the hammer. Gently hit the leaf sandwich with the round part of the hammer. Take care not to disperse the leaf sap from the midribs (*see* **Note 9**). Continue hitting until the tissues of the entire leaf collapses and the sap is completely absorbed onto the filter papers.
5. Collect the filter papers from the bag. Remove the paper towel and leaf debris from the surface of filter papers. Label the sample information on the filter paper using a pencil, if needed (*see* **Note 10**).
6. Completely air dry the filter papers at room temperature or in a hybridization oven set at 50 °C (*see* **Notes 8** and **11**).
7. Store the blotted filter papers in a plastic bag with desiccant at room temperature.

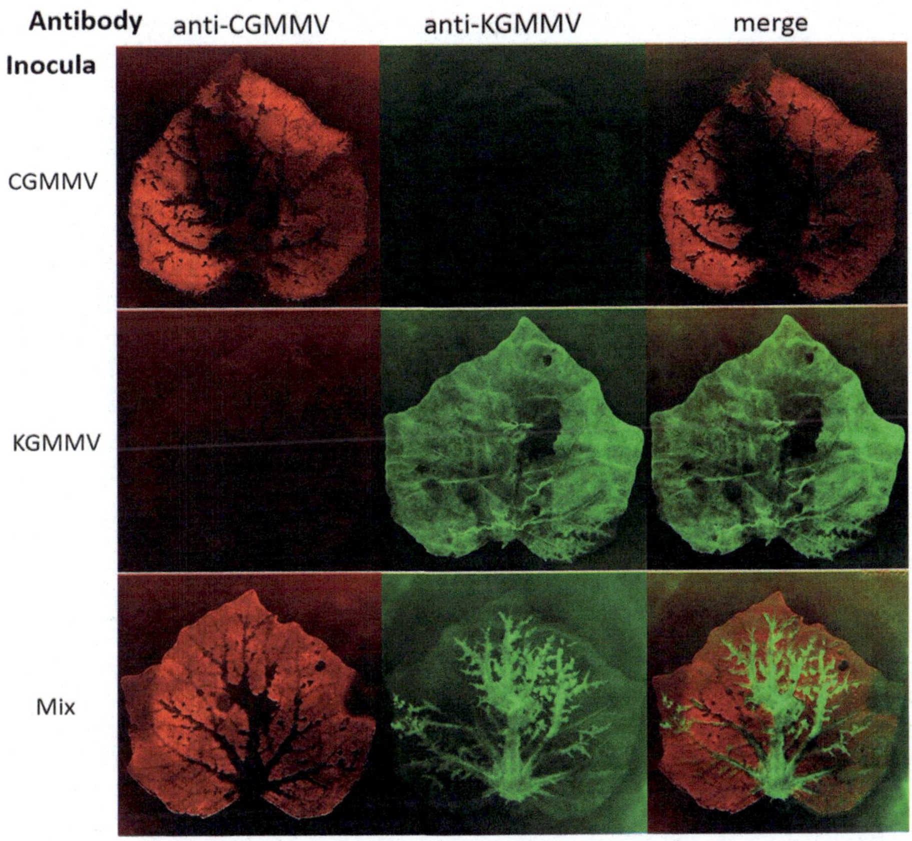

Fig. 4 Visualization of mutual exclusion by hammer blotting between cucumber green mottle mosaic virus (CGMMV) and kyuri green mottle mosaic virus (KGMMV). Cucumber leaves were inoculated with purified virions of CGMMV (upper panels), KGMMV (middle panels), or a mixture of the two viruses (bottom panels). At 3 weeks after inoculation, inoculated leaves were blotted onto two filter papers using hammer blotting. The two blots derived from one leaf were immunostained with antibodies against CGMMV (anti-CGMMV) and KGMMV (anti-KGMMV), respectively. The signals for CGMMV and KGMMV were pseudo-colored with red and green, respectively. The images for anti-KGMMV were mirror-reversed

8. Immunostain the filter papers following protocols described in Subheading 3.4.
9. Image processing: if needed, appropriately process the image using a graphics software (Photoshop, GIMP, etc.). The procedure includes adjusting the size of the image, tuning for correct brightness and contrast, pseudo-coloring, mirror-reversing, and overlaying (*see* Fig. 4).

3.4 Immunostaining of Blotted Filter Papers

Reaction conditions described in **steps 2–6** are general recommendations. Optimal conditions for each step must be empirically determined. During the following steps, the blotted surface of filter papers should always be kept facing upward.

1. To remove chlorophyll from the filter paper, pour chlorophyll removal buffer into a plastic container. Place the dried filter paper into the buffer and gently shake on an electric shaker. When the buffer turns green, discard the buffer and pour fresh wash buffer onto the sample. Repeat several times until most green color is removed.
2. Blocking: after completely removing the wash buffer, pour the blocking buffer and continue shaking for 30 min.
3. Preparation of primary antibody: while blocking, dilute the primary antibody at a 1000–10,000-fold dilution in a blocking buffer. Antibody concentration should be empirically determined. In general, prepare 5 mL of buffer for every 100 cm^2 paper. When multiple filter papers are treated in a container, prepare sufficient buffer solution to allow the papers to freely move.
4. For incubation with primary antibody, once the blocking is completed, completely discard the blocking buffer and add the primary antibody buffer to the container. Continue shaking for 2 h at room temperature. If desired, the papers in the buffer can be stored overnight at 4 °C.
5. Washing: pick the paper with tweezers, remove the buffer, and transfer the paper to a new container. Pour the blocking buffer over the paper. After gentle shaking for 5 min, discard the buffer by decanting it and pour in fresh buffer. Repeat the washing thrice. During this step, prepare the secondary antibody buffer, as described in **step 3**.
6. Incubation with secondary antibody: After washing, transfer the papers to a new container. Add the secondary antibody buffer and incubate for 2 h at room temperature with gentle shaking.
7. Washing: Discard the secondary antibody buffer. Wash the papers thrice with 1× TBS, until the buffer is free of skimmed milk.
8. Incubate the filter papers with alkaline phosphatase buffer for 5 min by gentle shaking. Prepare AP substrate buffer during incubation.
9. Transfer the filter papers to a new container with a flat bottom and spread them out in one layer. Remove air bubbles below the papers to prevent uneven staining.
10. Color development: Using a pipette, add AP substrate solution onto a filter paper (2 mL for every 100 cm^2). Incubate without agitation, avoiding light to prevent nonspecific backgrounds. A purple signal will appear within 5 min (*see* **Note 12**).
11. When sufficient color has developed, stop the reaction by rinsing the papers with deionized water. Place the papers on several layers of paper towels (*see* **Note 13**).
12. The image is recorded using a camera or a scanner. The filter papers can then be stored under dry conditions.

4 Notes

1. The inoculation method and apparatus described herein have been patented by Kyoto Biken Laboratories, Inc. (Japanese Patent No. 5938798).
2. According to our experience, mature leaves of *Arabidopsis thaliana*, tomato, pepper, cucumber, melon, watermelon, pea, and Japanese apricot are suitable for hammer blotting. Hammer blotting can also be applied to flower petals, stems, roots, or for an entire plant. Some species contain high amounts of phenolic compounds (e.g., *Chrysanthemum* spp.), which results in a brown pigmentation on the filter paper immediately after blotting. Blotting with infected positive and healthy negative controls should be performed under such conditions.
3. NBT/BCIP stock solution contains nitro blue tetrazolium chloride and 5-bromo-4-chloro-3-indolyl-phosphate, 4-toluidine salt, which are harmful to the skin. Gloves and goggles should be worn when using this solution. Follow the manufacturer's instructions.
4. Inoculating with a spray gun is also applicable to cucumber seedlings at the cotyledon stage. This method generally achieves higher infection rates (>95%) than manual inoculations, which requires rubbing with a glass rod or with fingers in the presence of carborundum (80–90%).
5. Wear goggles and latex gloves during the inoculation because spraying solution containing carborundum may harm your eyes or skin.
6. Handling large sheets of paper is more difficult than handling smaller ones.
7. Blot different plant parts within a leaf. To determine the appropriate leaf stage and sites for detection, use hammer blotting to preliminarily investigate the virus distribution in a plant. For pepper seedlings inoculated with the attenuated virus described in Subheading 3.1, we usually perform toothpick blotting with pepper seedlings on 4–5 true leaves 2 weeks after inoculation.
8. Incomplete drying of a blot may result in an imperfect removal of chlorophyll in Subheading 3.4, **step 1**.
9. If the dispersal of excessive sap is unwanted, remove the major veins using a razor blade.
10. Although filter papers can face either the adaxial and abaxial surfaces of a leaf, the latter tends to give better results. If blotting of a leaf sample is not sufficient, then freezing and thawing the leaves for a short time can provide a better blotting result. However, necrotic tissues will not be blotted.

11. Blotted filter papers can be stored for at least 3 months.
12. If color development is not sufficient, then the development can be extended up to ~3 h.
13. Take care not to wrinkle the papers while drying.

Acknowledgments

Part of this study was supported by a Grant-in-Aid for "The research project for utilizing advanced technologies in agriculture, forestry and fisheries" administered by the Ministry of Agriculture, Forestry, and Fisheries of Japan. We are grateful to Bo-Song Ryang and Naoko Nakamura for providing details on the usage of large-scale inoculation apparatus.

References

1. Oshima N (1981) Control of tomato mosaic disease by attenuated virus. Jpn Agric Res Q 14:222–228
2. Motoyoshi F, Nishiguchi M (1988) Control of virus diseases by attenuated virus strains. Comparison between attenuated strains of *Cucumber green mottle mosaic virus* and *Tobacco mosaic virus*. Gamma Field Symposia 27: 91–109
3. Nishiguchi M, Kobayashi K (2011) Attenuated plant viruses: preventing virus diseases and understanding the molecular mechanism. J Gen Plant Pathol 77:221–229
4. Ogai R, Kanda-Hojo A, Tsuda S (2013) An attenuated isolate of *Pepper mild mottle virus* for cross protection of cultivated green pepper (*Capsicum annuum* L.) carrying the L^3 resistance gene. Crop Prot 54:29–34
5. Sayama H, Sato T, Kominato M, Natsuaki T, Kaper JM (1993) Field testing of a satellite-containing attenuated strain of cucumber mosaic virus for tomato protection in Japan. Phytopathology 83:405–410
6. Schenk MF, Hamelink R, van der Vlugt RAA, Vermunt AMW, Kaarsenmaker RC, Stijger ICCMM (2010) The use of attenuated isolates of *Pepino mosaic virus* for cross-protection. Eur J Plant Pathol 127:249–261
7. Kosaka Y, Ryang BS, Kobori T, Shiomi H, Yasuhara H, Kataoka M (2006) Effectiveness of an attenuated *Zucchini yellow mosaic virus* isolate for cross-protecting cucumber. Plant Dis 90:67–72
8. Hull R (ed) (2014) A. Mild strain protection (cross-protection). Plant virology, 5th edn. Academic Press, London, p 853
9. Mandal B, Csinos AS, Martinez-Ochoa N, Pappu HR (2008) A rapid and efficient inoculation method for *Tomato spotted wilt tospovirus*. J Virol Methods 149:195–198
10. Akhter MS, Yaeno T, Yamaoka N, Nishiguchi M, Kobayashi K (2016) Hammer blot-mediated RNA extraction: an inexpensive, labor-saving method to extract RNA for plant virus detection. J Gen Plant Pathol 82:268–272
11. Komuro Y, Tochihara H, Fukatsu R, Nagai Y, Yoneyama S (1971) Cucumber green mottle mosaic virus (watermelon strain) in watermelon and its bearing on deterioration of watermelon fruit known as konnyaku disease. Ann Phytopathol Soc Jpn 37:34–42
12. Tochihara H, Komuro Y (1974) Infectivity test and serological relationships among various isolates of cucumber green mottle virus; some deduction of the invasion route of the virus into Japan. Ann Phytopathol Soc Jpn 40:52–58

Chapter 16

RNA Silencing-Mediated Apple Latent Spherical Virus Vaccine in Plants

Chunjiang Li, Noriko Yamagishi, and Nobuyuki Yoshikawa

Abstract

The apple latent spherical virus (ALSV), originally isolated from an apple tree in Japan, is a small spherical virus with a diameter of 25 nm and comprises a bisegmented, single-stranded RNA genome (RNA1 and RNA2) and three different capsid proteins (Vp25, Vp20, and Vp24). The virus can experimentally infect a broad range of plants including, not only model plants (*Arabidopsis thaliana* and *Nicotiana* species) but also economically important crops such as cucumber, soybean, tomato, fruit trees, and flowers. ALSV has been used as an effective plant virus vector for virus-induced gene silencing (VIGS) to assess gene functions because the virus infects most of the host plants without showing any symptoms and induces a uniform knockout phenotype in infected plants. Moreover, the VIGS persists throughout plant growth in infected plants. Here, we show that genetically engineered ALSV vectors (ALSV vaccines) containing a partial genome sequence of pathogenic viruses display a high degree of cross-protection against the challenge inoculation of the corresponding pathogenic viruses. Treatment effects can also be expected in virus-infected plants by subsequent inoculation with ALSV vaccine.

Key words Apple latent spherical virus (ALSV), Virus-induced gene silencing (VIGS), ALSV vaccine, Plant virus diseases, Cross-protection, Treatment

1 Introduction

RNA silencing is a sequence-specific mechanism of gene suppression that is highly conserved in the majority of eukaryotes including plants and animals. When an RNA virus infects a plant, dsRNA is synthesized through the action of RNA-dependent RNA polymerase (RdRp). This viral dsRNA is cleaved by Dicer-like (DCL) enzymes, which are the double-strand-specific RNase in plants, into small interfering RNA (siRNA) fragments 21–24 base pairs (nt) in length. Then, the siRNAs are incorporated into the RNA-induced silencing complex (RISC). RISC cleaves viral RNA with a base sequence complementary to that of the siRNA, thus suppressing viral infection [1, 2]. Furthermore, the silencing signal moves to neighboring cells and spreads throughout the entire plant [3, 4].

Kappei Kobayashi and Masamichi Nishiguchi (eds.), *Antiviral Resistance in Plants: Methods and Protocols*, Methods in Molecular Biology, vol. 2028, https://doi.org/10.1007/978-1-4939-9635-3_16,

This RNA silencing mechanism to prevent viral infection in plants has similarities with the immune system in animals, and an analogy can be drawn between dsRNA and antigens, siRNA and antibody molecules, and RISC and killer cells. Thus, RNA silencing plays an important role in suppressing viral infections in plants which have no immune system [5].

The apple latent spherical virus (ALSV), originally isolated from an apple tree in Japan, is a small spherical virus with a diameter of 25 nm, comprises a bisegmented, single-stranded RNA genome (RNA1 and RNA2) and three different capsid proteins [6]. Although apple tree is the only natural host of ALSV, it can infect a relatively broad range of hosts through artificial inoculation and result in latent infections without causing disease in model plants such as *Arabidopsis thaliana* and *Nicotiana tabacum*, and economically important crops (i.e., tomato, cucumber, soybean, apple, grapes, and citrus) [7–10].

We have previously generated an ALSV vector that can deliver a foreign gene based on an infectious cDNA clone of ALSV genome and used it to express exogenous genes in various herbaceous plants and apples, as well as to analyze gene function utilizing virus-induced gene silencing (VIGS) [11–16]. ALSV can efficiently induce systemic VIGS in various host plants [7–9, 17–19]. We considered the possibility of utilizing the viral characteristics of ALSV including its broad host range, latent infectivity, and ability to induce systemic VIGS as tools to prevent viral diseases. Namely, this idea involves introducing a portion of a pathogenic virus genome to the ALSV vector and using the genetically engineered ALSV as a viral vaccine (herein referred to as the ALSV vaccine) to prevent the pathogenic virus. If this idea becomes effective, ALSV vaccine could be prepared for numerous different viruses within a short time by merely altering the viral genome fragment to be inserted. In this chapter, we show that ALSV vaccines containing a partial genome sequence of pathogenic viruses display a high degree of protection against the challenge inoculation of the corresponding viruses [20, 21]. Treatment effects can also be expected in virus-infected plants by a subsequent inoculation with ALSV vaccine [21, 22].

2 Materials

2.1 Construction of DNA Clones to Prepare ALSV Vaccine

1. Infectious cDNA clones of ALSV-RNA1 (pEALSR1) and ALSV-RNA2 (pEALSR2mL5mR5XSB and pEALSR2mL5mR5-3′MN) (Fig. 1) [12, 21].
2. Bean yellow mosaic virus (BYMV) expressing the green fluorescent protein (GFP) (BYMV-GFP) derived from an infectious cDNA clone (pBIbG:FL) [23].
3. Zucchini yellow mosaic virus (ZYMV) expressing GFP (ZYMV-GFP) derived from an infectious cDNA clone (p35S-

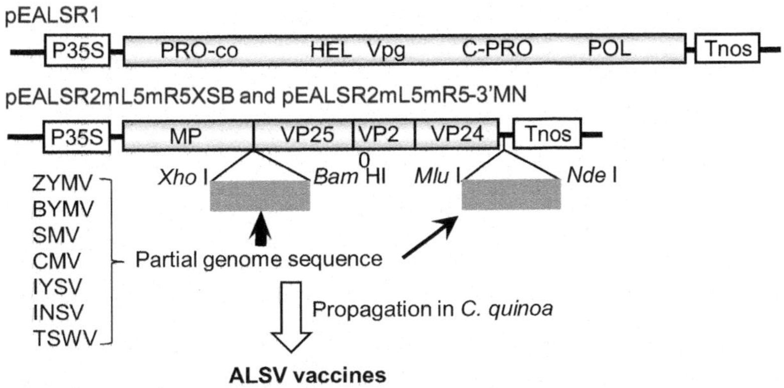

Fig. 1 Construction scheme for ALSV vaccines against ZYMV, BYMV, SMV, CMV, IYSV, INSV, TSWV. DNA fragments amplified from each virus genome using primer pares (Tables 1 and 2) were ligated to cloning site of pEALSR2mL5mR5 or pEALSR2mL5mR5-3′MN, and the plasmids used for inoculation to *Chenopodium quinoa* plants. The position and length of each fragment are shown in Tables 1 and 2

Z5vector + GFP-FL) of ZYMV isolate Z5-1 (kindly provided by Dr. T. Natsuaki).

4. Soybean mosaic virus (SMV, strain D).
5. Cucumber mosaic virus (CMV-Y) (kindly provided by Dr. H. Takahashi).
6. Three tospoviruses: Impatiens necrotic spot virus (INSV), Iris yellow spot virus (IYSV), and Tomato spotted wilt virus (TSWV). IYSV, INSV, and TSWV were kindly provided by Drs. S. Fuji, T. Natsuaki, and Y. Kosaka, respectively.
7. Primers for amplification of the genome fragments of target viruses in Table 1.
8. Reverse transcriptase: ReverTraAce (Toyobo Co., Ltd., Osaka, Japan).
9. PCR enzyme: ExTaq polymerase (Takara Co., Ltd., Tokyo, Japan).
10. Thermal cycler: TaKaRa PCR Thermal Cycler Dice® (Takara Co., Ltd., Tokyo, Japan).
11. DNA ligation kit: DNA ligation kit (Takara Co., Ltd., Tokyo, Japan).
12. Competent cells: *Escherichia coli* JM109 or DH 5α.
13. RNA extraction kit: TRI Reagent (Sigma-Aldrich Corporation, St. Louis, USA).
14. TE buffer:10 mM Tris–HCl, 1 mM EDTA (pH 8.0).
15. Midiprep kit: QIAGEN Plasmid Midi Kits (Qiagen, Tokyo, Japan).

Table 1
Primers used to construct ALSV vaccines against BYMV, ZYMV, CMV, SMV, INSV, IYSV, and TSWV

Virus (accession No.)	Primer	Sequence (5′-3′)[a]
BYMV (AB079887)	BYP3(−)-3059Mlu(+)	TACATACGCGT^{3298}TCAGGGGCTCAGATAAAAC^{3280}
	BYP3(−)-3298Nde(−)	TACATCATATG^{3059}GGCTATCCTCCTAGCTTCATT^{3077}
ZYMV (AB188115)	ZYP1-209Xho(+)	TACATCTCGAG^{209}GCAAGTAGTCGGGTTGGTATA^{230}
	ZYP1-418Bam(−)	TACATGGATCC^{418}TCCATTCGGACCCTTTTTCACC^{397}
	ZYHC-1082Xho(+)	TACATCTCGAG^{1082}GAAGTCCAGTTCTTTCAAGG^{1101}
	ZYHC-1261Bam(−)	TACATGGATCC^{1261}TCTACTAAGTTTCTCCCTAC^{1242}
	ZYP3-3269Xho(+)	TACATCTCGAG^{3269}GCGGCGACAAGGGAAGAAAAA^{3289}
	ZYP3-3457Bam(−)	TACATGGATCC^{3457}TGCGCCAGCCATGTACAAG^{3439}
	ZYCI-3662Xho(+)	TACATCTCGAG^{3662}GATAAGAGACTCACAATTGA^{3681}
	ZYPCI-3850Bam(−)	TACATGGATCC^{3850}TATTTCATTGGCCACAAAAG^{3831}
	ZYNIa-6701Xho(+)	TACATCTCGAG^{6701}AATGAAGGAGATTGCGGATT^{6720}
	ZYNIa-6901Bam(−)	TACATGGATCC^{6901}TCCCCACGCGATTTTGCTAG^{6882}
	ZYNIb-8321Xho(+)	TACATCTCGAG^{8321}CACACCGAGCTTTTGCAAGA^{8340}
	ZYNIb-8521Bam(−)	TACATGGATCC^{8521}TTGTTCAAAGAAGGTATCTT^{8502}
	ZYCP-8627Xho(+)	TACATCTCGAG^{8627}GGCTCCGGCTCAGGTGAGAAAA^{8648}
	ZYCP-8851Bam(−)	TACATGGATCC^{8851}AGAGGCGAATTGCTGATGAG^{8832}
SMV(AB100445.1)	SMVCP316Xho(+)	TACATCTCGAGG^{316}GTGTGGTTATGAATGGTT^{323}
	SMVCP519Bam(−)	TACATGGATCC^{519}AGAATTTCTCATCTCAATG^{501}
CMV (RNA1, D12537.1)	CM1a-179Xho(+)	^{179}AAGACAACTCATGAGCAACTCGAGGAGCAA^{208}
(RNA2, D12538.1)	CM1a-385Bam(−)	CGCGGATCC^{385}TAGACAATCGAGAGTTCCACA^{365}

(RNA3, D12499.1)	CM2a-313Xho(+)	CCGCTCGAG^{313}GATGAGTTTGTAACTTATGGT^{333}
	CM2a-522Bam(−)	CGCGGATCC^{522}TTCAAAACACTTCATGGTTCG^{502}
	CMCP-1494Xho(+)	CCGCTCGAG^{1494}AAAATAGACCGTGGGTCTTAT^{1514}
	CMCP-1733Bam(−)	CGCGGATCC^{1733}AGATGCAGCATACTGATAAAC^{1713}
INSV(AB109100)	IN:N-Bam(+)	TTAGGATCC^{2131}AGCTTTTGGACTACAAGCCTTT^{2152}
	IN:N-Xho(−)	TTTCTCGAG^{2341}GAAATAGCTTCACTAGTTCTGGCT^{2364}
IYSV(AF001387)	IY:N-Bam(+)	TTAGGATCC^{2361}GTCAGTGAGGATCTTGATAGT^{2381}
	IY:N-Xho(−)	TTTCTCGAG^{2561}ATGGCAAGTCTTGACATCTTT^{2581}
TSWV (D00645)	TS:N-Bam(+)	TTAGGATCC^{2080}CCCTTTAGCATTAGGATTGCT^{2100}
	TS:N-Xho(−)	TTTCTCGAG^{2280}ATCAGTGTTGTCTTGGCTATAT^{2301}

[a]Restriction sites are underlined and figures indicate the nucleotide positions of genomic RNA

2.2 Preparation of ALSV Vaccines

1. Quinoa plant: *Chenopodium quinoa* Willd.
2. Plasmid DNA clones from Subheading 3.1.
3. Inoculation buffers: 0.1 M Tris–HCl (pH 7.8), 0.1 M NaCl, and 5 mM $MgCl_2$.
4. RNA extraction kit: TRI Reagent (Sigma-Aldrich Corporation, St. Louis, USA).

2.3 Inoculation of ALSV Vaccines and the Challenge Inoculation with Pathogenic Viruses

1. *Nicotiana benthamiana* Domin.
2. Tobacco: *Nicotiana tabacum* L. 'Xanthi-nc'.
3. Cucumber: *Cucumis sativus* L. 'Aodai'.
4. Soybean: *Glycine max* L. 'Jack'.
5. Pea: *Pisum sativum* L. 'Matsushima Kinusaya'.
6. Eustoma: *Eustoma exaltatum* (L.) Salisb. 'Shalala blue'.
7. Inoculation buffers: 0.1 M Tris–HCl (pH 7.8), 0.1 M NaCl, and 5 mM $MgCl_2$ for ALSV and BYMV; 0.1 M phosphate buffer (pH 7.0) for CMV and ZYMV; 0.1 M phosphate buffer (pH 7.2) for SMV; 0.1 M phosphate buffer (pH 7.2), 0.01 M Na_2SO_3 for three tospoviruses.
8. Carborundum (Nacalai Tesque Inc., Kyoto, Japan).
9. Helios Gene Gun system (Bio-Rad, CA, USA).

2.4 Evaluation of Vaccine Effects

1. Microplates (Greiner Bio-One, Tokyo, Japan).
2. Coating buffer: 0.05 M carbonate buffer, pH 9.6.
3. PBST: The phosphate buffered saline with Tween®. 137 mM NaCl, 2.68 mM KCl, 8.1 mM Na_2HPO_4, 1.47 mM KH_2PO_4, pH 7.4, 0.05% (v/v) Tween 20, 2% (w/v) polyvinylpyrrolidone.
4. Primary antibody: Antisera against ALSV, BYMV, ZYMV, CMV, INSV, IYSV, and TSWV.
5. Secondary antibody: Anti-rabbit IgG-Alkaline Phosphatase linked antibody (Cell Signaling, Tokyo, Japan).
6. AP substrate: Substrate for alkaline phosphatase. 1 mg/mL *p*-nitrophenyl phosphate (Wako, Osaka, Japan) dissolved in 1 M diethanolamine buffer, pH 9.8, 0.5 mM $MgCl_2$.
7. A microplate reader (Bio-Rad, CA, USA).
8. VB-G25 fluorescent stereoscopic microscope with a GFP filter (Keyence, Osaka, Japan).
9. DP 70 digital camera (Olympus, Tokyo, Japan).

3 Methods

3.1 Construction of DNA Clones to Prepare ALSV Vaccine

1. Amplify DNA fragments of ZYMV and BYMV (Table 1) by ExTaq polymerase and primer pairs (Table 1) using p35S-Z5 + GFP-FL (ZYMV) and pIbG:FL (BYMV) as templates, respectively. Digest the PCR products with *Xho*I and *Bam*HI

or *Mlu*I and *Nde*I and then ligate them into pEALSR2mL5m-R5XSB or pEALSR2mL5mR5–3′MN (Fig. 1).

2. Extract total RNAs from plants infected with each virus using TRI Reagent to construct DNA clones for ALSV vaccines against SMV, ZYMV, CMV, INSV, IYSV, and TSWV (Tables 2 and 3), respectively. Amplify DNA fragment of each virus from total RNAs by RT-PCR using corresponding primer pairs (Table 1) and ligate them into pEALSR2mL5mR5 as shown in Fig. 1.
3. Artificially synthesize chimeric sequences of ZYMV and CMV genome fragments (150 bp each) and ligate them into pEALSR2mL5mR5 to prepare ALSV vaccines (ALSV-ZC300) against both ZYMV and CMV, respectively.

3.2 Preparation of ALSV Vaccines

1. Purify the plasmid DNAs constructed in Subheading 3.1 using a Midiprep kit according to the manufacturer's protocols and dissolve them in TE buffer at a concentration of 2 μg/μL and stored at −80 °C.
2. Mix purified DNA plasmids equally with pEALSR1 (2 μg/μL) and then use them to inoculate *C. quinoa* plants.
3. Homogenize symptomatic *C. quinoa* leaves with two to three volumes of inoculation buffer and use the crude sap for cross-protection tests (Fig. 2a).
4. Extract total RNAs from infected *C. quinoa* leaves using TRI reagent and use for inoculation of *Eustoma* by particle bombardment using a Helios Gene Gun system [22].

3.3 Inoculation of ALSV Vaccines and the Challenge Inoculation with Pathogenic Viruses

1. For ZYMV, inoculate both cotyledons of cucumber at the cotyledon stage with ALSV vaccines against ZYMV, then at 6 days after inoculation challenge-inoculate the first true leaf with an infectious clone of ZYMV (p35S-Z5 + GFP-FL) using a Helios Gene Gun system (Fig. 2a).
2. For BYMV, mechanically inoculate the third and fourth true leaves at the five-leaf stage of *N. benthamiana* with crude extracts of ALSV vaccines against BYMV (Table 2). Challenge-inoculate the fifth and sixth true leaves after 4 days, with crude extracts of *N. benthamiana* infected with BYMV-GFP.
3. For tospoviruses, inoculate the second, third, and fourth true leaves at the five-leaf stage of *N. benthamiana* with ALSV vaccine against INSV, INSV or TSWV. Challenge-inoculate the sixth, seventh, and eighth true leaves at the nine-leaf stage with the crude sap of the virus-infected *N. benthamiana* leaves by the carborundum method.
4. For CMV, inoculate the third to fifth leaves of tobacco at the five-leaf stage with ALSV vaccine against CMV, and challenge-inoculate the seventh to ninth leaves with the crude extracts of CMV-Y-infected tobacco leaves.

Table 2
Preventive effects of ALSV vector-based vaccines against potyvirus and cucumovirus infections

Target viruses (genus)	ALSV and ALSV vaccine	Gene regions of target virus inserted into ALSV vector (nt)		Preventive effect of ALSV vaccine			Curative effect of ALSV vaccine[c]
				Plants	Symptoms by target viruses[a]	Accumulation of target viruses[b]	
ZYMV (poty)	wtALSV	–	–	Cucumber	+++	+++	–
ZYMV (poty)	ALSV-Z:P1	P1	(210)	Cucumber	+/–	+	
ZYMV (poty)	ALSV-Z:HC-Pro	HC-Pro	(180)	Cucumber	+/–	+	
ZYMV (poty)	ALSV-Z:P3	P3	(189)	Cucumber	–	–	+
ZYMV (poty)	ALSV-Z:Cl	Cl	(189)	Cucumber	–	–	
ZYMV (poty)	ALSV-Z:Nla	Nla	(201)	Cucumber	–	–	
ZYMV (poty)	ALSV-Z:Nlb	Nlb	(201)	Cucumber	–	–	
ZYMV (poty)	ALSV-Z:CP	CP	(225)	Cucumber	–	–	
BYMV (poty)	wtALSV	–	–	*N. benthamiana*	–	+++	
BYMV (poty)	wtALSV	–	–	Pea	++	+++	–
BYMV (poty)	wtALSV	–	–	Eustoma	++	+++	–
BYMV (poty)	ALSV-BY:P3(–)	P3(–)	(240)	*N. benthamiana*	–	–	
BYMV (poty)	ALSV-BY:P3(–)	P3(–)	(240)	Pea	–	–	+
BYMV (poty)	ALSV-BY:P3(–)	P3(–)	(240)	Eustoma	–	–	+
SMV (poty)	wtALSV	–	–	Soybean	+++	+++	
SMV (poty)	ALSV-SM:CP	CP	(240)	Soybean	–	–	
CMV (cucumo)	wtALSV	–	–	Tobacco	+++	+++	
CMV (cucumo)	wtALSV	–	–	Cucumber	++	+++	

(continued)

Table 2 (continued)

Target viruses (genus)	ALSV and ALSV vaccine	Gene regions of target virus inserted into ALSV vector (nt)		Preventive effect of ALSV vaccine			Curative effect of ALSV vaccine[c]
				Plants	Symptoms by target viruses[a]	Accumulation of target viruses[b]	
CMV (cucumo)	ALSV-C:1a	1a	(188)	Tobacco	+/−	+/−	
CMV (cucumo)	ALSV-C:1a	1a	(188)	Cucumber	−	+	
CMV (cucumo)	ALSV-C:2a	2a	(210)	Tobacco	−	−	
CMV (cucumo)	ALSV-C:2a	2a	(210)	Cucumber	−	−	
CMV (cucumo)	ALSV-C:CP	CP	(240)	Tobacco	+	+	
CMV (cucumo)	ALSV-C:CP	CP	(240)	Cucumber	−	++	
ZYMV + CMV	wtALSV	−	−	Cucumber	+++	+++	
ZYMV + CMV	ALSV-ZC300	P3 + 2a	(300)	Cucumber	−	−	

[a]– ~ +++ symbols indicate relative intensities of symptoms and virus accumulations
[b]–; no symptoms and virus accumulation. Accumulation of target viruses was estimated by ELISA
[c]+ indicates that curative effect of ALAV vaccine was observed

5. In Eustoma plants, inoculate the first and second leaves at the six-leaf stage with ALSV vaccines against BYMV (Table 2), and challenge-inoculate the third and fourth leaves mechanically with the crude extracts of *N. benthamiana* infected with BYMV-GFP [21, 22]. For TSWV inoculate the fifth to eighth leaves at a ten-leaf stage with total RNAs from infected leaves with ALSV vaccines against TSWV using a Helios Gene Gun system. Challenge-inoculate the 11th to 16th leaves of Eustoma at the 16-leaf stage with leaf extracts of TSWV-infected *N. benthamiana* plants.
6. In pea plants, inoculate the first and second leaves at three-leaf stage with ALSV vaccines against BYMV (Table 2 and *see* **Note 1**) and challenge-inoculate the third and fourth leaves at the five-leaf stage mechanically with crude extracts of *N. benthamiana* infected with BYMV-GFP [21, 22].

Table 3
Preventive effects of ALSV vector-based vaccines against tospovirus infections

Target viruses (genus)	ALSV vaccine	Gene regions of target virus inserted into ALSV vector (nt)	Preventive effects of ALSV vaccine		
			Plants	Symptoms by target viruses[a]	Accumulation of target viruses[b]
INSV (tospo)	wtALSV	– –	*N. benthamiana*	+++	+++
INSV (tospo)	ALSV-IN:N	N (201)	*N. benthamiana*	+	+
IYSV (tospo)	wtALSV	– –	*N. benthamiana*	+++	+++
IYSV (tospo)	ALSV-IY:N	N (201)	*N. benthamiana*	+	+
TSWV (tospo)	wtALSV	– –	*N. benthamiana*	+++	+++
TSWV (tospo)	wtALSV	– –	Eustoma	+++	+++
TSWV (tospo)	ALSV-TS:N	N (201)	*N. benthamiana*	–	–
TSWV (tospo)	ALSV-TS:N	N (201)	Eustoma	–	–

[a]– ~ +++ symbols indicate relative intensities of symptoms and virus accumulations
[b]–; no symptoms and virus accumulation

7. In soybean plants, inoculate the cotyledons with ALSV vaccine against SMV (Table 2), and challenge-inoculate the first true leaves with the crude extracts of soybean leaves infected with SMV.

3.4 Evaluation of Vaccine Effects

3.4.1 Enzyme-Linked Immunosorbent Assay (ELISA)

1. Homogenize the inoculated leaves harvested at indicated time postinoculation in 400 μL coating buffer.
2. Transfer the 200 μL leaf sap into a well of the microplate and incubate the plate overnight at 4 °C or for 2 h at room temperature to coat the well surface with antigens.
3. After antigen coating, wash the wells of the plate with PBST three times (*see* **Note 2**).
4. Add 200 μL virus-specific primary antibody to each well of the plate. Incubate at RT for 1 h and wash the wells of the plate three times with PBST.
5. Add 200 μL secondary antibody to each well of the plate, and incubate at RT for 1 h and wash the wells three times with PBST.
6. Add 200 μL AP substrate to each well and incubate for 15–30 min at room temperature and measure the absorbance at 405 nm using an ELISA reader.

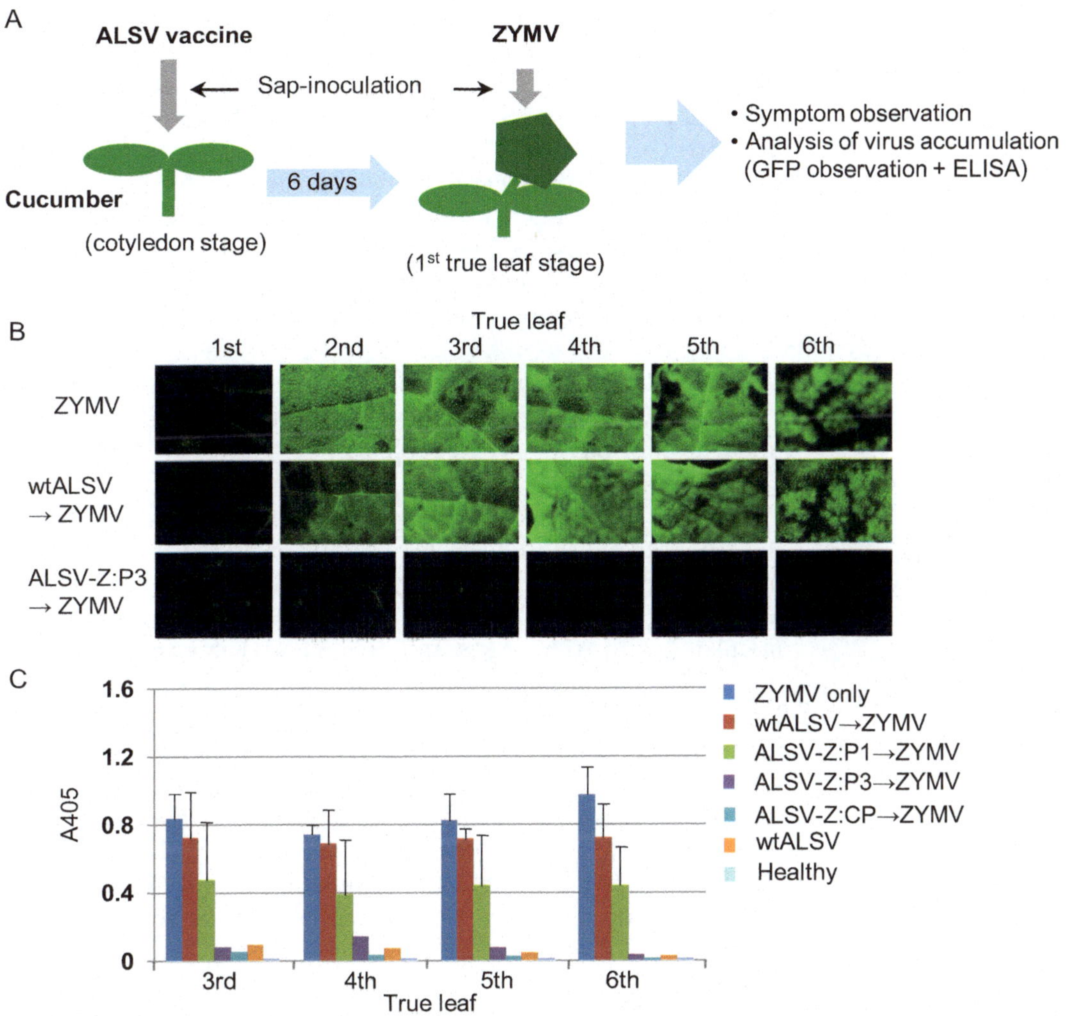

Fig. 2 Preventive effects of ALSV vaccines against ZYMV infection. (**a**) Protection tests for preventive effects of ALSV vaccine against ZYMV in cucumber plants. (**b**) GFP fluorescence showing the challenge-inoculated ZYMV distribution in cucumber plants pre-inoculated with ALSV vaccine (ALSV-Z:P3) at 14 days after ZYMV inoculation. (**c**) ELISA of ZYMV accumulation in the third to sixth true leaves of plants at 21 days after ZYMV inoculation

3.4.2 Observation of GFP Fluorescence

1. Harvest the plants inoculated with the virus harboring GFP gene at the desired time point and observe under the fluorescent stereoscopic microscope with a GFP filter.
2. Document the fluorescent images with a digital camera.

4 Outlook for the Practical Use of Plant Virus Vaccine

4.1 Preventive Effects of ALSV Vaccines

4.1.1 ZYMV

Approximately 12 days later after the challenge inoculation, mosaic symptoms developed on leaves inoculated only with ZYMV-GFP or challenge-inoculated with ZYMV-GFP virus after the inoculation with wild-type (wt) ALSV. In contrast, almost none of the plants firstly inoculated with each ALSV vaccine showed symptoms (Table 2). Observation of GFP fluorescence at 14 days after inoculation of ZYMV-GFP showed that no fluorescence observed in plants preinoculated ALSV vaccines (ALSV-Z:P3, ALSV-Z:Cl, ALSV-Z:Nla, ALSV-Z:Nlb, and ALSV-Z:CP) (Fig. 2b) (*see* **Note 3**). ELISA also indicated that the accumulation of ZYMV-GFP in leaves firstly inoculated with ALSV vaccines (ALSV-Z:P3, ALSV-Z:Cl, ALSV-Z:Nla, ALSV-Z:Nlb, and ALSV-Z:CP) was very low (Table 2 and Fig. 2c), suggesting ALSV vaccines induced strong cross-protection in cucumber plants against the challenge inoculation of ZYMV (*see* **Note 4**).

4.1.2 BYMV

All *N. benthamiana*, pea, and Eustoma plants preinoculated with ALSV-BY:P3(−) showed strong protection against the challenge inoculation of BYMV-GFP (Table 2).

4.1.3 SMV

The mosaic symptoms caused by SMV were observed on the second true leaf and subsequent leaves of plants inoculated with SMV only but not on the leaves of plants preinoculated with ALSV-SM:CP. SMV accumulation was not found in leaves of plants preinoculated with ALSV-SM:CP (Table 2 and data not shown).

4.1.4 CMV

Severe yellow mosaic symptoms were observed on leaves of cucumber plants inoculated with CMV only or plants with wtALSV followed by CMV inoculation (Table 2). In contrast, tobacco plants preinoculated with ALSV-C:1a or ALSV-C:2a showed preventive effects against the challenge inoculation of CMV-Y (Fig. 3a and Table 2) (*see* **Note 3**). No accumulation of CMV-Y was not detected in upper leaves of tobacco plants preinoculated with ALSV-C:2a (Table 2).

4.1.5 ZYMV and CMV

The cotyledons of cucumber were firstly inoculated with ALSV-ZC300, and then the first true leaf was challenge-inoculated with the mixture of ZYMV and CMV. Severe mosaic symptoms on leaves and dwarfing of plants were observed 21 days later, but not on the leaves of cucumber preinoculated with ALSV-ZC300. The accumulation of CMV in plants preinoculated with ALSV-ZC300 was quite lower compared to that of inoculated with wtALSV followed by ALSV-ZC300 (Table 2).

4.1.6 Tospoviruses (INSV, IYSV, and TSWV)

N. benthamiana plants inoculated with tospovirus alone or in combination with wtALSV developed the virus-induced necrosis and then yellowing on upper leaves, followed by shoot apex necrosis and complete withering (Table 3). However, the plants pre-inoculated with ALSV vaccines (ALSV-IN:N, ALSV-IY:N, and ALSV-TS:N) developed white spots on upper leaves, but no phenomenon of the withering of plants occurred (Table 3).

In Eustoma plants, preinoculation of ALSV-TS:N induced strong cross-protection against the challenge inoculation of

TSWV, and no apparent symptoms were observed in plants preinoculated with ALSV-TS:N. However, all plants without ALSV-TS:N eventually died (Table 3, Fig. 3b) [20]. ELISA indicated no accumulation in the upper leaves of plants preinoculated with ALSV-TS:N (Table 3) [20]. Surprisingly, ALSV containing NSs gene fragment of three tospoviruses did not show preventive effects (*see* **Note 1**).

4.2 Curative Effects of ALSV Vaccines

4.2.1 ZYMV

ZYMV-GFP was firstly inoculated into cotyledons of cucumber to test curative effects of ALSV vaccine. Five days later, the first true leaves of cucumber were postinoculated with ALSV-Z:P3. According to our results, no symptoms appeared on the leaves above the fourth true leaves of cucumber plants whereas severe mosaic symptoms were observed on the plants infected with ZYMV only. No GFP fluorescence was also observed in the fourth and subsequent true leaves of ALSV-Z:P3-inoculated plants. ELISA showed that the accumulation of ZYMV in plants postinoculated with ALSV-Z:P3 was also strongly suppressed (data not shown) [21].

4.2.2 BYMV

In pea plants, BYMV-GFP was firstly inoculated to first and second leaves and, after 5 days, the third and fourth leaves of pea seedlings were inoculated with ALSV-BY:P3(−). In the plants postinoculated with ALSV vaccine, both mosaic symptoms and GFP fluorescence were not observed on upper leaves above the seventh leaves whereas plants postinoculated with wtALSV developed clear mosaic symptoms and strong GFP fluorescence on upper leaves.

In Eustoma plants, the first and second pairs of seedling leaves were inoculated with BYMV-GFP by particle bombardment. Eustoma plants showing stunting after 3 weeks were postinoculated with ALSV-BY:P3(−) [22]. In Eustoma plants postinoculated with wtALSV, the stems remained stunted, and the growth of plants was not restored. In contrast, in plants postinoculated with ALSV-BY:P3(−), new shoots emerged from the stunted stem, the intermodal elongation was recovered, and no mosaic symptoms were detected on the leaves at 2 weeks after inoculation of ALSV-BY:P3(−). No GFP fluorescence from BYMV-GFP was observed from the new leaves of Eustoma plants postinoculated with ALSV-BY:P3(−).

5 Notes

1. The partial sequence (201 nt) of NSs gene of three tospoviruses was inserted into pEALSR2mL5mR5 and tested for ALSV vaccines (ALSV-IY:NSs, ALSV-IN:NSs, and ALSV-TS:NSs). Surprisingly, all ALSV vaccines containing NSs gene fragments did not show preventive effects against pathogenic viruses in *N. benthamiana* and Eustoma plants at all.

Fig. 3 (**a**) Symptoms of tobacco plants pre-inoculated with ALSV vaccines (ALSV-C:1a, ALSV-C:2a, and ALSV-C:CP) against the challenge inoculation of CMV-Y (9 days after CMV-Y inoculation). Preinoculation of ALSV vaccines in tobacco induced cross protection against CMV-Y. (**b**) Symptoms of Eustoma plants preinoculated with ALSV vaccines (TS:N and TS:NSs) against the challenge inoculation of TSWV (70 days after TSWV inoculation). Preinoculation of ALSV vaccine (ALSV-TS:N), but not ALSV-TS:NSs, induced strong cross protection against TSWV infection. H, uninoculated plant

2. The dilution factor of antibodies should empirically be determined.
3. Among ALSV vaccines containing seven different regions of the ZYMV genome, ALSV-Z:P1 or ALSV-Z:HC-Pro showed weak cross-protection compared to those by other ALSV vaccines (Table 2), indicating that the strength of cross-protection of ALSV vaccine depends on the genome region in ZYMV. The variation in cross-protection was also observed among ALSV vaccines containing fragments from different genome segments of CMV (Table 2).
4. The influence of the length of the introduced virus genome fragments on the effectiveness of cross-protection by ALSV vaccine was investigated. When ALSV vaccines with 225, 117, 66, or 33 nt in the CP region of ZYMV genome were compared to the effectiveness of cross-protection, we found

that the smaller the size of the introduced ALSV vaccine, the weaker the effects of cross-protection against ZYMV. We concluded that a virus genome fragments of 180 nt or longer are required for an effective vaccine.

Acknowledgments

This work was supported by the Program for Promotion of Basic and Applied Researches for Innovations in Bio-oriented Industry and Grants-in-Aid for Research and Development Projects for Application in Promoting New Policy of Agriculture, Forestry and Fisheries from the Ministry of Agriculture, Forestry and Fisheries, Japan.

References

1. Voinnet O (2005) Induction and suppression of RNA silencing: insights from viral infections. Nat Rev Genet 6:206–220
2. Zhang C, Wu Z, Li Y, Wu J (2015) Biogenesis, function, and applications of virus-derived small RNAs in plants. Front Microbiol 6:1237
3. Mlotshwa S, Voinnet O, Mette MF, Matzke M, Vaucheret H, Ding SW, Pruss G, Vance VB (2002) RNA silencing and the mobile silencing signal. Plant Cell 14(Suppl):S289–S301
4. Mermigka G, Verret F, Kalantidis K (2016) RNA silencing movement in plants. J Integr Plant Biol 58:328–342
5. Pumplin N, Voinnet O (2013) RNA silencing suppression by plant pathogens: defence, counter-defence and counter-counter-defence. Nat Rev Microbiol 11:745–760
6. Li C, Yoshikawa N, Takahashi T, Ito T, Yoshida K, Koganezawa H (2000) Nucleotide sequence and genome organization of *Apple latent spherical virus*: a new virus classified into the family *Comoviridae*. J Gen Virol 81:541–547
7. Igarashi A, Yamagata K, Sugai T, Takahashi Y, Sugawara E, Tamura A, Yaegashi H, Yamagishi N, Takahashi T, Isogai M, Takahashi H, Yoshikawa N (2009) *Apple latent spherical virus* vectors for reliable and effective virus-induced gene silencing among a broad range of plants including tobacco, tomato, *Arabidopsis thaliana*, cucurbits, and legumes. Virology 386:407–416
8. Yamagishi N, Yoshikawa N (2009) Virus-induced gene silencing in soybean seeds and the emergence stage of soybean plants with *Apple latent spherical virus* vectors. Plant Mol Biol 71:15–24
9. Yamagishi N, Yoshikawa N (2013) Highly efficient virus-induced gene silencing in apple and soybean by *Apple latent spherical virus vector* and biolistic inoculation. In: Becker A (ed) Virus-induced gene silencing. Methods in molecular biology, vol 975. Humana Press, New York, pp 167–181
10. Yamagishi N, Yoshikawa N (2016) A new plant breeding technique using ALSV vectors to shorten the breeding periods of fruit trees. In: James F (ed) Genetic engineering - an insight into the strategies and applications. InTech, London, pp 47–61
11. Li C, Yoshikawa N (2015) Virus-induced gene silencing of *N* gene in tobacco by apple latent spherical virus vectors. In: Uyeda I, Masuta C (eds) Plant virology protocols. Methods in molecular biology, vol 1236. Humana Press, New York, pp 229–240
12. Li C, Sasaki N, Isogai M, Yoshikawa N (2004) Stable expression of foreign proteins in herbaceous and apple plants using *Apple latent spherical virus* RNA2 vectors. Arch Virol 149:1541–1558
13. Li C, Yamagishi N, Kaido M, Yoshikawa N (2014) Presentation of epitope sequences from foreign viruses on the surface of apple latent spherical virus particles. Virus Res 190:118–126
14. Yamagishi N, Yoshikawa N (2011) Expression of *FLOWERING LOCUS T* from *Arabidopsis thaliana* induces precocious flowering in soybean irrespective of maturity group and stem growth habit. Planta 233:561–568
15. Yamagishi N, Kishigami R, Yoshikawa N (2013) Reduced generation time of apple seedlings to within a year by means of a plant virus

vector: a new plant-breeding technique with no transmission of genetic modification to the next generation. Plant Biotech J 12:60–68

16. Yamagishi N, Sasaki S, Yamagata K, Komori S, Nagase M, Wada M, Yamamoto T, Yoshikawa N (2011) Promotion of flowering and reduction of a generation time in apple seedlings by ectopical expression of the *Arabidopsis thaliana FT* gene using the *Apple latent spherical virus* vector. Plant Mol Biol 75:193–204
17. Kon T, Yoshikawa N (2015) An effective and convenient method for the delivery of apple latent spherical virus (ALSV)-based vectors into plant cells by agroinoculation. In: Mysore KS, Senthil-Kumar M (eds) Plant gene silencing. Methods in molecular biology, vol 1287. Humana Press, New York, pp 191–199
18. Sasaki S, Yamagishi N, Yoshikawa N (2011) Efficient virus-induced gene silencing in apple, pear and Japanese pear using *Apple latent spherical virus* vectors. Plant Methods 7:15–25
19. Yaegashi H, Yamatsuta T, Takahashi T, Li C, Isogai M, Kobori T, Ohki S, Yoshikawa N (2007) Characterization of virus-induced gene silencing in tobacco plants infected with apple latent spherical virus. Arch Virol 152:1839–1849
20. Taki A, Yamagishi N, Yoshikawa N (2013) Development of apple latent spherical virus-based vaccines against three tospoviruses. Virus Res 176:251–268
21. Tamura A, Kato T, Taki A, Sone M, Satoh N, Yamagishi N, Takahashi T, Ryo BS, Natsuaki T, Yoshikawa N (2013) Preventive and curative effects of *Apple latent spherical virus* vectors harboring part of the target virus genome against potyvirus and cucumovirus infections. Virology 446:314–324
22. Satoh N, Kon T, Yamagishi N, Takahashi T, Natsuaki T, Yoshikawa N (2014) *Apple latent spherical virus* vector as vaccine for the prevention and treatment of mosaic diseases in pea, broad bean, and Eustoma plants by *Bean yellow mosaic virus*. Viruses 6:4242–4257
23. Takahashi T, Sugawara T, Yamatsuta T, Isogai M, Natsuaki T, Yoshikawa N (2007) Analysis of the spatial distribution of identical and two distinct virus populations differently labeled with cyan and yellow fluorescent proteins in coinfected plants. Phytopathology 97:1200–1206

INDEX

A

Acetosyringone 3, 5, 16, 21, 87, 89, 105, 109, 157, 162, 173, 235, 251
Agrobacterium tumefaciens 2–4, 15, 86, 93, 94, 99, 104, 156, 159–161, 171, 173, 177, 178, 233, 238, 240, 250
Agroinfiltration 2, 3, 5, 6, 8, 16, 23, 87, 89, 168, 179, 235, 240
Antibody 6, 8, 24, 75, 117, 121, 124, 172, 178, 252, 257, 263, 265, 267–270, 274, 278, 282, 286
Antiserum 258
Antiviral RNAi 215, 216, 218, 222, 225, 226
Apple latent spherical virus (ALSV) 273–287
Arabidopsis thaliana .... v, 4, 7, 12, 27–77, 86, 98, 145–147, 168, 216, 218, 219, 234, 244, 271, 274
Argonaute (AGO) 215, 216, 232
Artificial microRNA (amiRNA) 231–244

B

Bean yellow mosaic virus (BYMV) 274, 275, 278, 279, 281, 284, 285
Begomovirus 168
Binary vector 2, 4, 5, 13, 87–89, 94, 154–157, 159–161, 171, 176, 177, 249, 255
Bioinformatics .... v, 186, 187
Biolistic bombardment 124, 129, 130, 139–142
BY-2 116–121
BY-2 protoplast lysate (BYL) 116–121

C

Calmodulin-like protein 153
Carborundum 67, 74, 105, 236, 242, 252, 256, 263, 264, 266, 267, 271, 278, 279
Cauliflower mosaic virus (CaMV) 100, 140, 234, 249
cDNA
- library 124–127, 133, 134, 138
- synthesis 88, 126, 130, 131, 249

Cell-free .... v, 115–121
Cellulase 116, 146
Cetyltrimethylammonium bromide (CTAB) 31, 69, 70
Cleaved amplified polymorphic sequence (CAPS) 70–72
Coat protein (CP) 4–7, 23, 24, 75, 98–101, 103, 104, 107–111, 124, 140, 286
Competent cells 102, 137, 154, 159, 169, 171, 173, 233, 249, 250, 275
Computational pipeline 219, 223
Cotyledons 130, 255, 257, 263, 266, 271, 279, 282, 284, 285
CRISPR/Cas9 .... 153–165, 167–182
Cucumber green mottle mosaic virus (CGMMV) 261, 263, 267, 269
Cucumber mosaic virus (CMV) 7, 12, 28, 67, 74, 75, 77, 154, 216–218, 220–223, 225–227, 261, 275, 278, 279, 284, 286
Cucumis sativus 278

D

Defense 1, 27–29, 67, 68, 74, 77, 86, 93, 94, 115, 123, 153, 215, 248
Derived CAPS (dCAPS) 70–73
Dicer-like (DCL) endonuclease 248
DIG 124, 174–175, 181
DNA
- extraction 20, 21, 30, 31, 69, 70, 173, 179, 251, 255
- fragments 20, 105–107, 111, 133, 162, 174–175, 179–181, 253, 254, 275, 278, 279
- methylation 248
- polymerase 14, 31, 70, 102, 103, 107, 108, 111, 112, 169, 249, 253

DNase
- RNase-free 88, 249

Double-stranded RNA (dsRNA) 215, 216, 232, 248, 273, 274

E

Electroporation 3, 4, 15, 21, 103, 104, 106, 109, 112, 177, 248, 254
ELISA reader 248, 257, 282
Epitope tag
- c-Myc 2, 6–8, 16, 22, 23
- FLAG 2, 16, 22, 117, 120, 121
- HA 2, 7, 23

Kappei Kobayashi and Masamichi Nishiguchi (eds.), *Antiviral Resistance in Plants: Methods and Protocols*, Methods in Molecular Biology, vol. 2028, https://doi.org/10.1007/978-1-4939-9635-3,

Eschrichia coli (E. coli)....14, 20–22, 99, 102, 108, 109, 121, 126, 128, 131, 133–135, 137, 138, 142, 158, 169, 171, 173, 176, 180, 233, 238, 240, 249, 254
Eustoma exaltatum....278
Evacuolation....116–121

F

Far-western screening....123–143
Fasta....189, 190, 197, 199, 200, 203, 205, 206, 208, 210, 212
Fastq....192–195, 197, 202, 203, 212
Fluorescence microscope....130, 142, 147

G

Gel electrophoresis
  agarose....15, 31, 72, 73, 103, 105, 108, 111, 155–158, 160, 162, 172–174, 181, 224, 249, 254
  polyacrylamide....67, 72–74, 138, 178
Geminivirus....153, 167–182
Gene
  expression....2, 22, 89, 90, 92, 124
  silencing....v, 85–94, 232, 248, 274
Genotyping....28, 31, 68, 70–72, 74, 75, 219, 222, 227
Glass rods....252, 265, 267, 271
β-Glucuronisase (GUS)....249
Glutathione Sepharose....128, 135, 143
Glutathione S-transferase (GST)....124, 128, 134–136, 138, 143
Glycine max....278
Green fluorescent protein (GFP)....92, 124, 139, 140, 142, 147, 150, 240, 243, 262, 274, 278, 283–285
Guide RNA (gRNA)....154, 155, 157, 159, 168–170, 172, 174–176, 178, 179, 182

H

Hairpin RNA....249, 250, 253, 254
Hammer blot....110, 263–265, 267–269, 271
Homolog....6, 7, 99, 204, 226, 227, 247, 248, 253
Host
  defense....248
  factors....86, 92, 94, 123, 145, 146, 247–258
  gene....v, 86, 88, 90, 92, 94, 168, 205, 247
Housekeeping gene....258
Hypersensitive reaction (HR)....98, 99
Hypocotyls....255, 257

I

Immunoblotting....121, 178, 263
Immunoprecipitation....24, 117, 120, 121
Impatiens necrotic spot virus (INSV)....275, 278, 279, 284, 285
Infection....v, 2, 15, 21, 23, 28, 67, 74, 85, 88, 90, 92, 99–109, 115, 116, 123, 140, 142, 153, 168, 199, 203, 205, 215, 217, 221, 222, 225–227, 232, 247, 252, 257, 261–263, 267, 271, 273, 274, 283, 286
Infectious clone....129, 140, 172, 178, 279
Inoculation....v, 74, 90, 92–94, 98, 99, 105, 109, 110, 150, 222, 236, 242, 243, 252, 256, 261–272, 274, 275, 278, 279, 281–286
Internal control....7, 150, 256, 258
Inverted repeats (IR)....248, 254
Iris yellow spot virus (IYSV)....275, 278, 279, 284, 285

K

Kanamycin....3, 4, 14, 15, 20–22, 87, 89, 102, 104, 106, 156–158, 163, 169, 171, 176–178, 233–235, 238, 240, 250–252, 254–256
Kyuri green mottle mosaic virus (KGMMV)....263–265, 267, 269

L

LeTH1....248, 253, 254, 256, 257
LeTH2....248
LeTH3....248

M

Macerozyme....146
Melon....167, 247, 261, 271
Meropenem....251
Microtiter plate....252, 256
Micro-Tom....249, 250
Miracloth....221, 248, 257
Mitogen-activated protein kinase (MAPK)....92
Movement protein....100, 123–143
Multiplication....116, 248
Mutagenesis....28, 97–112, 154
Mutant T-DNA insertion....71, 77, 218, 222, 223

N

Negative strand....220, 248
Next generation sequencing (NGS)....185, 186, 191, 212
Nicotiana Sp.
  N. benthamiana....v, 2, 3, 5, 6, 15, 21, 22, 85, 130, 142, 168, 171, 172, 174–175, 177–180, 182, 232, 235, 240, 263, 266, 278, 279, 281, 284, 285
  N. sylvestris....12, 98, 99, 104, 109, 110
  N. tabacum....98, 274
Northern blot....150, 217, 219, 220, 223–225, 235, 236, 241–243

NtTOM1 248
NtTOM3 248
Nucleotide-binding site and leucine-rich repeat (NB-LRR) ....v, 1, 2, 11–24, 27, 74, 97–112, 115

O

Oligonucleotides 13, 16, 22, 24, 31, 107, 126, 130, 131, 154, 155, 157–159, 182, 220, 225, 233, 236, 238, 242–244
Oligo(dT) primer 249, 253

P

Pathogen....1, 12, 29, 76, 97–99, 145
Pectolyase 116
Pepper mild mottle virus (PMMoV)....4–8, 23, 98–112, 261–263
Percoll....117–119, 121
Perl
  script....188, 189, 193, 199, 200, 202, 203, 206, 207, 210, 213
Petiole....256
Petri dishes 14, 15, 102, 104, 126, 127, 129–131, 133, 136, 142, 146, 150, 163, 171, 177, 248, 250
Pimers 129
Pisum sativum....278
Plant boxes 171, 248, 251, 255
Plasmids 2, 4, 7, 20–22, 99, 102, 103, 105–109, 111, 112, 124, 125, 129, 137–141, 146–150, 154, 157–159, 161, 168–170, 176, 177, 180, 182, 233, 236, 238, 249, 250, 253–257, 275, 278, 279
Polyethylene glycol (PEG) 126, 130, 146, 147, 149, 218, 221, 236
Polymerase chain reaction (PCR)
  error-prone 101, 103, 108, 112
Post-inoculation 110, 222, 256, 258, 282, 285
Post-transcription....248
Potato virus X....11
Primers 13, 14, 19, 20, 31, 70–72, 77, 88, 90, 91, 94, 101, 103, 105–108, 111, 112, 137, 155, 160, 162, 163, 169, 170, 172, 173, 176, 177, 179–182, 212, 250, 251, 253–256, 275, 278, 279
Protein kinase....92, 124, 128
Protoplasts....v, 94, 116–121, 145–147
Protospacer adjucent motif (PAM)....154, 155, 174–175
Python....188

R

Random mutagenesis....97–112, 154
Recessive virus resistance....247
Recombinant protein....120, 125, 128, 134, 135, 139
Replication ....v, 115–121, 123, 124, 145, 146, 168, 223, 226, 227, 247–258
Replication proteins....100, 115, 116, 120, 121
Resistance
  breakdown 98
Restriction enzymes 14, 31, 70–72, 76, 102, 154, 155, 157, 159, 165, 169, 173, 174, 179, 180, 182
Reverse transcriptase 91, 105, 111, 172, 177, 249, 253, 275
R gene
  L....4, 7, 12–14, 22, 99
  N....11, 98
  N′....4, 5, 7, 8, 12, 98, 99
  RCY1 ....7, 74, 77
  Rx....11
  Tm-1....98, 116, 120, 121
rgs-CaM....153–165
Rifampicin....3, 4, 15, 87, 89, 104, 157, 159, 161, 163, 171, 173, 177, 178, 235, 238, 240, 250
RNA
  degradation....248
  extraction....87, 90, 105, 111, 172, 213, 219, 235, 236, 240, 249, 253, 275, 278
  interference....168, 215–227
  silencing....v, vi, 27, 86, 153, 185, 247–258, 273–287
RNA-dependent RNA polymerase (RdRp)....117, 119, 120, 216, 217, 273
RNA-induced silencing complex (RISC)....215, 248, 273, 274
RNA interference (RNAi)....168, 215–227, 248
RNase
  DNase-free....102
RNase inhibitor....90, 91, 117, 249, 253
Rolling circle amplification (RCA)....173–175, 180, 181

S

SAMtools....188, 208
Seedlings 21, 89, 109, 163, 172, 177, 218, 221–223, 227, 249, 252, 253, 256, 262–264, 266, 271, 285
Signaling....v, 1, 2, 27–77, 85–94, 153, 278
Skim(med) milk 172, 178, 265, 270
Small interfering RNA (siRNA)....27, 205, 215, 216, 220, 224–226, 231, 248, 273, 274
Small RNA (sRNA)....v, 27, 185–213, 225, 231–245
Sodium dodecyl sulfate-polyacrylamide gel electrophoresis (SDS-PAGE) 6, 138, 143, 172
Sodium hypochlorite 156, 161, 250
Solanum lycopersicum 249
Southern blot analysis....173–175, 180, 181
Soybean mosaic virus (SMV)....275, 278, 279, 282, 284
35S RNA promoter....249

Synthetic trans-acting small interfering RNA (syn-tasiRNA) 231, 232, 234, 236, 237, 243, 245

T

T7 endonuclease I (T7EI) 173, 179, 180
Tobacco 2, 98, 153, 167, 168, 221, 247, 262, 278, 279, 284, 286
Tobacco rattle virus (TRV) 86–90, 92, 94, 168, 169, 172, 174–176, 178, 179, 182, 189, 192, 201, 204, 213
Tobamovirus 2, 4, 5, 7, 8, 17, 18, 24, 97–99, 247, 252, 261, 263
TOM1 .... 247, 249, 253, 254
TOM3 247
Tomato 97, 98, 116, 120, 167, 168, 247–255, 257, 261, 271, 274
Tomato mosaic virus (ToMV) 116, 117, 119–121, 124, 128, 139, 140, 142, 248, 252, 256, 257
Tomato spotted wilt virus (TSWV) 245, 275, 278, 279, 281, 284–286
Tomato yellow leaf curl virus (TYLCV) 168–170, 172, 174–176, 178–181
Toothpick 89, 99, 105, 109, 110, 137, 240, 263, 264, 267, 271
Transformation
  bacteria 7, 112, 161
  plant 87–89, 94
Transgene 6, 77, 251, 255
Transgenic
  tomato 97, 98, 248, 251, 252, 255
Translational inhibition 248

U

Ultracentrifuge 218, 221

V

Vectors .... v, 2, 4, 5, 7, 8, 20–22, 86, 124, 125, 127, 132, 133, 147, 157, 159, 160, 165, 172–178, 182, 223, 232–234, 236, 238, 240, 243, 244, 249, 250, 274
Vermiculite 248, 255, 256
Virus
  accumulation 93, 217, 225, 232, 243, 247, 258
  antibody 252, 263, 282
  antiserum 258
  infection .... v, 28, 67, 74, 90, 92, 98, 99, 115, 123, 139, 153, 185–213, 215, 227, 242, 247, 252, 262, 273, 274, 279
  inoculation .... v, 74, 94, 150, 222, 236, 242, 243, 262, 263
  resistance .... v, 1–8, 11–24, 97–112, 167–182, 247, 248, 256, 258
  suspension 221, 256
  titer 181
  vectors 86, 90, 94, 169
Virus-derived small RNA (vsRNA) 185, 186, 193, 195, 197–205, 213
VirusDetect 203, 204, 213
Virus-induced gene silencing (VIGS) 85–94, 274

W

Western blot 3, 4, 6, 8, 24, 124, 150, 178, 243

Z

Zucchini yellow mosaic virus (ZYMV) 261, 274, 275, 278, 279, 283–286

Printed by Printforce, the Netherlands